Springer Geophysics

For further volumes:
http://www.springer.com/series/10173

Alexander Nickolaenko · Masashi Hayakawa

Schumann Resonance for Tyros

Essentials of Global Electromagnetic Resonance in the Earth–Ionosphere Cavity

Springer

Alexander Nickolaenko
Usikov Institute for Radio Physics
and Electronics
National Academy of Sciences
of the Ukraine
Kharkov
Ukraine

Masashi Hayakawa
Advanced Wireless Communications
Research Center
The University of Electro-Communications
Chofu, Tokyo
Japan

ISBN 978-4-431-54357-2 ISBN 978-4-431-54358-9 (eBook)
DOI 10.1007/978-4-431-54358-9
Springer Tokyo Heidelberg New York Dordrecht London

Library of Congress Control Number: 2013949720

Printed on acid-free paper

Springer is part of Springer Science+Business Media (www.springer.com)

Preface

This book is devoted to global electromagnetic resonance, which occurs in the cavity bounded by the conducting ground and the lower edge of the ionosphere. Radio waves in the frequency range of a few Hz can travel around the globe and return to the starting point, so that global resonance becomes possible. This phenomenon was predicted by W. O. Schumann in 1952. More than 50 years have passed since that time, and Schumann resonance has become a recognized branch of radio science that is used in the global sensing of thunderstorm activity and of the lower ionosphere on planetary scales.

In this book we direct our main attention to the properties of Schumann resonance, its detection, typical receiving equipment, and the arrangement of an observatory. We describe the interpretation of Schumann resonance data and outline geophysical information obtained from the records. We also demonstrate recent applications of the resonance phenomenon. A short description is given of other resonances that occur in the adjoining frequency bands: the Alfvén ionospheric resonance (frequencies between 0.1 and 5.0 Hz) and the transverse resonance (basic frequency 1.7 kHz). We also mention the latest reports on the detection of these resonances on board satellites in the ionosphere.

The history of radio science and radio engineering started more than 100 years ago. Natural electromagnetic radiation from lightning strokes was applied at that time as a prototype for the first man-made transmitters. Later, with the development of electronic devices, radio waves appeared with sinusoidal modulation and were used in radio transmissions and radio location. After the discovery of the ionosphere, great attention was directed to the development of shortwave radio communication systems. Higher and higher frequencies were used during and after World War II. Further development of radio science was characterized by revisions of applications of different frequency bands including very high frequencies. The latter are widely used for cellular radio communications. A "return" was made to the band of very low frequencies (VLF: from 3 to 30 kHz) and to the extremely low frequencies (ELF: from 3 Hz to 3 kHz). The low-frequency bands exploit radio wave propagation within the spherical non-conducting gap between the reflecting ionosphere and the ground, so-called sub-ionospheric propagation.

VLF waves were and are used for radio navigation and for delivering the signals of standard time. The existence of global electromagnetic resonance was experimentally confirmed 50 years ago. Interest in these waves was initially conditioned by possible applications of such long radio waves in global communications with submarines. In the 1960s–1980s, ELF radio propagation was studied predominantly from the point of view of military applications, so that interest in these radio waves decreased after the end of the Cold War.

It was understood in the 1990s that Schumann resonance is an efficient tool for studying two interesting objectives: the first is global warming, which might be assessed through the monitoring of global lightning activity. The second goal was the global location of the "red sprites" (optical phenomena in the mesosphere), which are associated with extremely powerful lightning strokes. A promising step was taken when it was demonstrated that some unusual ELF radio signals are observed in association with seismic activity. The most recent observational results indicate that properties of low-frequency waves depend on space weather. Moreover, the radio waves penetrate from the Earth–ionosphere cavity right into space, the upper ionosphere.

Thus, observational data indicate that the Schumann resonance frequency band is a natural (and free of charge) instrument for the remote sensing and monitoring of many interesting phenomena from a single or a few ground-based observatories. The advantage of such a method is in its "passivity": only the receiving equipment is applied while the radio signals are supplied by natural sources of radiation. Thus, the studies we describe are ecologically clean. The disadvantage of such an approach is connected with the same feature: owing to its natural origin, the solution of inverse electromagnetic problems becomes rather complicated since we do not know much about the source of particular events or radiation.

The signals we address originate from lightning discharges. Therefore, one can establish the effective characteristics of global lightning activity from the records and simultaneously deduce the properties of the medium where the propagation took place.

Apparently, some material in the present book overlaps with the contents of earlier books on Schumann resonance. We have tried to describe such facts in the most explicit and compact way. However, substantial new information had accumulated since the time of our latest publication of a special monograph on Schumann resonance in 2002. We tried to include all the new results and present them from the point of view of possible geophysical applications.

We express our sincere thanks to the colleagues with whom we published the results used in this book and with those who took part in discussing the data collected here. Our special thanks are directed to those who kindly gave us permission for presenting and taking advantage of their results in this book. These individuals should be mentioned in particular: Drs. A. V. Shvets, E. I. Yatsevich, A. Y. Schekotov, P. P. Belyaev, G. G. Belyaev, M. Füllekrug, M. J. Rycroft,

V. C. Mushtak, C. Price, O. Pechony, D. L. Jones, A. C. Fraser-Smith, F. Simões, M. Parrot, D. D. Sentman, K. Ohta, Y. Hobara, M. Sekiguchi, and V. C. Roldugin.

Finally, the Authors would like to thank their wives, Irina and Noriko, for their unceasing support. Additional thanks are due to Ms. Yuko Ozawa and Ms. Yuko Watanabe for their great efforts in the editorial work, and we are also grateful to Mr. Thomas R. Walker for his English correction.

Kharkov, Ukraine — Alexander Nickolaenko
Tokyo, Japan — Masashi Hayakawa

Contents

Chapter 1
Introduction

In this introductory chapter we describe general properties of the Earth–ionosphere cavity, outline the relationship of the global electromagnetic resonance to other natural low frequency radio signals, briefly describe the early history, and mention essentials of the signal measurements.

1.1 The Atmosphere as a Waveguide

Our planet is a well-conducting sphere covered by a thin dielectric atmosphere. The thickness of the air shell is small in comparison with the Earth's radius $a \approx 6{,}400$ km. The 30 km air slab includes about 99 % of the atmosphere mass. Although the remaining air or its small portion plays an important role in electromagnetic phenomena. The conductivity of air becomes noticeable at altitudes of a few tens of kilometers above the ground, and it increases by six orders of magnitude or more when entering the ionosphere and space plasma.

A rapid increase of atmospheric conductivity allows for treating the air slab as a dielectric layer positioned between two relatively good conductors. The lower one is the ground, and the upper one is the ionospheric plasma. Thus, a spherical cavity is formed where the radio waves propagate in different frequency bands. The ionosphere plasma density controls the upper frequency of the sub-ionospheric radio propagation. The ionosphere becomes transparent when radio frequency exceeds 10–20 MHz. The lower frequency limit is equal to zero: electromagnetic waves propagate in the Earth–ionosphere cavity starting from arbitrary small frequencies. At zero frequency (direct current), one speaks about the spherical Earth–ionosphere capacitor. The upper 'electrode' (the ionosphere) carries the +250 kV potential relative to the ground, and the electrostatic field of about 100 V/m is observed near the ground. This field is regarded as the fair weather field, which can be observed all around the globe in the local fair weather conditions.

It is possible to regard the Earth–ionosphere cavity as a waveguide, cavity resonator or capacitor. Any particular term (and the formal treatment) depends on

A. Nickolaenko and M. Hayakawa, *Schumann Resonance for Tyros*,
Springer Geophysics, DOI: 10.1007/978-4-431-54358-9_1,

the frequency or on the electromagnetic wavelength λ. We speak about the Earth–ionosphere waveguide when the Earth's radius substantially exceeds the wavelength: $a \gg \lambda$. The oscillations become quasi-electrostatic in the opposite case $a \ll \lambda$, for which the relevant electric field might be described by the process of charging - discharging the Earth–ionosphere capacitor. The term 'electromagnetic resonator' is applied when the wavelength is comparable with the Earth's circumference $\lambda \approx 2\pi \cdot a = 40$ Mm (1 Mm = 1,000 km).

It seems that the idea of global electromagnetic resonance appeared immediately after the discovery of the ionosphere in the beginning of 20th century. However, the formal solution for the problem of global resonance was published by Schumann in (1952a, b, c, 1957; Schumann and König 1954), and the global electromagnetic resonance is referred to as Schumann resonance (hereafter abbreviated as SR). The history of early explorations of SR is presented in a detailed paper by Besser (2007). A series of works on modern SR studies was published in the same issue of Radio Science (2007 42, No. 2).

To predict the global electromagnetic resonance, Schumann not only considered the Earth as a spherical celestial body, but also adequately described the Earth–ionosphere cavity as a kind of radio device. Such an approach is customary nowadays, like the sights of the globe from the space that we meet everyday in the TV weather forecasts.

It looks like that Yugoslavia born American scientist Tesla (1905) was the first who considered the globe as a radio device. He tried to excite the electric oscillations of the whole planet (e.g. Bliokh and Nickolaenko 1986). At the end of 19th—the beginning of 20th century, intense studies started on radio waves. Natural electric discharges (lightning strokes) were used as the sources of electromagnetic radiation in the 1895 experiments by Popov AS, Markoni G used the spark transmitters for radio communications, which are in fact a small copy of a lightning stoke. Half-a-century later, Schumann noted that radio emissions from lightning strokes might be the major source of electromagnetic oscillations in the global cavity.

The first internationally recognized proof of detection of the global electromagnetic resonance was presented by Balser and Wagner in (1960, 1962a, b). The number of publications on SR has rapidly increased since that time. Similar to many other physical phenomena, the military applications were driving these studies. The resonance signals were intended for radio communications with submarines in the World Ocean (e.g. special issue of IEEE Trans 1974, Com-22, No. 4). After the completion of ELF (extremely low frequency) radio links, the interest in SR faded out. It re-appeared again in 90 s when Williams (1992, 1994) demonstrated a connection of the SR amplitude with the global temperature.

SR monitoring allows for studying the natural sources of electromagnetic radiation, so that it is a tool for passive remote sensing of the global lightning activity. Resonance parameters (peak amplitudes , peak frequencies, and the width of resonance lines) also reflect propagation conditions on the global scale, i.e., the general properties of the lower ionosphere, which is out of reach for other frequencies.

It was found that powerful lightning strokes can cause the luminous structures above the thunder clouds, the so-called 'red sprites ', 'blue jets', 'blue starters', etc. [see the latest review by Hayakawa et al. (2012)]. The details might be found in the collective monograph edited by Füllekrug et al. (2006). These strong discharges usually generate the well-known ELF transients or Q–bursts that are the discrete pulsed radio signals propagating in the frequency band of SR. Such powerful strokes might be responsible for the 'slow tail' atmospherics that occupy the frequency band above the SR (Sentman and Wescott 1993; Boccippio et al. 1995; Sentman et al. 1995; Sukhorukov and Stubbe 1997; Pasko 1998; Füllekrug and Reising 1998; Füllekrug et al. 1998; Marshall et al. 1998, 2005; Huang et al. 1999; Rodger 1999; Williams 2001; Hobara et al. 2001, 2006; Price et al. 2002, 2004, 2007; Yair et al. 2003, 2009; Chern et al. 2003; Neubert et al. 2005; Greenberg et al. 2007; Hayakawa et al. 2011, 2012). Thus, the idea appeared of the global ELF radiolocation of the powerful lightning strokes which can modify the middle atmosphere. The ELF occupies the frequency range from 3 Hz to 3 kHz. The SR oscillations are usually observed at frequencies below 50 Hz, while the slow tail atmospherics are detected above 300 Hz.

It was demonstrated that seismic and pre-seismic activity are often reflected in the low frequency electromagnetic activity at the ground surface (e.g. Hayakawa and Fujinawa 1994; Hayakawa 1999; Hayakawa and Molchanov 2002). We must mention, for example, the unusual SR signals in Japan associated with the earthquakes (EQs) in Taiwan (Ohta et al. 2002). Elementary considerations show that seismic electromagnetic radiation should be sought at the lowest frequencies, since it travels through the conducting soil from the underground seismic source into the air [e.g. Molchanov and Hayakawa (2008)].

The global resonance is conditioned by an interference of the direct and antipodal waves. Constructive and destructive interference is realized when these waves have comparable amplitudes. The condition is true in the SR frequencies, as the wave attenuation factor does not exceed a few parts of dB/1,000 km. The ionosphere reflections acquire the noticeable amplitude in ambient night conditions, and the transverse resonance becomes visible. This resonance is relevant to radio waves trapped between the ground and the lower ionosphere boundary, so that waves move vertically between the upper and lower walls (Lazebny et al. 1988; Nickolaenko and Hayakawa 2002). The major distinction of these two resonances is that the SR has a global nature: the radio wave "multiply circles the Earth". The transverse resonance is a local phenomenon, in which the wave is trapped within the cavity in the vicinity of the source and observer.

1.2 Typical Characteristics of the Lower Ionosphere

The atmosphere conductivity depends on the time of day, as the air ionization arises from the solar ultraviolet radiation. The nighttime ionization, especially at the lower ionosphere regions E and D, is conditioned by the galactic background.

Ionosphere is a highly variable medium in time and space, and it depends on the current solar condition, etc. We mention the major characteristics of the lower ionosphere (Alpert 1990), and the detailed information might be found in special literature.

Typical data is collected in Table 1.1. The first column here shows the height above the ground in km. The second and third columns contain the typical electron density N_e (electrons in cubic centimeters) at the middle latitudes in ambient day and night conditions. The fourth column presents the density of neutral atmosphere N_0 (in cm^{-3}), and the fifth column depicts the effective collision frequency ν_{eff} of electrons (it primarily depends on the electron—neutral particle collision). The last three columns present the height profiles of the electron gyro-frequency ω_H (in s^{-1}), plasma frequency ω_0 (in s^{-1}), and the air conductivity σ_0 (in S/m).

Table 1.1 demonstrates that the electron content of the lower ionosphere rapidly increases with altitude. One may also note considerable changes of the ratio of the electron gyro-frequency to the effective collision frequency. This ratio indicates whether the plasma is considered to be isotropic or anisotropic. The ratio is approximately 0.2 at the 60 km altitude, and it becomes equal to 5 when the altitude reaches 80 km.

SR was predicted more than 60 years ago. Since that time it has become a popular and continuously advancing branch in the environmental studies. There are many publications on the global electromagnetic resonance and ELF radio propagation. We are not able to discuss or even mention all of them. Naturally, it would be desirable to cite all published papers, however, this is impossible. We give, in advance, our apologies to authors whose works were omitted as it was really impossible to include them all in a book of limited size.

Table 1.1 Typical parameters of vertical profile of atmosphere

Height h km	Electron density N_e 1/cm^3		Neutral density N_0 1/cm^3	Electron collision frequency ν_{eff} s^{-1}	Electron gyro-frequency $\omega_H \times 10^{-6}\ s^{-1}$	Plasma frequency $\omega_0 \times 10^{-5}$ s^{-1}	Air conductivity σ_0 S/m
	Day	Night					
0	–	–	2.7×10^{19}	–	8.73	–	–
10	–	–	2.7×10^{18}	–	8.7	–	–
20	–	–	1.5×10^{18}	–	8.66	–	–
30	–	–	3.2×10^{17}	–	8.62	–	–
40	–	–	8.1×10^{16}	–	8.59	–	–
50	1–10	–	2.5×10^{16}	1×10^8	8.55	0.6 –1.8	8.8×10^{-10}
60	70	–	7×10^{15}	5×10^7	8.51	4.8	4.1×10^{-9}
70	90	10	2×10^{15}	1.4×10^7	8.48	5.4	1.5×10^{-7}
80	110	50	3×10^{14}	1.7×10^6	8.44	6	1.8×10^{-6}
90	1,000	500	6×10^{13}	7.6×10^5	8.4	18	3.8×10^{-5}
100	4,000	2,000	1×10^{13}	1.7×10^5	8.37	36	6.7×10^{-4}

The major attention will be directed to the following aspects:

1. General demands on the SR observatory, positioning of the equipment, general description of antennas and receivers, compatibility of channels, calibration of measurements, data acquisition.
2. When interpreting observational data, we use only the compulsory equations and model data. Detailed descriptions of relevant formalism can be found in the references.
3. We prefer to use 'effective' or 'engineering' models. We will try to discuss details of the signal processing and finding information on the global lightning activity on the diurnal and seasonal scales. We will outline a possible connection of SR with the global temperature, address unusual resonance signals, impact of the space weather and of seismic activity on the SR records.
4. We will use the simple and compact time domain solutions and demonstrate their excellent agreement with observations.
5. To demonstrate peculiarities of obtaining geophysical information, we use original and published data of various kinds.

1.3 Schumann Resonance (SR) Among Natural Radio Signals

The global electromagnetic resonance was predicted by Schumann (1952a, b, c, 1957). Together with the concept of resonance, the idea was spelled out in these works stating that the major field source must be the planetary lightning activity. The narrow dielectric interface separates the conducting ground and the ionosphere. It is less than 100 km in height, while the Earth's radius is about 6,400 km. Besides, the cavity height is small in comparison with a wavelength; the latter is comparable with the equator circumference of 40,000 km. This is why fields in a three-dimensional Earth–ionosphere cavity become the functions of a single variable—the angular distance from the source to an observer. By exploiting the small height of the cavity, Schumann obtained his famous formula for the eigenfrequencies of the Earth–ionosphere cavity $(ka)^2 = n(n + 1)$ where $k = \omega / c$ is the free space wave number, a is the Earth's radius, c is the light velocity in vacuum, $\omega = 2\pi f$ is the angular wave frequency, and n is the mode number (the number of spherical harmonic). The product $n(n + 1)$ is the so-called separating constant of the Legendre polynomials, which proves the spherical form of the cavity. The above relation is readily transformed into the well-known SR formula for the cavity with perfectly conducting boundaries:

$$f_n = \frac{c}{2\pi a} \sqrt{n(n + 1)} \tag{1.1}$$

By substituting the relevant numbers n into Eq. (1.1), one obtains the following succession of resonance frequencies: $f_1 = 10.6, f_1 = 18.3, f_1 = 26.0$ Hz, etc.

The results were dubious in the first experiments aiming at the detection of global electromagnetic resonance (Schumann and König 1954): An attempt was unsuccessful to find more or less durable sinusoidal trains of resonance frequencies in the natural terrestrial radio noise. As it became clear later, the technique applied would hardly be a success due to the low quality factor of the Earth–ionosphere cavity resonator. The existence of resonance was proven in the experiments by Balser and Wagner (1960, 1962a, b), who analyzed the power spectra of natural radio noise in the frequency band from a few Hz to a few tens of Hz. SR was observed as a series of peaks in the power spectrum at the frequencies around 8, 14, 20, 26 Hz, etc. The computer signal processing was applied in their particular measurements: the Fourier transformation was performed of the 10 min samples of record. It is remarkable that the algorithm of the fast Fourier transform (FFT) was not known yet at that time.

Since the frequency step was relatively large, the parameters of resonance peaks were found by fitting the spectral lines with the help of Lorentzian curves. This curve is regarded as the Cauchy curve in mathematics, and it is widely used in the functions of complex argument. A particular function $U(f) = \frac{-i}{f - f_1 - if_2}$ has a pole at the point $z_0 = f_1 + if_2$ in the plane of the complex argument z. When the argument z is frequency, the Fourier conjugate of the $U(f)$ function is the damping in time sinusoid $\exp\{i2\pi(z_0)t\}$. One may readily obtain the following relation for the $U(f)$ spectrum: $U(f) = \frac{f_2}{f_1^2}\frac{1}{(x-1)^2+\Delta^2} - i\frac{1}{f_1}\frac{x-1}{(x-1)^2+\Delta^2}$. Here $x = \frac{f}{f_1}$ is the dimensionless variable, and $\Delta = \frac{f_2}{f_1} = \frac{1}{2Q}$. The quantity $f_1 = \mathrm{Re}\{z_0\}$ is regarded as a peak frequency, and Q is the quality factor. The real part of the complex Cauchy function $U(f)$ is the Lorentz curve :

$$y(x) = y_0 + \frac{A}{(x - x_0)^2 + \left(\frac{x_0}{2Q}\right)^2} \tag{1.2}$$

where y_0 is a constant, A is the resonance amplitude, x_0 is the resonance frequency, and Q is the quality factor of resonance oscillations, which is connected with the width of resonance line.

Thus, the resonance frequencies, amplitudes and Q-factors were found for the first time. We know that the typical resonance amplitude in the vertical electric field component is around 0.3–1 mV/m/Hz$^{1/2}$, peak frequencies are equal to 8, 14, 20, 26 Hz, etc., and the Q-factors occupy an interval from 4 to 5.

In a completely inaccurate, “home made” manner one may assert that a Q-factor shows the amount of turns around the globe that the radio wave is able to go. Of course, the larger the Q-factor is, the longer is the duration of response to a pulsed excitation in the time domain. However, the popular phrase “turns around the globe” contradicts the physics, because resonance waves do not go around the world. Indeed, imagine a point vertical lightning stroke placed at the pole for convenience. It generates the circular wave expanding in time similarly to circles in a pond diverging from the thrown stone. Our 'planetary pond' is not flat, but a sphere. Therefore, a wave initially expands, it reaches the equator and starts to converge toward the other pole (the antipode of the source). In a period of about 1/16 s the wavefront will merge into a 'splash' at the antipode and afterwards, a 'new' wave starts to expand from the source antipode. After $\sim 1/8$ s, the circular wave returns and provides a field burst at the source point. The process is repeated afterwards. A remote observer detects the direct wave first coming from the source. In time, the wave from the antipode will arrive regarded as its antipodal wave. Later the wave appears coming from the source again (the round-the-world wave), etc. It is clear now that ELF waves do not 'circle the globe'. Instead, they arrange a kind of 'push–pull' regime coming from two opposite directions one after another. In this context, the Q-factor shows how many noticeable "pushes" come to the observatory.

The observation of SR is not a problem. One must have relevant field sensors, appropriate receivers, and devices that perform signal analysis. We will briefly describe all of these. The natural radio signal is a noise, and this is an important feature. To detect resonances, one has to analyze the spectra, i.e., to perform the Fourier transform of the time domain records. Choosing a particular spectral technique plays an important role. It was found experimentally that the well-outlined resonance peaks emerge in the power spectra as a result of considerable averaging over the spectral samples obtained in the 'parallel' spectral analysis, i.e. when the same time realization is processed by a 'bank' of filters. The 'serial' spectral analysis that was widely applied to the man-made narrow band signals was inefficient for the SR studies. The serial analysis uses a single narrow band filter with a peak frequency slowly varying in time in a given band. Such filtering was unproductive for the terrestrial ELF radio noise: the same waveform should be fed to the set of filters tuned to different frequencies, and spectral amplitudes must be obtained simultaneously, in parallel. We must remark that modern spectrum analyzers utilize the parallel scheme. The spectra of separate time segments fluctuate from one to another. This is why individual spectra should be averaged for obtaining stable output data. This instability is rarely mentioned in literature.

Successful observations of SR spectra in the city are rare owing to the high level of industrial interference. A field site is usually positioned in the rural area at a distance from power supply and communication lines. We consider general demands on the SR observatory in the next section.

We present in Fig. 1.1 the typical amplitude spectra of natural ELF radio signals observed in the magnetic field component at the Karimshino observatory, Kamchatka, Russia (geographic coordinates; 52.94°N and 158.25°E). This particular record was made in 2000. Three spectral functions are presented in two columns of plots, each function corresponding to a separate magnetic field component: H, D, or Z. These are standard notations for the components of geomagnetic field. The horizontal component in the plane of magnetic meridian is called the H–component. The vertical field component is Z–component (the OZ axis is directed downward). The horizontal magnetic field, which is perpendicular to the magnetic meridian, is regarded as D–component. The frequency in Hz is

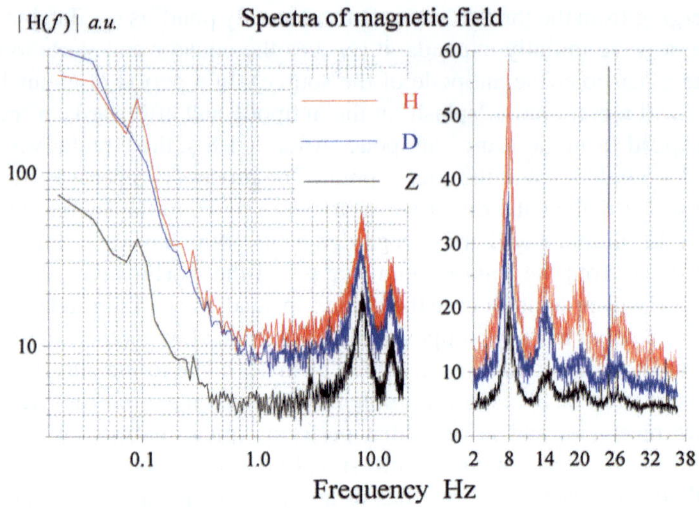

Fig. 1.1 Typical experimental ULF – ELF spectra of natural electromagnetic signal, the magnetic field components recorded at the Karimshino observatory, Kamchatka, Russia. The *left* plots show amplitude spectra in the ULF–ELF band, and the *right* plots show the SR range

plotted along the abscissa in this figure. The effective amplitude of magnetic field spectrum is shown on the ordinate in arbitrary units. The effective amplitude is the square root of the average power spectrum of every field component:

$$A^{(i)}_{\text{eff}}f = <|\,\text{H}^i(f)\,|^2>^{1/2}.$$

The data acquisition system (DAS) included the three induction coil magnetometers (see Sect. 1.4), a common calibration generator, scaling amplifiers, connecting cables and the three channel 24-bit analog-to-digital device (ADC). The sensors were the magnetic induction coils with the high permeability amorphous permalloy ribbon core (see Fig. 2.8d). Shielded coils and the low noise amplifiers were fixed in the hermetic fiberglass tube. The DAS was connected through the serial port with the PC, and the GPS time stamps were used. The DAS had the characteristics listed in Table 1.2. It worked in the frequency band from 10^2 to 10^{-2} Hz with the sampling frequency of 150 Hz. The amplitude response of antenna grew linearly with the frequency from 0.01 to 0.02 Hz, and it remained practically constant at frequencies above 4 Hz. The record, in time, was subdivided into the pieces of 54.6 s duration (the number of samples is equal to 8,192). Every fragment was processed by the FFT code. Afterwards, the 24 individual power spectra were averaged thus providing the effective power spectrum from which the spectral amplitudes were obtained. The spectra shown in Fig. 1.1 present the digital record of the three magnetic field components of approximately 20 min duration.

The left survey plots in Fig. 1.1 are shown against the logarithmic frequency axis. The right plots present the SR spectra, which are shown versus the linear

Table 1.2 Characteristics of data acquisition system (DAS) of Karimshino observatory

Characteristic parameter	Value
Frequency band, Hz	0.003–40.0
Threshold sensitivity, pTHz	
$f = 0.01$ Hz	16
$f = 0.1$ Hz	1.6
$f = 1.0$ Hz	0.16
$f > 10$ Hz	<0.02
Conversion function, Hz*V/nT	
$f < 4$ Hz	0.4
$f > 4$ Hz	1.6
Output voltage, V	±5, ($R_{out} = 10$ K, $C_{out} = 10$ nF)
Supply voltage, V	±5–7
Supply current, mA	<15
Sensor weight, kg	<6.0
Total weight without common cable, kg	20
Dimensions of one sensor, mm	Ø 60 × 630
Operating temperature range C	−30 + 50

frequency scale. The left plots clearly indicate a place of the SR among the other low frequency electromagnetic phenomena. One can see that amplitudes within the ultra low frequency (ULF) band exceed those of the SR oscillations by an order of magnitude in the quiet geomagnetic conditions. Sometimes, geomagnetic pulsations are observed with the periods between a second and a few tens of seconds. In this case, the characteristic peaks appear in the spectra at frequencies between 0.01 and 1 Hz, and we call these geomagnetic pulsations (Pc). Signals of ionosphere Alfvén resonance (IAR) are observed in the frequency band from 1 to 10 Hz during the local night (Polyakov 1976; Polyakov and Rapoport 1981; Belyaev et al. 1987, 1989, 1990, 1999a, b). The IAR signal was absent in particular spectra of Fig. 1.1, however, there is a weak Pc1 geomagnetic pulsation visible around the frequency of 0.1 Hz in the spectra of the $H_H(f)$ and $H_Z(f)$ field components.

The right plots correspond to the 2–37 Hz frequency band. We clearly observe four SR peaks here plus a less distinctive fifth mode. The relevant peak frequencies are equal to 8, 14, 20, 26, and 32 Hz. Spectrum of the vertical magnetic field component also contains resonance peaks. This behavior contradicts with the theoretical expectations, since at ELF only the TEM (transverse electromagnetic) radio waves propagate having the non-zero vertical electric and the horizontal magnetic field components. Probably, the resonance structure in the vertical magnetic field arises from a 'tilt' of the magnetic field vector, which was caused by geological irregularities in the Kamchatka crust. As a result, the local 'electromagnetic vertical line' deviates from the normal to the ground surface. The real distribution of currents in the underlying soil depends on the conductivity of the rocks, which is highly irregular within the seismically active Kamchatka region.

SR spectra in Fig. 1.1 contain a typical narrow band interference at a frequency of about 25 Hz (approximately one-half of the industrial 50 Hz frequency).

The amplitude of this mysterious signal is time dependent. It is present at many observatories throughout the world, and it might surpass the SR amplitude.

Resonance spectra are always structured, as Fig. 1.1 demonstrates. Fine structures did not vanish after averaging over 20 min. On the other hand, the number of averaged spectra was not great since we had to use relatively durable data fragments of 54.6 s for obtaining an appropriate frequency resolution of about 0.01 Hz. Such a resolution was necessary for resolving the Pc1 geomagnetic pulsation in the spectrum, and the fine structure of SR spectra is discussed in the following section.

1.4 Natural Stabilizations of SR Spectral Estimates

The stable spectral estimates are obtained after averaging the power spectra accumulated in the time intervals from a few to 10 min. The process of natural stabilization is illustrated by plots in Fig. 1.2. We show here a succession of

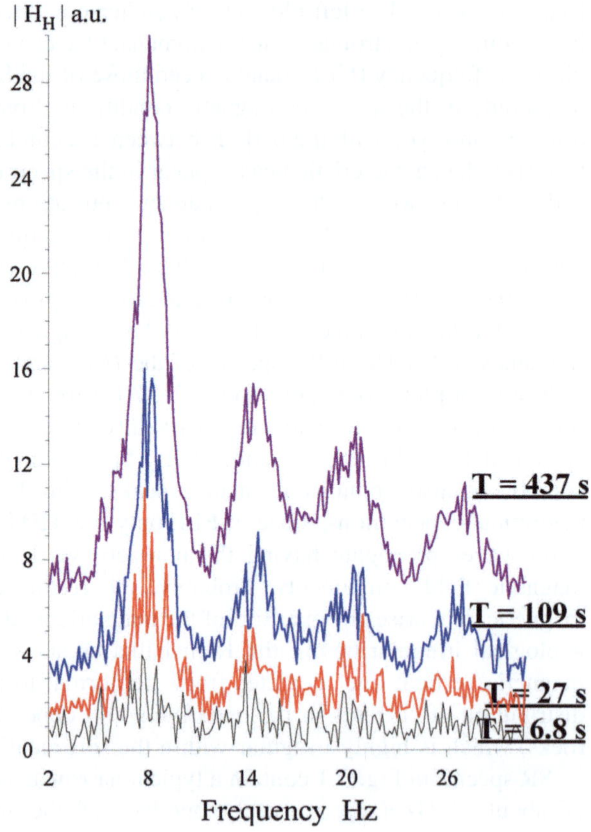

Fig. 1.2 Natural stabilization of SR spectra in the H horizontal magnetic field component. The numbers to the *right* from the plots denote the accumulation time *T*. The smallest one being 6.8 s corresponds to the durations of elementary data segment processed by the FFT algorithm. The largest time corresponds to the sum of 64 elementary power spectra

amplitude spectra. The power spectra of segments 6.8 s long were accumulated and the relevant amplitude spectrum was obtained as a square root of the intensity spectrum. We show the spectra of H-field component of the record presented in Fig. 1.1. However, the FFT procedure is applied to 1,024 points now, so that the elementary time segment is shorter, 6.8266 s long.

Figure 1.2 depicts four amplitude spectra that were obtained similarly. First, the power spectra were computed for every segment of the record, afterwards, these spectra were summed. The spectrum of effective amplitude is the square root of the accumulated power spectrum. This is a customary procedure of preliminary data processing in SR monitoring (e.g. Belyaev et al. 1999a, b). Particular plots of Fig.1.2 correspond to amplitude spectra accumulated with 1, 4, 16, and 64 elementary successive realizations.

Experimental SR spectra Spectra are always structured for both electric and magnetic field components. Such an irregularity seems customary if we recall that natural ELF radio signal is a composition of random electromagnetic pulses arriving from the lightning strokes distributed all over the globe. Positions of particular discharges, the arrival times, and the pulsed amplitudes are stochastic. The average rate of pulses is about 100 events per second. Therefore, the resonance peaks are hardly recognized in the individual spectrum shown in Fig. 1.2 by the lower line: only 600–700 pulses took part in the formation of this spectrum. One may expect the spectral smoothing in the process of accumulation (averaging). Data processing verifies these expectations, but the stabilization occurs not so fast as it might be expected from the statistical central limiting theorem. This theorem states that for the normal statistical process, the relative deviations of the averages in N tests decreases as N. The slower stabilization might be caused by the Poisson nature of the pulse flux, which decelerates the process of "normalization". Besides, a definite role might be played by relatively rare intense pulses from the powerful lightning strokes regarded as Q-bursts or ELF transients. These arrive at a rate of one event in a minute (Jones and Kemp 1971).

It is clear that the accumulation of individual spectra Spectra noticeably reduces the fine structure of SR spectrum, but it does not remove it completely. One can model the spectra shown in Fig. 1.2. We have used for this purpose the formal solution of the ELF radio propagation problem in the time domain. We describe this solution in Chap. 9 when discussing the modern Q-burst observations. Here, we depict a series of spectra of a synthetic radio signal composed from the model pulses arriving from independent random strokes of lightning (Nickolaenko and Hayakawa 2002).

1.5 Stabilization of SR Spectra in the Model Time Domain Record

By using the formal expressions for the pulsed response, we synthesized the ELF noise directly in the time domain and then process this 'signal' in the same way as the real SR records are processed (see e.g. Belyaev et al. 1999a, b). Such a numerical experiment allows for demonstrating the features of natural stabilization following from the random nature of the signal.

We initially generated a relatively long (longer than 10 min) time domain realization of the vertical electric field component that is a superposition of discrete random independent pulses.

$$E(t) = \sum_k A_k \cdot e_k(t - t_k) \tag{1.3}$$

Casual amplitudes of pulses A_k were normally distributed, the pulse arrival times t_k corresponded to Poisson law, and the random waveform of a pulse e_k was a function of coordinates of the k-th stroke having the stochastic longitude and latitude. Different generators of quasi-random numbers were used in the computations. In particular, the first one produced a uniform distribution that was applied to model the source–observer distance. The second one generated the normally distributed numbers used for modeling the stroke amplitude, and the third one provided the random variable with the exponential distribution—the mutual time delay of individual pulses in the succession (1.3). It was accepted that the pulsed flux has a rate of 50 events per second, the source–observer distance is uniformly distributed in the interval from 5 to 15 Mm, and the normal distribution has an average value of 15 and the standard deviation of 40, which corresponds to the measurements of amplitudes of vertical lightning strokes (Nickolaenko and Hayakawa 2002). The sampling frequency of the $E(t)$ realization was 204.8 Hz, which was equal to the sampling rate in the records at the Lehta observatory, Karelia, Russia (Belyaev et al. 1999a, b).

Spectral processing was performed in the following way. Data segments of the 10 s length (2,048 real quantities) were transformed by the FFT standard procedure. Then, separate power spectra were averaged, whose relevant results are shown in Fig. 1.3. The lower scalloped curve demonstrates the spectrum obtained in the first 10 s realization. The next, averaged, power spectra were vertically shifted to facilitate their comparison. The figure N to the right of a curve denotes the number of individual spectra (of the 10 s realizations) involved in the averaging. The smoothing of spectral estimates is clearly seen. SR peaks emerge in the spectra when the accumulation includes ensembles greater than $N > 16$.

The behavior of model spectra is similar to that observed experimentally, as seen in Fig. 1.2. It is clear that data accumulation should include the interval not less than 3 min. During this time interval, about 7,500 pulses arrived at the observatory, which is an enormous number from the viewpoint of central limiting theorem. Thus, our modeling showed that rather slow stabilization results from the

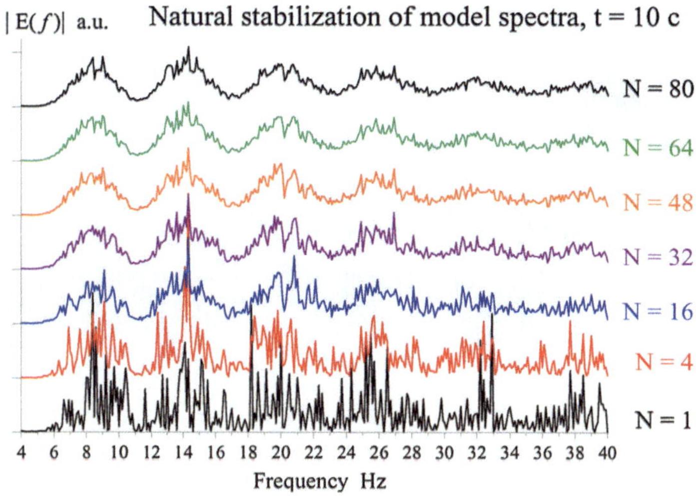

Fig. 1.3 Natural stabilization in the spectra of a model SR time domain record

Poisson nature of the pulse flux. By chance, the accumulation over 10 min interval was used in the first successful experiments by Balser and Wagner (1960), and thus provided obtaining the stable spectral estimates.

We noted that elementary considerations do not explain the necessary time intervals of spectral accumulation. In the model realization we used the Poisson process with the pulse rate of 50 events per second, so that there are about 3,000 pulses arriving every minute. This value itself seems to be quite sufficient for providing the stable spectral estimates. A detailed treatment and the model computations confirm that a combination of three random functions in the source distribution (in time, in space, and in amplitude) increases the necessary ensemble of averaged spectra by an order of magnitude. The model spectra in Fig. 1.3 demonstrate that the fine structure of spectra does not vanish completely even when accumulation time exceeds 10 min. We will see in Chap. 4 that resonance spectra remain structured even after the 1 h accumulation. Simultaneously, the accumulation times from 3 min and greater provide reasonably smooth spectra with distinct SR pattern. An exception was the spectra of the global electromagnetic resonance recorded at high latitudes (Egeland and Larsen 1968). The fine structure was not reduced noticeably with accumulation time, and this was probably explained by a rather poor ensemble of the source–observer distances when the field-site is positioned in the vicinity of the pole.

Finally, spectra of Fig. 1.3 have a feature revealing their artificial origin. The modeling we used does not provide the quasi-static background over the frequency axis, which is always present in the experimental spectra. Model spectral density reduces almost to zero between the resonance peaks, and this property allows for more accurate estimates of the resonance quality factors (Q factors) . In all other aspects, the model spectra are similar to the experiment, and the computations

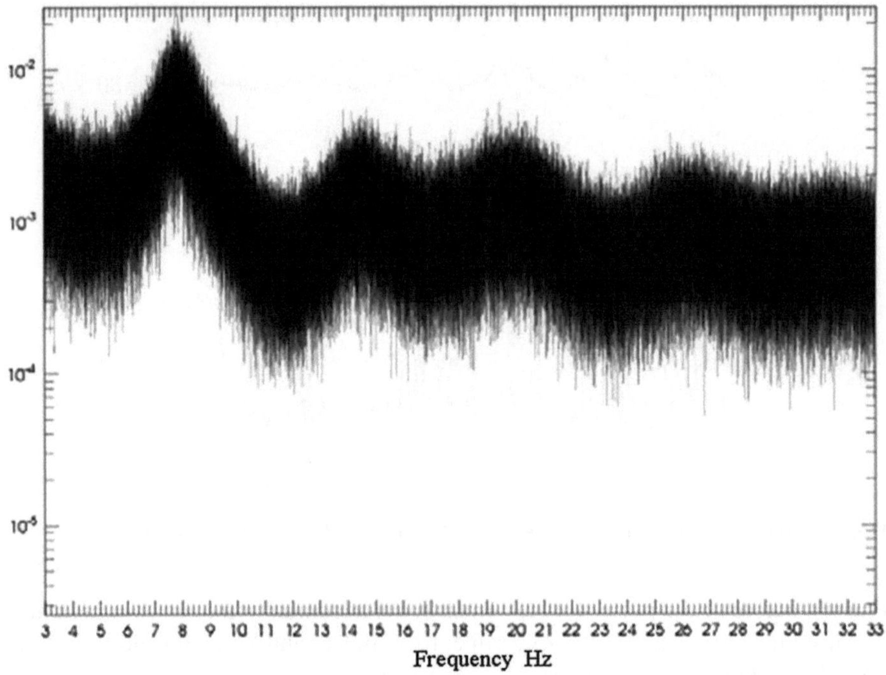

Fig. 1.4 An example of experimental power spectra of a continuous long SR record (adopted from Neska and Sátori 2006)

allow for a better understanding of the behavior of real signals. A synthetic radio signal might be helpful in the laboratory tests of equipments, because the detection of natural radio signals in the urban environment is hampered.

The fine structure is seen in Fig. 1.3, which arises from the interference of separate pulses in the complex spectrum. The impact of inter-pulse interference decreases when substantial averaging is applied. Since the natural ELF radio signal is a random process, its Fourier transform (the spectrum) is also a random function of frequency. An increase in the duration of time domain record does not make the spectral estimates stable (Blackman and Tukey 1958; Sveshnikov 1968). According to the localization principle of Fourier transform, an increase in duration T of the time domain record simply raises the spectral resolution $F = 1/T$. Physically this means that spectral densities become independent, provided that their frequencies are separated by F or a greater amount.

This property of Fourier transform was clearly demonstrated by Neska and Sátori (2006) who presented the spectrum of very durable realization of natural ELF signal recorded at the Polish sub-polar station Hornsund, Spitzbergen (geographic coordinates: 77 N and 15.5 E). We adopted their spectrum in Fig. 1.4. One can easily recognize the resonance modes in the spectrum of radio noise, but the spectrum looks like a noisy strip rather than a resonance line.

To stabilize the spectrum estimates, one has to apply the averaging procedure. This goal can be achieved by applying the spectral windows convolving the "raw" spectrum and a special window function (Blackman and Tukey 1958; Marple 1987). The windowing procedure averages the closest spectral densities and thus exploits the principle of localization. Since the adjacent spectral densities are independent, averaging over the frequencies becomes an equivalent to averaging over ensemble of realizations (Sveshnikov 1968), however, the frequency resolution is reduced as a result. The idea might be clearly illustrated by using the spectrum of Fig. 1.4. Obviously, the averaging over the frequencies will reduce the random spectral deviations at adjoining frequencies, and a habitual SR pattern will appear.

In the standard data processing, the power spectra are averaged, and each spectrum is computed for a fragment of record of the standard duration. The majority of modern spectrum analyzers works in this specific way. The averaging of individual spectra reduces random fluctuations arising from the inter-pulse interference: these patterns do not repeat from one sample to another. The SR peaks are continuously replicated in all the segments, and therefore, the smooth resonance lines emerge.

References

Alpert Ya L (1990) Space plasma, vols 1 and 2, Cambridge University Press, Cambridge-New York-Melbourne-Sydney

Balser M, Wagner CA (1960) Observation of Earth-ionosphere cavity resonances. Nature 188:638–641

Balser M, Wagner CA (1962a) Diurnal power variations of the Earth-ionosphere cavity modes and their relationship to worldwide thunderstorm activity. J Geophys Res 67:619–625

Balser M, Wagner CA (1962b) On frequency variations of the Earth-ionosphere cavity modes, J Geophys Res, 67(4):081–4083

Belyaev PP, Polyakov SV, Rapoport VO, Trakhtengertz VY (1987) Discovery of the resonance spectrum structure of atmospheric electromagnetic noise background in the range of short-period geomagnetic pulsations. Dokl Akad Nauk SSSR 297:840–846

Belyaev PP, Polyakov SV, Rapoport VO, Trakhtengertz VY (1989) Experimental investigations of resonance structure in the spectrum of atmospheric electromagnetic noise in the band of short-period geomagnetic pulsations. Izvestiya VUZov, Radiofizika 32(6):663–672 (in Russian)

Belyaev PP, Polyakov SV, Rapoport VO, Trakhtengertz VYu (1990) The ionospheric Alfvén resonator. J Atmos Terr Phys 52:781–788

Belyaev PP, Bösinger T, Isaev SV, Kangas J (1999a) First evidence at high latitudes for the ionospheric Alfvén resonator. J Geophys Res 104(4):305–317

Belyaev GG, Schekotov AYu, Shvets AV, Nickolaenko AP (1999b) Schumann resonances observed using Poynting vector spectra. J Atmos Solar-Terr Phys 61:751–763

Besser BP (2007) Synopsis of the historical development of Schumann resonances, Radio Sci, 42, RS2S02, doi:10.1029/2006RS003495

Blackman RB, Tukey JW (1958) The measurement of power spectra. Dover Publications, New York

Bliokh PV, Nickolaenko AP (1986) Global electromagnetic resonances, Priroda (Nature), 4: 3–15, (in Russian)

Boccippio DJ, Williams ER, Heckman SJ, Lyons WA, Baker IT, Boldi R (1995) Sprites, ELF transients and positive ground strokes. Science, 269: 1088–1091

Chern JL, Hsu RR, Su HT, Mende SB, Fukunishi H, Takahashi Y, Lee LC (2003) Global survey of upper atmospheric transient luminous events on the ROCSAT-2 satellite. J Atmos Solar-Terr Phys 65:647–659. doi:10.1016/S1364-6826(02)00317-6

Egeland A, Larsen TR (1968) Fine structure of the Earth-ionosphere cavity resonances. J Geophys Res 73:4986–4989

IEEE Trans, Com-22, No.4 (1974) Special issue of the 'Sanguine' project

Füllekrug M, Fraser-Smith AC, Reising SC (1998) Ultra-slow tails of sprite associated lightning flashes, Geophys Res Lett, 25(3): 495–3498

Füllekrug M, Reising SC (1998) Excitations of Earth-ionosphere cavity resonances by sprite-associated lightning flashes, Geophys Res Lett, 25(4): 145–4,148

Füllekrug M, Mareev EA, Rycroft MJ (eds) (2006) Sprites, Elves and Intense Lightning Discharges. NATO Science Series, Springer, Dordrecht

Greenberg E, Price (2007) Diurnal variations of ELF transients and background noise in the Schumann resonance band, Radio Sci, 42, RS2S08, doi:10.1029/2006RS003477

Hayakawa M, Fujinawa Y (eds) (1994) Electromagnetic phenomena related to earthquake prediction, TERRAPUB, Tokyo, p 677

Hayakawa M (ed) (1999) Atmospheric and ionospheric electromagnetic phenomena associated with earthquakes, TERRAPUB, Tokyo, p 996

Hayakawa M, Molchanov OA (eds) (2002) Seismo electromagnetics; lithosphere—atmosphere—ionosphere coupling, TERRAPUB, Tokyo, 477

Hayakawa M, Nickolaenko AP, Shvets AV, Hobara Y (2011) Recent studies of Schumann resonance and ELF transients, In: Wood MD (ed) Lightning: properties formation and types. Nova Science Publications, New York, 39–71

Hayakawa M, Hobara Y, Suzuki T (2012) Lightning effects in the mesosphere and ionosphere. In: Cooray V (ed) Lightning electromagnetics. Institute of Engineering and Technology, UK, pp 611–646

Hobara Y, Iwasaki N, Hayashida T, Hayakawa M, Ohta K, Fukunishi H (2001) Interrelation between ELF transients and ionospheric disturbances in association with sprites and elves. Geophys Res Lett 28:935–938

Hobara Y et al (2006), Location and electrical properties of sprite-producing lightning from a single ELF site, in "Sprites, elves and intense lightning discharges", NATO Sci Ser, Ser. II. V.225. In:Füllekrug M, Mareev EA, Rycroft MJ (eds), p 211–235, Springer, Dordrecht, Netherlands

Huang E, Williams E, Boldi R, Heckman S, Lyons W, Taylor M, Nelson T, Wong C (1999) Criteria for sprites and elves based on Schumann resonance observations, J Geophys Res, 104:16943–16964

Jones D Ll, Kemp DT (1971) The nature and average magnitude of the sources of transient excitation of Schumann resonances. J Atmos Terr Phys 33:557–566

Lazebny BV, Nickolaenko AP, Rafalsky VA, Shvets AV (1988) Detection of transverse resonances of the Earth-ionosphere cavity using average spectra of VLF atmospherics. Geomagnetism and Aeronomia 28:329–330 (in Russian)

Marple SL Jr (1987) Digital spectral analyses with applications, Prentice-Hall, Englewood Cliffs, New Jersey

Marshall IH, Hale LC, Croskey CL, Lyons WA (1998) Electromagnetics of sprite and elve-associated sferics. J Atmos Solar-Terr Phys 60:771–786

Molchanov OA, Hayakawa M (2008) Seismo electromagnetics and related phenomena: history and latest results. TERRAPUB, Tokyo 189

Marshall RA, Inan US, Neubert T, Hughes A, Sátori G, Bor J, Collier A, Allin TH (2005) Optical observations geomagnetically conjugate to sprite-producing lightning discharges, Ann Geophys, 23 (6): 2231–2237

Neska M, Sátori G (2006) Schumann resonance observation at Polish polar station at Spitzbergen and geophysical observatory in Belsk. Przeglad Geofizyczny 51(3–4):189–198 (in Polish)
Neubert T, Allin TH, Blanc E, Farges T, Haldoupis C, Mika A, Soula S, Knutsson L, van der Velde O, Marshall RA, Inan U, Sátori G, Bor J, Hughes A, Collier A, Laursen S, Rasmussen IL (2005) Coordinated observations of transient luminous events during the Euro Sprite 2003 campaign. J Atmos Solar-Terr Phys 67:807–820. doi:10.1016/j.jastp.2005.02.004
Nickolaenko AP, Hayakawa M (2002) Resonance in the Earth-ionosphere Cavity. Kluwer Academic Publishers, Dordrecht, Boston, London, p 380
Ohta K, Umeda K, Watanabe N, Hayakawa M (2002) Relationship between ELF magnetic fields and Taiwan earthquakes. In: Hayakawa M, Molchanov OA (eds) Seismo electromagnetics: lithosphere–atmosphere–ionosphere coupling. TERRAPUB, Tokyo, pp 233–237
Pasko VP, Inan US, Bell TF, Reising SC (1998) Mechanism of ELF radiation from sprites. Geophys Res Lett 25:3493–3496
Polyakov SV (1976) On the properties of the ionospheric Alfvén resonator, KAPG Symposium on Solar-Terrestrial Physics, 3, Nauka, Moscow, pp 72–73
Polyakov SV, Rapoport VO (1981) Ionospheric Alfvén Resonator. Geomag Aeronomy 21:816–822
Price C, Asfur M, Lyons W, Nelson T (2002) An improved ELF/VLF method for globally geolocating sprite-produced lightning, Geophys Res Lett, 29 (3), doi: 10.1029/2001GL013519
Price C, Greenberg E, Yair Y, Sátori G, Bor J, Fukunishi H, Sato M, Israelevich P, Moalem M, Devir A, Levin Z, Joseph JH, Mayo I, Ziv B, Sternlieb A (2004) Ground-based detection of TLE-producing intense lightning during the MEIDEX mission on board the space shuttle Columbia. Geophys Res Lett 31(20):L20107. doi:10.1029/2004GL020711
Price C, Pechony O, Greenberg E (2007) Schumann resonance Schumann resonance in lightning research. J Light Res 1:1–15
Rodger CJ (1999) Red sprites, upward lightning, and VLF perturbations. Rev Geophys 37:317–336
Radio science (2007), 42(2), (special issue on Schumann resonance)
Schumann WO (1952a) Über die strahlungslosen Eigenschwingungen einer leitenden Kugel, die von einer Luftschicht und einer Ionosphärenhülle umgeben ist, Zeitschrift und Naturfirschung, 7a: 149–154
Schumann WO (1952b) Über die Dämpfung der elektromagnetischen Eigenschwingnugen des Systems Erde—Luft—Ionosphäre, Zeitschrift und Naturfirschung, 7a: 250–252
Schumann WO (1952c) Über die Ausbreitung sehr Langer elektriseher Wellen um die Signale des Blitzes, Nuovo Cimento, 9:1116–1138
Schumann WO, König H (1954) Über die Beobactung von Atmospherics bei geringsten Frequenzen. Naturwiss 41:183–184
Schumann WO (1957) Elektrische Eigenschwingnugen des Systems Erde—Luft—Ionosphäre. Z Angew Physik 9:373–378
Sentman DD, Westcott EM (1993) Observations of upper atmosphere optical flashes recorded from an aircraft, Geophys Res Lett, 20, 2857–2860
Sentman DD (1995) Schumann resonances. In: Volland H (ed) Handbook of atmospheric electrodynamics, 1: 267–298, CRC Press, Boca Raton, London, Tokyo
Sukhorukov AI, Stubbe P (1997) On ELF pulses from remote lightning triggering sprites, Geophys Res Lett, 24(13): 1639–1642
Sveshnikov AA (1968) Applied methods of the theory of random functions, Moscow, Nauka Publ, 463, §42
Tesla N (1905) The transmission of electrical energy without wires as a means of furthering world peace. Electrical World and Engineer, January 7: 21–24
Williams ER (1992) The Schumann resonance: a global tropical thermometer, Science, 256, 1184–1188
Williams ER (1994) Global circuit response to seasonal variations in global surface air temperature. Mon Weather Rev, 122:1917–1929

Williams ER (2001) Sprites, elves, and glow discharge tubes. Phys Today 41:41–47
Yair Y, Price C, Levin Z, Joseph J, Israelevitch P, Devir A, Moalem M, Ziv B, Asfur M (2003) Sprite observations from the space shuttle during the Mediterranean Israeli dust experiment (MEIDEX). J Atmos Solar-Terr Phys 65:635–642. doi:10.1016/S1364-6826(02)00332-2
Yair Y, Price C, Ganot M, Greenberg E, Yaniv R, Ziv B, Sherez Y, Sátori G (2009) Optical observations of transient luminous events associated with winter thunderstorms near the coast of Israel. Atmos Res 91(2–4):529–537

Chapter 2
Choosing a Site and Positioning of Equipments

General demands are described that should be satisfied by a field-site. Typical allocation is shown of the antennas, grounding, receivers, calibration and data acquisition system at a site. The block diagram is given of a standard SR receiver and its characteristics are briefly discussed. The classical design, the equivalent circuit, and gain are given of the vertical electric antenna. We discuss the impact of the metallic carrying construction on the antenna effective height, and the role of antenna mechanical vibration is demonstrated. Effective characteristics are addressed of magnetic induction coil antennas as well. The electric and magnetic antennal collocations are demonstrated at SR observatories. Practical parameters of antennas are listed and examples are given of a typical SR receiver including the notch filter. The material is illustrated by photo views of real field sites and equipment.

2.1 Demands on the Field Site and Equipments

Industrial activity of man and urban environment cause many kinds of ELF interference. First of all, we should mention any radiation from the power supply lines at the frequency of 50 (or 60) Hz, the induced signals from telephone lines, telex, and other communication lines, the leakage currents from the grounding of industrial objects, the control signals of electrified rail roads, etc. The man-made signals have high amplitude and are random in nature. These features impede the SR records and force the experimenters to position their observatories in the rural areas at considerable distance from cities and large settlements.

The following general requirements should be satisfied when choosing a field site:

- An observatory should be positioned as far as possible from the power supply lines. All non-working electric lines and devices should be removed from the site. Though this demand may disagree with the setting of computers and the data acquisition system (DAS).

A. Nickolaenko and M. Hayakawa, *Schumann Resonance for Tyros*,
Springer Geophysics, DOI: 10.1007/978-4-431-54358-9_2,

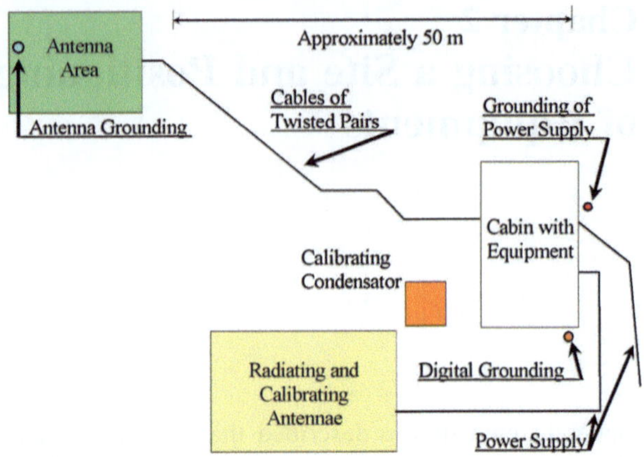

Fig. 2.1 Position of the equipment at a typical SR observatory

- The battery power supply is preferred, however, it remains expensive and causes problems in the long-term measurements, because the re-charging or replacement of batteries becomes necessary.
- When it is impossible to avoid the power supply lines, the equipment must be positioned at the 'dead end' of the line. Antennas must be placed at the largest possible distance along the straight line that continues the feeding wire, as seen in the diagram in Fig. 2.1.
- It is obligatory to use the battery supply in the antenna preamplifiers. The symmetric low impedance outputs should be preferred, and the signal to the main receiver or to the DAS should be fed via twisted shielded pair.
- Antenna grounding must be as close to antennas as possible. The main receiver is usually placed in a room or in a cabin, and its grounding is placed nearby this building. The separate 'digital' grounding is desirable. It is important that grounding wires and the currents in the soil should not form the closed loops, e.g., the cable shielding must be grounded at a single point only.
- The flat open places should be preferred with the well-conducting homogenous soil. Trees, bushes and other plants (if any) must be at a sufficient distance from the antenna. The grass should be mowed, so that antenna area becomes a 'football' field.

Antenna vibrations occupy the second place in the interference list after the man-made signals. Indeed, the electric and magnetic field sensors are submerged into outer static fields. The first one is the electrostatic fair weather field being an amplitude of $\sim$100 V/m. The second one is the geomagnetic field of about 0.5 Gauss. The minute mechanical vibrations of the antennas cause a huge voltage at the receiver input. For example, a mechanical change of the effective height of the vertical electric antenna of 0.1 mm in the outer electrostatic field of 100 V/m will cause the voltage of 10 mV. Such an interference exceeds the typical SR signal by an order of magnitude. To reduce the vibration interference, the mechanical oscillations should be damped, and the resonance frequencies of antenna construction should lie beyond the working frequencies. The latter is granted by antenna design.

The magnetic field sensors are buried in the ground for reducing vibrations. As a rule, the magnetic antenna is placed in a wooden or concrete box filled by the dry sand. Sometimes, antennas are covered by pokes with dry sand. The box is placed at the massive concrete base. The construction must have a heavy hermetic cover with the edge at the ground level, which reduces the wind impact.

Summarizing, the observatory should be positioned at a remote place, far away from the industrial activity and the roads (electric trains, cars, tractors), even from the pedestrians, whose steps inevitably cause the ground trembling. The above listed elements were taken into account in Fig. 2.1. Here, the cabin with the DAS is found in the lower right corner of diagram. The power supply is fed to this building. Two grounding points are arranged nearby: one for the electric equipment and the other for the digital devices. The power supply transformers should be removed as far from the observatory as possible and the feeder should be a symmetric shielded cable (see below). Antennas are positioned in the left upper corner of diagram, far away from the power supply line along its continuation by the straight line. The shelter for magnetic antennas is found there, the accumulators, and the mast for the electric antenna. Here is the point of antenna grounding. The symmetric outputs of antenna preamplifiers are connected via twisted pairs with the receiver. The calibrating system is also shown in the diagram of Fig. 2.1 (see Chap. 3).

A typical block diagram of SR receiver is shown in Fig. 2.2. The equipment records the vertical electric and two orthogonal horizontal magnetic field components. The analog part of receiver contains three channels. Each field sensor is connected to the preamplifier, which has an individual power supply. We show in the diagram the scaling amplifiers and the notch filters. These latter are necessary for reducing the industrial interference at 50 Hz (60 Hz) and its harmonics. This is done when the natural waveforms are used. For example, when the components of Poynting vector are measured in the time domain. The notch filters are not present in many modern receivers. Instead, the high-resolution 24-bit analog–digital converters (ADCs) are applied, and the signal is recorded as it is in the time domain. Industrial interference is removed later during the digital signal processing. The application of 24-bit ADC guarantees the appropriate accuracy of the resonance record. Indeed, the observatories are usually positioned at places with low levels of the man-made interference. Let the industrial interference exceed the SR signal by 60 dB (by a factor of 1,000 or about 1 V/m in the amplitude). Then, one binary bit (of 24) recorded by the PC is reserved for the signal sign, and two more bits are left to avoid overloading when the signal amplitude abruptly increases. Ten binary digits are reserved for the industrial interference. As a result, we have eleven bits left for the SR signal. In other words the natural radio signal is sub-divided into $\sim$2,000 levels. Such an accuracy exceeds that of the SR records by the analog devices. Filtering of the industrial noise becomes unnecessary, provided that the analog time domain records are not directly used at the observatory.

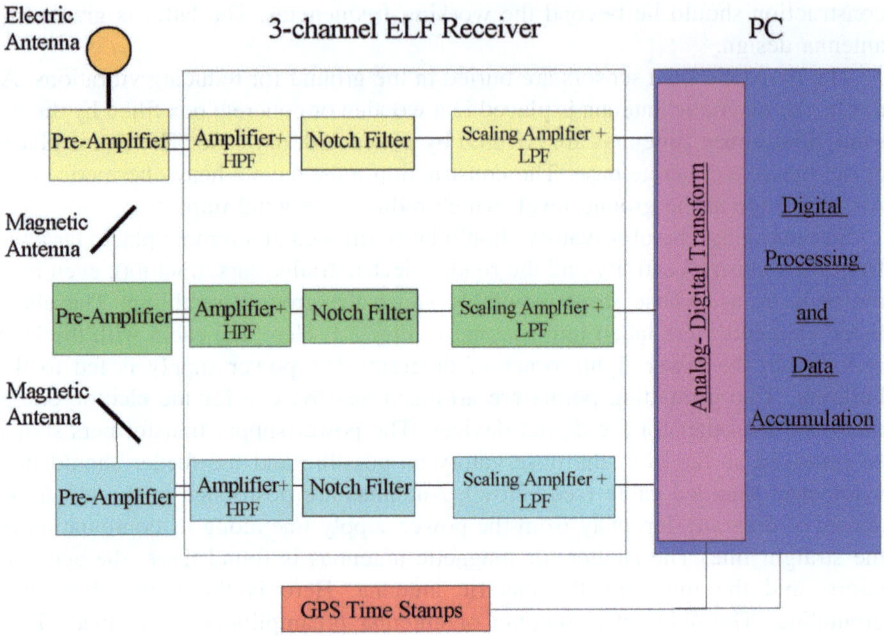

Fig. 2.2 Block diagram of a SR receiver

Characteristics of all receiver channels must be identical. We list typical parameters of receivers and briefly describe them.

1. The 'through gains' of magnetic and electric channels must be identical. Mutual deviations in the frequency dependence of gains should not exceed a few percents in amplitude and a few degrees in the phase. For this purpose, a phase rotator is used in the input circuit of electric channel (Belyaev et al. 1999), otherwise the correct estimates of the Poynting vector become problematic in the time domain.
2. The equipments including the field sensors must be regularly calibrated.
3. The 24-bit ADC should be preferred, as these allow for avoiding the analog notch filtering and thus increase the accuracy of the records. In this case, the natural signal passes through a high-pass filter, a scaling amplifier and the low-pass filter that forms the bandwidth which depends on the sampling frequency.
4. Data collected by different observatories must be compatible. Unfortunately, this property remains just a dream.
5. The software of observatories should be unified, so that digital records are compatible without additional efforts (one more dream).
6. Modern computers successfully perform both the data acquisition and the preliminary processing. The GPS time stamps should be used in the record. This makes the collected data synchronous and coherent.

7. The sound cards of PC might be used in the DAS. However, the spectral roll-off is usually present at the frequencies below 15 Hz due to the frequency response of the sound card. This feature must be compensated in the signal processing.

2.2 Design of Vertical Electric Antenna

The ground might be considered as a perfect conductor at low frequencies, so that only the vertical electric and the horizontal magnetic fields exist at the ground surface. A vertical electric ELF antenna is an electrode of a given self-capacitance elevated by a carrying construction. The most popular electrode is the ball antenna, which is the spherical active electrode (see Fig. 2.3a). A particular design was introduced by Ogawa et al. (1966a, b), and we reproduce its major details in Fig. 2.3a. The antenna spherical active electrode is separated from the iron carrying pipe by the Teflon insulator. The sphere is made of two parts—the covering hemisphere and a 'skirt' defending the insulator from dirt, rain, snow, etc. The antenna preamplifier (not shown) is positioned inside the sphere, and it is connected to the inner surface of the active electrode. All connecting wires are placed inside the iron pipe. At present, the ball, disk, rod, wire, or bodies of other forms are widely used in measurements of electric field in the range from the ultra low (ULF) to very low frequencies (VLF) (see e.g. Sátori et al. 1996; Williams et al. 1996; Belyaev et al. 1999; Price et al. 1999; Füllekrug and Constable 2000; Hobara et al. 2000a, b; Nakamura et al. 2000).

The antenna equivalent circuit was also introduced by Ogawa et al. (1966b) together with its gain in the frequency domain (see Fig. 2.3b and c). The equivalent circuit of all vertical electric antennas does not depend on the particular form of active electrode: It might be the ball antenna, the metallic tower (Balser and Wagner 1960), the horizontal wire extended over the meadow (Rycroft 1965), a pair of parallel plates or the capacitor antenna (Jones and Kemp 1970), a cylindrical barrel (Bliokh et al. 1971, 1977a), the disk antenna (Füllekrug 2010), etc. The only distinction is in the way of how the self-capacitance C_0 is reached of the antenna active electrode.

The input voltage of the antenna preamplifier is found from the following equation (see the equivalent circuit in Fig. 2.3b):

$$U_{IN} = E \cdot h_{eff} \cdot G_{EA} \tag{2.1}$$

where

$$G_{EA} = \frac{i\omega\, RC_0}{1 + i\omega\, R(C_0 + C_1)} \tag{2.2}$$

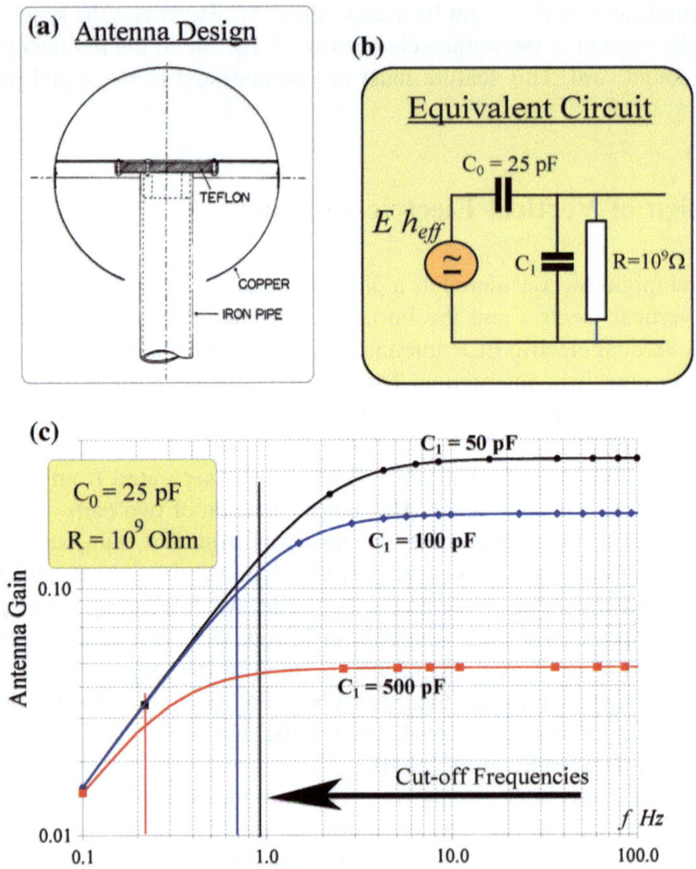

Fig. 2.3 Equivalent circuit and frequency response of the vertical electric antenna

Here E is the outer vertical electric field in V/m; h_{eff} is the antenna effective height in meters; C_0 is the self-capacitance of active electrode; C_1 and R are the input capacitance and the input resistance of antenna preamplifier correspondingly.

The lower cut-off frequency of an electric antenna depends on the product $R\,(C_0 + C_1)$, and the high frequency gain is a constant equal to the ratio $C_0/(C_0 + C_1)$. It is clear that the input resistance of preamplifier must be very high for realistic C_0 values. Three different frequency characteristics are shown in Fig. 2.3c, each corresponding to the same self-capacitance $C_0 = 25$ pF (the spherical active electrode of 55 cm diameter), the constant input resistance $R = 10^9$ Ohm, and a set of the input capacitances of $C_1 = 50$, 100, and 500 pF. The signal frequency in Hz is shown along the abscissa on logarithmic scale, and the logarithm of antenna gain is shown along the ordinate. Vertical lines denote the cut-off frequencies for different values of C_1, being correspondingly 2.1, 1.3, and

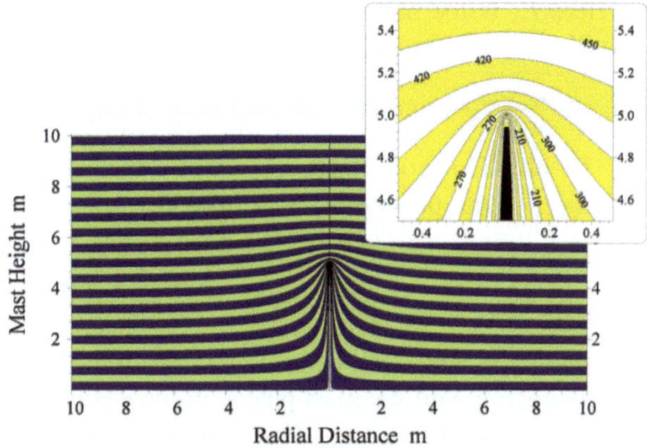

Fig. 2.4 Equipotential lines around the vertical conducting mast 5 m high

0.3 Hz. It is clear that the cut-off frequency reduces with the increase in input capacitance. However, the overall antenna gain simultaneously decreases.

The antenna input voltage is the product of the effective height h_{eff} and the amplitude of the incident radio wave E (about 1 mV/m). It is clear that the effective height depends on the elevation of the active electrode, especially when the carrying constriction is made of dielectric with permittivity of about 1. This is why many experimentalists prefer the dielectric carrying construction (made out of high voltage insulators). However, the initial design and the consequent practice of measurements showed that non-conducting masts and dielectric pipes are not very convenient in the exploitation. They become dirty rather quickly being covered with dust, remnants of the raindrops, web, snow, icicles, etc. The surface resistance of dielectric is reduced, and a regular cleaning is necessary for maintaining the constant antenna gain and its effective height. Application of the metallic masts removes such a problem. Fears of conducting construction are based on the fact that a metal body 'elevates' the lines of low potential above the ground toward the antenna active electrode (see Fig. 2.4). This is true, but we show that the effect has a minor impact on the antenna effective height.

The calibration of vertical electric antenna was always a challenge (see, e.g. Watt 1967; Clayton et al. 1973). We evaluate the influence of metallic mast with the help of the following model (Nickolaenko and Hayakawa 2002). Let the mast be a half of a prolate spheroid with the small horizontal half-axis m_b and a large vertical half-axis m_a. The ground potential is equal to zero. The antenna size is small in comparison with the wavelength, therefore we can apply the classical solution of the electrostatic problem: the electric potential around the conducting spheroid (Stratton 1941; Landau and Lifshits 1957):

$$U = -E_S z \left\{ 1 - \frac{\ln\sqrt{\frac{1+s}{1-s}} - s}{\ln\sqrt{\frac{1+e_t}{1-e_t}} - et} \right\} \tag{2.3}$$

Here E_S is the outer static electric field; x and y are the horizontal Cartesian coordinates of the current point $M(x, y, z)$, z is the height of this point; $s = \sqrt{\frac{(m_a)^2 - (m_b)^2}{(m_a)^2 + \xi}}$, and the spheroid eccentricity is equal to $e_t = \sqrt{1 - \left(\frac{m_b}{m_a}\right)^2}$.

The elliptical coordinate ξ is found by using the radius $\rho = \sqrt{x^2 + y^2}$ and altitude z of the point $M(x, y, z)$ from the following equation:

$$\frac{\rho^2}{m_b^2 + \xi} + \frac{z^2}{m_a^2 + \xi} = 1 \tag{2.4}$$

Figure 2.4 shows the central cross-section of the potential distribution around the conducting vertical mast placed in the electrostatic field. The radial distance is shown on the abscissa in meters, and the ordinate shows the altitude over the ground also in meters. The mast has the height $m_a = 5$ m, and its diameter at the ground is 10 cm ($m_b = 5$ cm). One may observe the elevation of the zero equipotential line together with all other lines of equal potential. The disturbed zone is extended horizontally by approximately the mast height, and the field becomes practically regular at altitudes of the doubled mast height.

The inset in Fig. 2.4 shows the vicinity of the mast top in detail. Equipotential lines are shown here with a step of 30 V/m. The electrostatic field is equal to 100 V/m (the regular fair weather field). The markers of 210, 270, 300, 420, and 450 V/m denote the position of the relevant equipotential lines. One may note that the equipotential line of about 3.5 m passes at the height of about 10 cm above the mast top. Hence, the antenna active electrode will have the effective height of 3.5 m when placed at this point, while its geometric elevation will be 5.1 m. Obviously, an ascent of equipotential lines is partially compensated by their 'compression' around the mast top. Therefore, the antenna effective height increases with the mast height.

Let us consider this variation in detail. We choose the point right above the mast top $M(0, 0, m_a + T)$, as seen in Fig. 2.5. The dependence shown in this figure characterizes the field concentration and corresponds to the cross-section of Fig. 2.4 at $\rho = 0$.

Figure 2.5 illustrates a rapid increase of potential with altitude above the tip of metallic mast. To show it in more detail, we used the inset in this figure. The elevations T range from 1 to 14 cm there. The outer electrostatic field was assumed to be 100 V/m. Thus, the elevation $T = 2$ cm corresponds to the effective height $h_{eff} = 2.5$ m, and elevation $T = 10$ cm is an equivalent of the 3.5 m height.

An active electrode placed at the 10 cm elevation above the tip of the 5 m mast has the effective height of 3.5 m, provided that the form of this electrode repeats the equipotential surface. The effective height of such an electrode might be computed by using the following relation:

$$h_{eff} = \frac{U}{E} = (m_a + T)\left[1 - \frac{\frac{1}{2}\ln\frac{2m_a + T}{T} - \frac{m_a}{m_a + T}}{\ln\frac{2m_a}{m_b} - 1}\right] \tag{2.5}$$

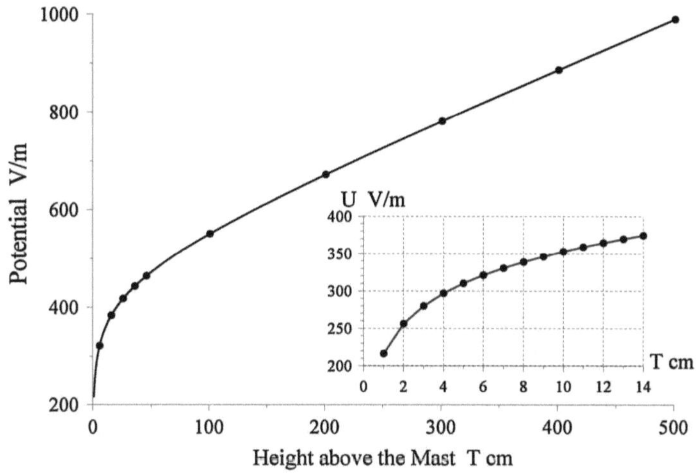

Fig. 2.5 Potential of an antenna active electrode as a function of its height T above the top of vertical conducting mast

or with the help of approximate formula:

$$h_{eff} \cong (m_a + T)\left[1 - \frac{\frac{1}{2}\ln\frac{2m_a}{T} - 1}{\ln\frac{2m_a}{m_b} - 1}\right] \tag{2.6}$$

Equations (2.5) and (2.6) show that the antenna effective height grows practically linearly with the elevation T for the fixed mast height $m_a = const$, provided that $m_a \gg T$. The growth is somewhat decelerated owing to the 'depolarization' factor $\ln(2m_a/m_b) - 1$, as seen in Fig. 2.5.

The effective height versus the mast height m_a is plotted in Fig. 2.6 for a few of fixed elevations T. We assumed, for simplicity, that the active electrode is positioned right above the mast top. The form of this electrode is coincident with the surface of equal potentials. Thus we can describe the effective height by potential of the point $M(0, 0, m_a + T)$.

The abscissa indicates the mast height in meters, and the ordinate depicts the antenna effective height for three particular size of insulator $T = 10$, 20, and 40 cm. The major plots and the upper inset demonstrate in what way the 'compression' of potential lines around the mast tip compensates the 'elevation of the ground' by the conducting mast. The antenna effective height is practically a linear function of the mast length. The upper inset demonstrates in detail the impact of the mast shorter than 5 m. The lower inset shows the elevation effect for the 'capacitor antenna' installed on the metallic mast. The term 'capacitor antenna' was introduced by Jones and Kemp (1970) who applied the flat cylindrical capacitor mounted on the tripod. They used such an antenna for the absolute

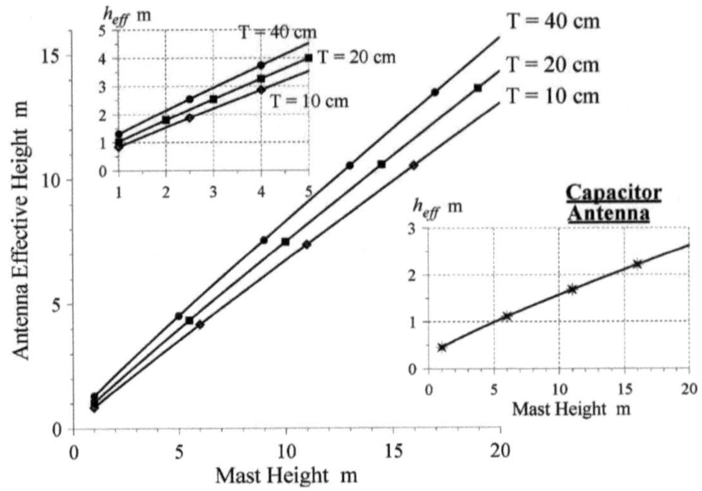

Fig. 2.6 Effective height of vertical electric antenna versus the mast height for different thickness of insulator *T*

measurements of the time varying vertical electric field by assuming that the effective height remained equal to the capacitor gap regardless the tripod height.

Exact computations of the field around such a construction are difficult. When evaluating the influence of the carrying mast on the capacitor antenna, we assume that the electrodes are coincident with a pair of equipotentials above the mast. The lower plate corresponds to the equipotential of $T = 10$ cm, and the upper one is the surface of $T = 40$ cm (see the inset in Fig. 2.4). Figure 2.4 demonstrates that the capacitor plates will be practically flat if their horizontal extension is not very large, say, their diameter does not exceed 60 cm. The effective height of the solitary capacitor antenna is 30 cm, and we will see whether it remains constant in the presence of the mast. The lower inset in Fig. 2.6 indicates that the effective height exceeds this value. The mast 1 m high increases the effective height of capacitor antenna to 46 cm. It becomes equal to 53 cm when the mast is 1.5 m high. The effective height is doubled when the elevation reaches 2 m. The mast of 10 m increases the effective height of the capacitor antenna up to 1.5 m.

The disk and capacitor antennas applied in measurements have flat electrodes, and these modify the potential distribution in the space. However, the compression of equipotential lines persists, and the effective height will noticeably increase, but to a smaller extent than Fig. 2.6 predicts.

It follows from our analysis that there are two competing mechanisms associated with the metallic carrying construction. On one hand, the conducting mast elevates the zero potential. On the other hand, the equipotentials are concentrated above the mast tip. Our modeling and measurement show that the second effect overcomes the first one: an increase of the mast length causes the overall growth of antenna effective height.

Equation (2.5) additionally allows for evaluating the interference amplitude caused by the mechanical vibrations of electric antenna. When the carrying construction is trembling, the position of active electrode varies cyclically against the potential lines. The electric voltage appears at the frequency of mechanical oscillations as a result, and its amplitude might be readily evaluated in the following way. If the gap between the mast tip and the active electrode is a function of time $T(t) = T + \delta T(t)$ where $\delta T(t)/T \ll 1$, the effective height will vary in respect to a median value, and the voltage will appear directly proportional to the $\delta T(t)$ variations.

$$\delta U(t) = \frac{\delta\, T(t)}{2\, T} \frac{m_a\, E_0}{\ln \frac{2m_a}{m_b} - 1} \tag{2.7}$$

By accepting the vibration amplitude of $\delta T(t) = 0.01\,\mathrm{mm}$, $m_a = 5$ m, $m_b = 5$ cm, $T = 10$ cm, and $E_0 = 100$ V/m, we obtain that the induced voltage is equal to $\delta U(t) = 6.9$ mV. This amplitude is in accord with observations, and the estimate itself demonstrates the role of antenna vibrations. We show in Fig. 2.7 the amplitude spectrum of radio noise recorded on board the research vessel "Professor Zubov" in motion (Nickolaenko 1995). Here, the frequency in Hz is plotted on the abscissa and the spectrum of the signal at the antenna output is shown on the ordinate in mV·m^{-1}·Hz$^{-1/2}$. The position of antenna at the navigation deck of ship is demonstrated by the photo in the inset. One may see that the active electrode of ELF antennas was a 'wide rod with a conical skirt'.

The wide band noise appeared from the deck vibrations around the antenna mounting, and it reduced in the drift. This kind of interference played the dominant role in the sea-borne measurements: it noticeably exceeded the typical level of the SR signal (0.3–1 mV·m^{-1}·Hz$^{-1/2}$) and the level of the ship power supply interference. We must admit that interference of the 50 Hz frequency was rather small on board the "Professor Zubov" research vessel. Its amplitude did not surpass a few mV·m^{-1}·Hz$^{-1/2}$, which is at least two orders of magnitude lower than the customary level at the ground-based observatories. Such a small interference was conditioned by a symmetric power supply: the electric energy was fed by the symmetric lines carrying the couple of 110 V anti-phase voltages instead of the single phase of 220 V. Besides, the power supply cables were hidden inside the steel pipes, which served as an additional shielding.

Electric antennas must be designed to move the mechanical frequencies of the construction outside the frequency band of future measurements. This means at ELF that a carrying construction should not be very rigid and the antenna active electrode must be massive enough.

On the other hand, antenna vibrations might be used in measurements of the static electric field. The fair weather field is detected as a rule with the help of so called "field-mills" (Ogawa 1985; MacGorman and Rust 1998). This gadget exploits a rotating capacitor of varying capacitance, the "butterfly" capacitor. Its

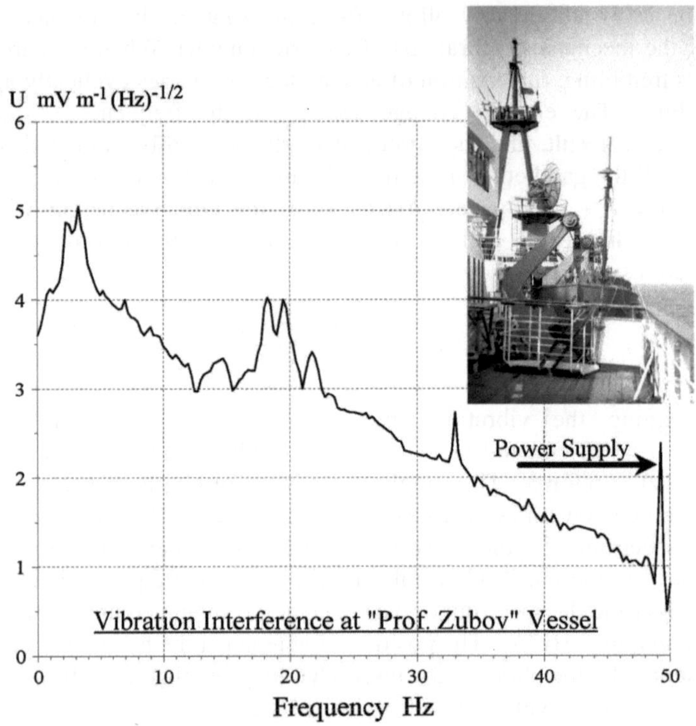

Fig. 2.7 Amplitude spectrum of vibration interference recorded by the vertical electric antenna on board the "Professor Zubov" research vessel in motion with the cruise speed. Inset shows positioning of the antenna at the navigation deck

moving plates are periodically exposed to the outer electrostatic field, and its output voltage varies in time having the amplitude proportional to the measured field. The trembling antennas might be an alternative field sensor of the fair weather field (Nickolaenko 1990).

The effective height of a vertical electric antenna increases with its elevation regardless of the conductivity of the carrying construction. The mast might be made of an insulator or a metal. If we want to increase the signal amplitude at the receiver input, we must raise the active electrode, and the effective height will increase almost linearly with the elevation. The particular value of the effective height must be measured experimentally since only the highly idealized models might be used in computations. The procedure of antenna calibration at the site is based on the application of radio signals of the known amplitude, for example, the signals of VLF radio transmitters or the signals generated at the site (see Chap. 3).

2.3 Design of Magnetic Antenna

Prior to discussing the magnetic antennas, we show a few working ELF antennas in Fig. 2.8. An electric antenna is shown in Fig. 2.8a. Its active electrode is a combination of a thick rod and the 'skirt' covering the Teflon insulator at the box with the preamplifier. Junction with the metallic mast is also visible. Such a construction proved its efficiency in the records in the sea. Figure 2.8b shows the vertical electric antenna of Tel Aviv University positioned at the Mitspe Ramon observatory in the Negev Desert—Israel (courtesy of Dr. Colin Price). The active electrode is the aluminum sphere of $\sim$60 cm diameter. The construction is mounted at a couple of high voltage ceramic insulators. The antenna preamplifier is placed at the antenna basement, and the connecting wires are going inside the insulator. Figure 2.8c shows two orthogonal magnetic antennas of Tel Aviv University. These antennas were taken out and placed on the boxes where they are usually buried. Each antenna is $\sim$2 m long. Figure 2.8d shows the compact ELF magnetic antenna in the hands of its designer G.G. Belyaev. This antenna became rather small due to application of the amorphous permalloy in its core. Such antennas were used in measurements at the Lehta observatory, Karelia, and Karimshino, Kamchatke, Russia (Belyaev et al. 1999).

At ELF, the horizontal magnetic fields are usually detected by the induction coils. Such an antenna is a solenoid of many turns winded on the ferromagnetic core with a high magnetic permeability. Antennas have the sensitivity angular pattern in the 8–form (lemniscata). Therefore, one has to use a couple of orthogonal antennas to detect the signals arriving from arbitrary directions. The ferromagnetic core antennas are used in the SR band, however, the air core coils were also applied by Polk (1969, 1982). We describe only general properties of magnetic antennas, and a detailed description might be found in the monograph by Burrows (1978).

The voltage induced in the magnetic antenna coil is the product of the time derivative of the outer magnetic induction B and the effective area of antenna winding S_{eff}. The design of magnetic antennas must provide the greatest possible effective area, and this is a real challenge. We show below the simple estimates for the effective area. The area S_{eff} depends on the geometry of its core (the length and the diameter), the magnetic material of the core, number of turns, and the type of winding. The latter controls the average cross-section of a turn and the mutual capacitance of turns. This capacitance combined with the antenna inductance governs the undesired resonances in the magnetic antenna sensitivity.

The number of turns N of a typical ELF magnetic antenna reaches 100,000. The winding is split into $\sim$10 sections set at the cover of the ferromagnetic core. The core is approximately 1 m long and is a few centimeters in diameter. It is made out of ferrite or the isolated permalloy leafs firmly placed in the plastic wrapping. Such a construction has the nickname of "the log" (see Fig. 2.8).

Effective parameters of magnetic antennas did not change essentially during the last 30 years in spite of great progress in the magnetic materials. In particular, the

Fig. 2.8 ELF antennas: (**a**) vertical electric antenna for measurements in the sea, (**b**) and (**c**) vertical electric and horizontal magnetic antennas at their concrete boxes at the Mitspe Ramon observatory, Israel (courtesy of Dr. C. Price), (**d**) the small SR magnetic antenna in hands of its designer G.G. Belyaev

size of the inductive field sensors did not change crucially. One of the smallest antennas is shown in Fig. 2.8. It has the core of amorphous permalloy with the relative permeability of a few millions. The construction is 80 cm long and has the mass of 6 kg (Belyaev et al. 1999).

The effective area of the magnetic antenna is equal to (Burrows 1978):

$$S_{eff} = N\,\pi\,\mu_{eff}\,m_b^2 \tag{2.8}$$

Here N is the number or turns. It is assumed that the core has the form of the prolate spheroid with the axes $m_a \gg m_b$, so that its effective permeability is equal to μ_{ff}. The latter is smaller than the permeability of the core material μ:

$$\frac{1}{\mu_{eff}} = \frac{1}{\mu} + n^{(x)} \tag{2.9}$$

The depolarization coefficient $n^{(x)}$ depends on the geometry of the core:

$$n^{(x)} = \frac{1-e_t^2}{2e_t^3}\left(\ln\frac{1+e_t}{1-e_t} - 2e_t\right) \cong \left(\frac{m_b}{m_a}\right)^2\left(\ln\frac{2m_a}{m_b} - 1\right) \tag{2.10}$$

$$n^{(y)} = n^{(z)} = \frac{1}{2}\left(1 - n^{(x)}\right) \tag{2.11}$$

The spheroid eccentricity is equal to $e_t = \sqrt{1 - \left(\frac{m_b}{m_a}\right)^2}$.

It is known that a ferromagnetic cylinder "draws in" the outer magnetic field lines. The volume from which the field is "sucked in" increases when the core becomes longer (its orientation is parallel to the outer field). The depolarization coefficient characterizes the portion of the field lines that were left outside the core. One readily obtains:

$$\mu_{eff} = \frac{\mu}{1+\mu n^{(x)}} \cong \left(\frac{m_a}{m_b}\right)^2 \frac{1}{\ln\left(\frac{2m_a}{m_b}\right) - 1} \tag{2.12}$$

and

$$S_{eff} \cong N\,\pi\,(m_a)^2 \frac{1}{\ln\left(\frac{2m_a}{m_b}\right) - 1} \cong N\,\pi\,(m_a)^2\frac{1}{3} \tag{2.13}$$

These formulas are approximate, and they are valid when $n^{(x)} \gg 1/\mu$. For the fixed geometry, the growth in antenna effective area caused by an increase of the material permeability will slow down when the μ value exceeds 1,000–2,000. In this case the antenna sensitivity is limited by the core geometry. This limitation explains the relative invariance of the magnetic antenna design. The majority of antennas described in the literature have the core 1–2 m long and a diameter of a few centimeters. Indeed, one has to apply the extremely long and slim cores for accomplishing the increasing permeability of magnetic material. Simultaneously, the core should not be extra thin, otherwise, the saturation might arise in the magnetic material when the core is parallel to the geomagnetic field. Besides, the thin antenna meets the problems of soundness and vibrations. To avoid saturation, one has to use the core a few centimeters in diameter. Thus, the antenna core must be a few tens meters long to realize the high permeability of the material. Such construction becomes expensive and inconvenient to use.

It follows from Eq. (2.13) that the effective area of a magnetic antenna with ferromagnetic core is approximately equal to that of an air loop antenna having the diameter equal to the great semi-axis of the core m_a. This equation is useful for the preliminary estimates of a future antenna size, while the detailed exact computations might be made afterwards by using rigorous and more complicated equations from Burrows (1978).

It is worth mentioning the mutual position of magnetic antennas at an observatory. The core of magnetic antenna modifies the incident field in the volume of a radius equal to the doubled length of the core. Therefore, two antennas might modify the angular pattern and frequency characteristics of each other. The simplest way for avoiding interaction is separating the antennas by a distance exceeding their double lengths. In the case of a closer positioning, the cross orientation is helpful with antennas placed one above the other. One may recall that crossed air loops are used in the direction finders. In case that the crossed positioning is impossible, the antennas might form the letter T. The configuration in a form of letter Γ is slightly worse, as the antenna interaction starts to depend on the wave arrival angle, and the directivity (angular pattern) might deviate from the symmetric 8-shape. Besides, additional phase shifts might appear between the channels.

2.4 Antenna Preamplifiers

The major demand on antenna preamplifier is in matching the output resistance of antenna and the input resistance of ELF receiver. In the SR frequency band, the input impedance of vertical electric antenna preamplifier must be equal to or exceed 10^9 Ohm. This is achieved by applying the electronic tubes or the MOS (metal–oxide–semiconductor) transistors, or the microchips with such transistor elements. We depict in Fig. 2.9a variant of relevant circuit. This particular preamplifier worked in the SR monitoring at the Lehta observatory, Karelia, Russia (Belyaev et al. 1999). The first stage here is a high resistance operational (differential) amplifier, and the second stage is a scaling linear amplifier having the symmetric low resistance output. ELF radio signal is fed from the preamplifier to the ADC positioned in the laboratory building through the shielded twisted pair approximately 200 m long.

The vertical electric antenna at the Lehta observatory was fixed at the polyethylene tube of 0.4 m diameter and of ~2.5 m height. The preamplifier and its battery power supply were placed inside the tube at its bottom. Photo in Fig. 2.10 presents a general view of the antenna area in the swampy land of Karelia. The big box in the front left part of this photo is the shelter for two orthogonal magnetic antennas (one of them was shown in Fig. 2.8d), their preamplifiers and the battery

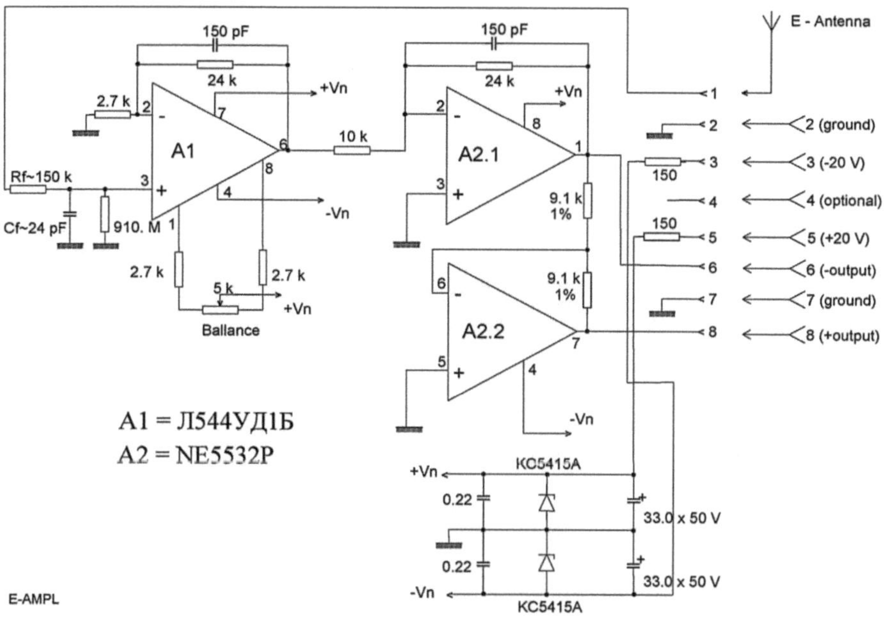

Fig. 2.9 Typical vertical electric antenna preamplifier (courtesy of Dr. A.Yu. Schekotov)

power supply. The vertical electric antenna is positioned about 10 m further, across the drainage groove. The cables connecting outputs of antenna preamplifiers with the DAS in laboratory were buried and therefore they are not seen in the photo.

The magnetic field sensor has as a rule the symmetric winding so that its output is also symmetric, otherwise its directivity pattern might deviate from the symmetric 8-shape outline. The input of magnetic antenna preamplifier must be symmetric, as we show in Fig. 2.11a. Similar preamplifiers were used in measurements at the Nakatsugawa observatory, Japan (Ohta et al. 2001, 2006; Hobara et al. 2011). Here, three orthogonal magnetic field components are recorded. Another simple low noise ELF receiver is shown in Fig. 2.11b. The description and explanations could be picked at the site: www.vlf.it/inductor/ulpreampl_schema.gif. In fact, the first integrated circuit in Fig. 2.11b is the preamplifier, and the subsequent stages are the notch 2-T bridge filters tuned to the basic (50 Hz) and third (150 Hz) harmonics of man-made interference. The circuit of low-pass filter with the 50 Hz cut-off is shown in the lower part of Fig. 2.11b.

Circuits shown in Figs. 2.9 and 2.11 should be treated as mere examples demonstrating the major principles of ELF design. Owing to the permanent and rather fast development of electronic devices, the novel chips appear with the better characteristics. These advanced elements should be used in the measurements. Though the principle should be followed: the antenna preamplifier must match the antenna with the shielded twisted pair cable connected to the receiver, a

Fig. 2.10 Antenna positions at the Lehta observatory (courtesy of Dr. A.Yu. Schekotov)

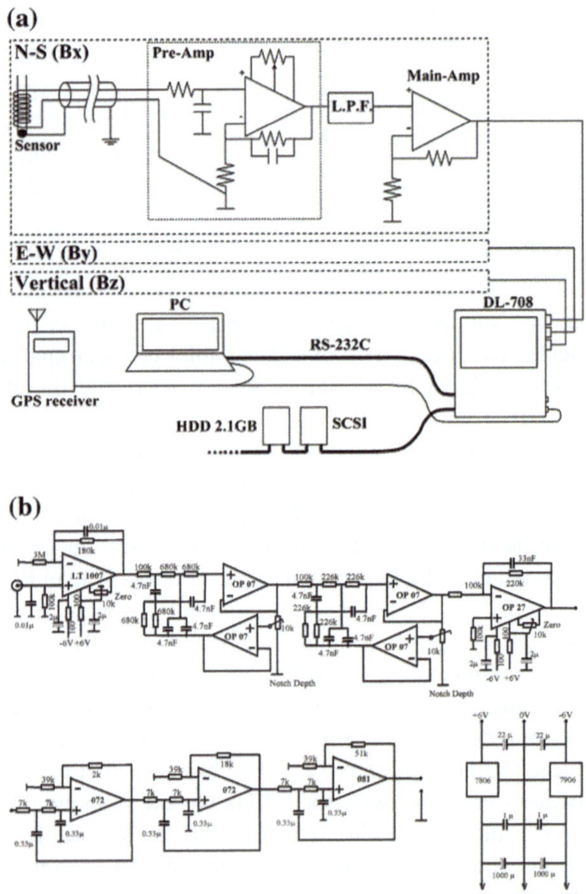

Fig. 2.11 Typical circuits of magnetic antenna preamplifier

scaling amplifier should be used that provides the signal amplitude appropriate for the ADC, and the low-pass filter must complete the receiver. The cut-off frequency depends on the sampling frequency: in accordance with the Nyquist theorem, the first should not exceed one-half of the second. Sometimes, the analog notch filters are added to the scaling amplifier, but this is rarely done at present. Such filters become redundant when the DAS uses the 16-bit or 24-bit ADC. The high resolution ADC allow for filtering the industrial interference in the computer, by using the codes, and the results obtained are of a higher quality.

We must note that the application of modern multi-channel sound cards proves to be rather practical sometimes. These boards must have four identical channels at least, a symmetric input combined with sufficiently high input impedance and sensitivity, and the sampling frequency must be high enough. In this case, a PC really becomes an element of receiving equipment.

References

Balser M, Wagner CA (1960) Observation of Earth-ionosphere cavity resonances. Nature 188:638–641

Belyaev GG, Schekotov AYu, Shvets AV, Nickolaenko AP (1999) Schumann resonances observed using Poynting vector spectra. J Atmos Solar-Terr Phys 61:751–763

Bliokh PV, Bormotov VN, Kontorovich VM, Nickolaenko AP, Sapogova NA, Shulga VF, Filippov YF (1971), On the degeneracy lift in the spherical Earth-ionosphere cavity. Preprint of Institute of Radio-Physics and Electronics of Ukrainian Acad. of Sci, No. 10, Kharkov (in Russian)

Bliokh PV, Nickolaenko AP, Filippov YF (1977a) Global electromagnetic resonances in the earth-ionosphere cavity. Naukova Dumka, Kiev, pp 199. (in Russian)

Burrows ML (1978) ELF communications antennas. Peter Peregrinus, London

Clayton MD, Polk C, Etzold H, Cooper WW (1973) Absolute calibration of antennas at extremely low frequencies. IEEE Trans Ant Prop AP-21:514–523

Füllekrug M, Constable S (2000) Global triangulation of lightning discharges. Geophys Res Lett 27:333–336

Füllekrug M (2010) Wideband digital low-frequency radio receiver. Meas Sci Technol 21:015901, 9. doi:10.1088/0957-0233/21/1/015901

Hobara Y, Iwasaki N, Hayashida T, Tsuchiya N, Williams ER, Sera M, Ikegami Y, Hayakawa M (2000a) New ELF observation site in Moshiri, Hokkaido Japan and the results of preliminary data analysis. J Atmos Electr 20(2):99–109

Hobara Y, Yamaguchi H, Akinaga Y, Watanabe T, Koons HC, Rieder JL, Hayakawa M (2000b) New ULF/ELF measurement in Seikoshi, Izu peninsula, Japan. Abstracts and Program of International Workshop on Seismo Electromagnetics. The University of Electro-Communications, Chofu-city, Tokyo, September 19–22, p 9

Hobara Y, Harada T, Ohta K, Sekiguchi M, Hayakawa M (2011) A study of global temperature and thunderstorm activity by using the data of Schumann resonance observed at Nakatsugawa, Japan. J Atmos Electr 32(2):11–119

Jones DL, Kemp DT (1970) Experimental and theoretical observations of Schumann resonances. J Atmos Terr Phys 32:1095–1108

Landau LD, Lifshits EM (1957) Electrodynamics of continuous media. Gostekhizdat, Moscow (in Russian)

McGorman DR, Rust WD (1998) The electrical nature of storms. Oxford University Press, Oxford

Nakamura T, Okugumo M, Sasaki H (2000) Simultaneous observation of ELF/VLF electromagnetic waves on the seashore and seabed. Abstracts and Program of International Workshop on Seismo Electromagnetics. The University of Electro-Communications, Chofu-city, Tokyo, Sep 19–22, 2000, p 137

Nickolaenko AP (1990) Sensor of the vertical electric field in the atmosphere. Author's certificate No.1626208, State Committee for Inventions and Discoveries of the USSR, Class G01R29/12, Moscow, USSR (in Russian)

Nickolaenko AP (1995) ELF/VLF propagation measurements in the Atlantic during 1989. J Atmos Terr Phys 57:821–831

Nickolaenko AP, Hayakawa M (2002) Resonance in the Earth-ionosphere Cavity. Kluwer Academic Publishers, Dordrecht, p 380

Ogawa T, Tanaka Y, Miura T, Yasuhara M (1966a) Observations of natural ELF and VLF electromagnetic noises by using ball antennas. J Geomagn Geoelectr 18:443–454

Ogawa T, Tanaka Y, Miura T (1966b) On the frequency response of the ball antenna for measuring ELF noise signals. Spec Contr Geophys Inst Kyoto Univ 6:9–12

Ogawa T (1985) Fair weather electricity. J Geophys Res 90:5951–5960

Ohta K, Umeda K, Watanabe N, Hayakawa M (2001) ULF/ELF emissions observed in Japan, possible associated with the Chi–Chi earthquake in Taiwan. Nat Hazards Earth System Sci 1:37–42

Ohta K, Watanabe N, Hayakawa M (2006) Survey of anomalous Schumann resonance phenomena observed on Japan in possible association with Earthquakes in Taiwan. Phys Chem Earth 31:397–402

Polk C (1969) Relation of ELF noise and Schumann resonances to thunderstorm activity. In: Coronoti SC, Hughes J (eds) Planetary electrodynamics, vol 2. Gordon and Breach, New York, pp 55–83

Polk C (1982) Schumann resonances. In: Volland H (ed) Handbook of atmospherics, vol 1. CRC Press, Bocca Raton, pp 111–178

Price C, Finkelstein M, Starobinets B, Williams E (1999) A new Schumann resonance station at the Negev desert for monitoring global lightning activity. In: Proceedings of 11th international conference on atmospheric electricity, Guntersville, Alabama, June 7–11, pp 695–697

Rycroft MJ (1965) Resonances of the earth-ionosphere cavity observed at Cambridge, England. J Res NBS Radio Sci. 69D:1071–1081

Sátori G, Szendröi J, Verö J (1996) Monitoring Schumann resonances—I. Methodology. J Atmos Terr Phys 58:1475–1482

Stratton JA (1941) Electromagnetic theory. McGraw-Hill, New York

Watt AD (1967) VLF radio engineering. Pergamon Press, Oxford

Williams E, Wong C, Boldi R, Lyons W (1996) Dual Schumann resonance methods for monitoring global lightning activity. In: Proceedings of 10th international conference on atmospheric electricity, Osaka, Japan, June 10–14 1996, pp 704–707

Chapter 3
Calibrating the Antennas

We address the procedure of a 'through' antenna and receiver calibration in this chapter. Electric and magnetic channels are calibrated individually, and we separately describe the relevant sets of devices. The necessary formulas are given to describe a calibrating capacitor, the gnome and radiating antennas, Helmholz rings, and the accuracy of calibration. Procedures are addressed of calibration at the field-site. A regular calibration is desirable of the receiver channels together with the field sensors. Such a procedure exploits the known 'standard' field applied to the antennas. Calibration is usually performed between the measurement sessions, and its results are stored and used in the data processing. We explain why the accurate calibration is hard to do at the field-site, so that root mean square error of the procedure is about 10 %. Attention is given also to the symmetry of angular patterns of magnetic field sensors.

3.1 Calibration of Vertical Electric Antennas

An absolute through calibration of the vertical electric field channel is a special task, since the antenna usually is a few meters high. We describe the "antenna substitution" procedure. The idea is based on application of the local tunable signal generator, which is connected to a radiating antenna through a high voltage transformer. The antenna-radiator is positioned at a distance of 30–50 m from the receiving antenna or from the place where this antenna will be installed. First of all, the effective height of the 'gnome' antenna (a reduced copy of the receiving antenna) is found by placing it into the flat calibration capacitor, which provides the known field at a given frequency. Then, the calibrated gnome is positioned at the place of the real antenna. The preamplifier of gnome is connected to the receiver, and the calibration signal is recorded from the radiating antenna. Thus, one can establish the amplitude of radiated field at the point where the real antenna will be settled. After fixing the amplitude of incident calibrating field, we remove the gnome and raise the real antenna, which is imposed now into the known sinusoidal field. This allows for establishing the effective height of the operational

A. Nickolaenko and M. Hayakawa, *Schumann Resonance for Tyros*, Springer Geophysics, DOI: 10.1007/978-4-431-54358-9_3, vertical electric antenna and obtaining the 'through gain' of the electric channel including the antenna and the data acquisition system.

Such a scheme is arduous. Therefore, calibration might be performed, say, once a few months or after the repair of preamplifier or other equipment. The routine tests of the system gain might be performed by the regularly turning on the calibrating signal generator. The latter could be a special computer code. Everyday tests might be performed automatically aiming at the frequency dependence of the through gain and its phase. The complete calibration is desirable after the equipment failure and its repair, and the measurements go on afterwards.

A calibrating system of the classical vertical electric antenna includes the gnome antenna, a calibration capacitor, a radiating antenna, a standard generator of ELF signals (or a computer code), and the wide band high voltage transformer. We will describe these elements one by one.

A few words should be mentioned about the capacitor antenna. Such an antenna has a relatively small height, which is close to the gap between the upper and the lower electrodes, i.e., $h_{eff} \approx$ 20–30 cm. Therefore it is possible to measure this effective height directly in the calibrating capacitor. However, the capacitor antenna must be elevated over the ground to reduce the impact of grass and other small vegetation. As we already know, positioning of a capacitor antenna at a metallic mast or a tripod increases its effective height. To evaluate its value, one has to initially place the antenna at the working point on the ground, switch on the radiating antenna, and to measure the signal amplitude. Afterwards, the capacitor antenna is raised to its working position, and again the calibrating signal is measured. The ratio of these two amplitudes indicates the extent to which the effective height had increased due to the antenna elevation.

Gnome antenna is a reduced copy of receiving ELF vertical electric disk antenna. It has the effective height of a few decimeters. The gnome is a disk of ~40 cm diameter elevated by an insulator over the horizontal metal base. Thus, it is similar to a capacitor antenna. One can apply a section of the PVC (Polyvinyl chloride) pipe as the insulator. An antenna preamplifier of the gnome antenna is the same as of the working antenna. A possible design of the gnome antenna is

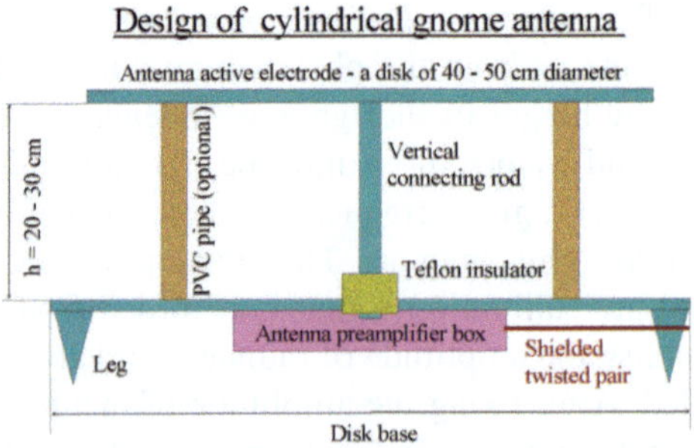

Fig. 3.1 Conventional design of the gnome antenna that might be easily calibrated in the capacitor

shown in Fig. 3.1. The antenna with dimensions shown in this figure has a self-capacitance of $C_0 \approx 10$ pF, and its frequency response becomes flat at frequencies above 10 Hz. Construction of a capacitor receiving antenna is similar to that of the gnome, but it is only somewhat greater in size. In particular, its diameter is about 1 m, which provides the self-capacitance of active electrode $C_0 \approx 28$ pF. Indeed, $C_0 = D/\pi$ is measured in centimeters while 1 cm = 0.9 pF.

Calibrating capacitor is formed by two parallel plates. The construction must satisfy a few demands. Since the gnome antenna must be portable, its diameter is not greater than 0.5 m or so, and hence its height must be 20–30 cm. When the conducting body of a gnome antenna is inserted into the capacitor, it distorts the initially uniform field of the calibrating capacitor. To make the disturbances small, one has to apply the capacitor of the height exceeding that of the gnome by a factor of 4 or greater. The capacitor height is usually about 1.5–1.7 m. We must remark that a gnome antenna modifies the initially uniform field of calibrating capacitor practically in the same way as it does in the open air when submerged into the field of ELF radio waves.

The lower plate of calibrating capacitor is made from the metal sheets. To lighten the upper electrode, it is made as a square network of horizontal wires extended by the steel springs with the 10 by 10 cm step. Each wire of the network is fixed to the frame of the angular profile, and its one end is pulled by a spring. The carrying frame is positioned at the top of four vertical rigid dielectric poles (PVC pipes), as seen in Fig. 3.2. The horizontal dimension of the construction is chosen in such a way that the electric field at its center is uniform with the predetermined accuracy. In other words, the edge effect of plates and the impact of dielectric poles must be small enough. We evaluate the edge effect by using the well-known solution for the half-infinite capacitor.

Let a two-dimensional (2D) capacitor be formed by two half-infinite parallel plates. Each plate is extended along the X-axis from zero to minus infinity (see Fig. 3.3). The plates coincide with the half-planes $y = \pm\, d$ and carry the potentials $+V$ and $-V$. correspondingly. This 2D electrostatic problem is solved with the help of the conform transformation of a semi-infinite strip in the complex plane $(x + iy)$ into the upper half plane of the complex variable $(u + iv)$. The solution is of the following form (Morse and Feshbach 1953; Fuks and Shabat 1964):

$$x = \frac{d}{\pi}\,(\rho\,\cos\psi + \ln\rho) \tag{3.1}$$

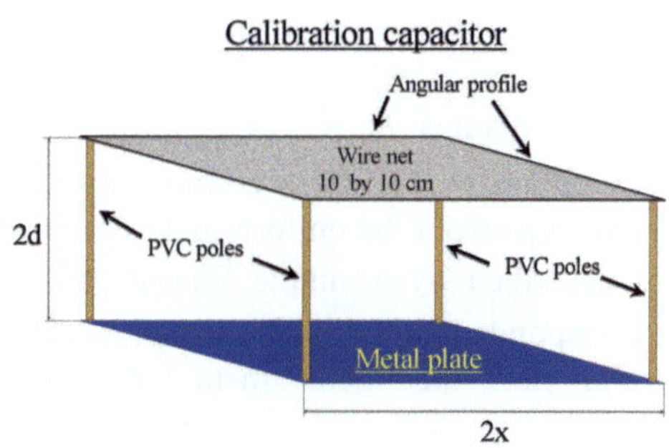

Fig. 3.2 Calibrating capacitor

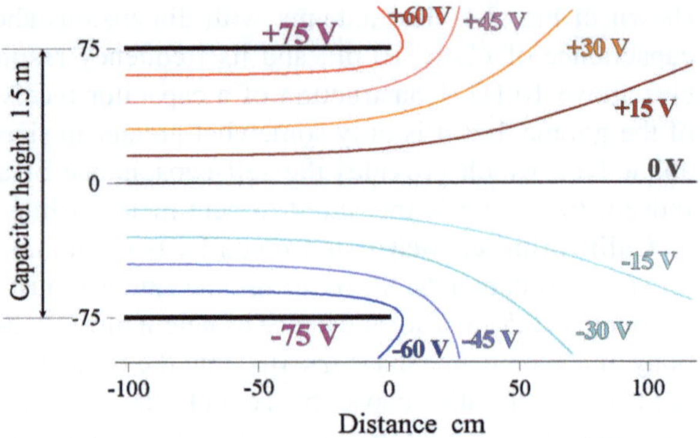

Fig. 3.3 Lines of equal potential at the rim of a half-infinite flat capacitor

$$y = \frac{d}{\pi}(\rho \ \sin\psi + \psi) \tag{3.2}$$

where $\rho = \exp\left(\frac{\pi u}{V}\right)$ and $\psi = \frac{\pi\, v}{V}$.

Equations (3.1) and (3.2) describe parametrically the distribution of electrostatic field at the rim of a capacitor, which we show in Fig. 3.3. The points with coordinates $x(\rho)$ and $y(\rho)$, corresponding to $\psi = const$, occupy the same equipotential line. The condition $\psi = \pm\pi$ corresponds to the capacitor electrodes: $x = d\ (\ln\rho - \rho)/\pi$ and $y = \pm d$.

Figure 3.3 depicts the spatial distribution of the lines of equal potential near the edge of a half-infinite 2D capacitor. The zero potential line is parallel to the abscissa, and it is the axis of symmetry of the field distribution. Coordinates of the plates were chosen to be $y = \pm$ 75 cm, and they are shown by black thick horizontal lines. The upper plate carries the +75 V potential, and the lower one is charged to −75 V. The capacitor height is 1.5 m, so that the electrostatic field has the 100 V/m amplitude. The field equipotential lines are plotted in Fig. 3.3 with a step of 15 V. One can note that the potential distribution inside the capacitor becomes practically uniform at a distance about 1 m from its edge. The same is valid for the field lines.

Curves $x(\psi)$ and $y(\psi)$ found from Eqs. (3.1) and (3.2) for ρ = const describe the electric field lines of the capacitor. The vertical component of electric field is found from the following equation.

$$E_y = \frac{V}{d}\frac{1}{\sqrt{\rho^2 + 2\rho\ \cos\psi + 1}} \tag{3.3}$$

One obtains $E_y = \frac{V}{d(1+\rho)}$ for $\psi = 0$. This is the field at median point of capacitor $y = 0$ and $x = d\ (\ln\rho - \rho)/\pi$. We are interested in the normalized field deviations from the uniform field $E_U = V/d$ of an infinite capacitor. The deviation is described by a simple formula $\frac{\delta E}{E_U} = -\frac{\rho}{1+\rho}$. The internal part of capacitor corresponds to small values of ρ, therefore the following approximate relation is valid for the field disturbance caused by the border of electrodes

$\frac{\delta E}{E_U} \cong -\rho = -\exp\left(-\frac{\pi x}{d}\right)$. By assuming that the relative field disturbance is 1 %, we obtain that $x = 4.61\ d/\pi \cong 1.47\ d$. It is easy to see that distance to the edge must be $x \cong 1.1$ m when the capacitor height is $d = 2 \times 0.75 = 1.5$ m. The real capacitor is of the square form, and each of its four rims disturbs the field. Thus the deviations might reach 4 % when we account for all the edges. We conclude that the 4 % accuracy would be guaranteed when the capacitor 1.5 m high has the horizontal size of 2.2 m. We evaluated the minimal appropriate width of a capacitor of a given height. Now, we can formulate a heuristic rule: the necessary width of calibration capacitor must exceed its height by the factor of about 1.5.

Calibration accuracy depends on many additional factors. These are the accuracy of the plate positioning (they must be strictly parallel), the impact of dielectric poles carrying the upper plate, the distance between the wires of the upper plate, the field disturbance caused by introducing the gnome antenna into the capacitor, etc.

The above estimates indicate that calibration of a small antenna 0.5 m high with the 4 % standard deviation or less is achieved when the capacitor has the size of 3 × 3 × 2 m. An increase in accuracy demands significant efforts. One will have to derive the more accurate relations and to exploit the high precision experimental equipment (see e.g. Burrows 1978). This objective lies beyond the goals of our book.

Radiating antenna is a metal disk of ~1 m diameter set at an insulator placed at the top of a carrying metallic pipe 3–4 m high. The high voltage signal of ~10 kV amplitude is fed to the disk by an isolated wire inside the mast. The signal generator and the transformer should be placed in the vicinity of radiating antenna to avoid any possible resonance in the transformer-feeding cable circuit.

A standard distance between the receiving and radiation antennas is equal to 30–50 m. The calibrating electric field from the radiating antenna is the near-zone quasi-electrostatic field. It depends on the distance R from antenna as R^{-3}. This field might be treated as appropriately uniform at great distances from the radiating antenna, and so, the accuracy of calibration increases with the distance R. The vertical electric field produced by a radiating antenna is found from the following simple formula:

$$E = \frac{h_R U C_0}{2\pi\varepsilon_0 R^3} \tag{3.4}$$

Here h_R is the height of radiating antenna in meters, U is the antenna voltage in V, C_0 is the capacitance of radiating antenna in pF, R is the distance from radiating antenna in m, and $\varepsilon_0 = (36\ \pi \times 10^9)^{-1}$ F/m is the dielectric permittivity of vacuum. The capacitance of a remote disk electrode of the diameter D is equal to $C_0 = D/\pi$ cm, while 1 cm = 0.9 pF.

The following operations are executed when calibrating the vertical electric antenna:

1. The gnome antenna is connected to the receiver electric channel and it is calibrated in the capacitor, so that its effective height H_{GA} becomes known.
2. The gnome antenna is installed at the place of the receiving antenna.
3. The calibrating signal of fixed amplitude (about 10 kV) is fed to the radiating antenna, and the voltage U_{GA} is registered at the output of electric channel of the receiver.
4. The operational receiving antenna is returned to its position, its output is connected to the receiver instead of gnome, and the output voltage of receiver U_{RA} is measured at the same frequency and for the same voltage of radiating antenna.
5. The effective height of receiving antenna is calculated by using the obvious relation:

$$h_{eff} = H_{GA} \frac{U_{RA}}{U_{GA}} \tag{3.5}$$

The major imprecision of such a procedure is evaluated by the quantity $3\frac{h_{eff}}{R}$. It arises from the inaccuracy with which we know the distance between the radiating and receiving antenna: we have in mind distance deviations between the point of installation of one antenna and the active electrode of the other. There is also a definite error in the amplitude of the voltage at the radiating antenna, which we neglect. We also put aside the problem of obtaining the sinusoidal high voltage waveform at low frequencies.

3.2 Calibration of Horizontal Magnetic Antennas

The field sensors with ferromagnetic core are calibrated by the known outer field produced by a solenoid (Huang et al. 1999) or the Helmholz rings (Nickolaenko and Hayakawa 2002). The ferromagnetic core of an antenna collects the outer field from a zone of the diameter approximately equal to the length of the core. Hence, the calibrating facility must provide (with the given accuracy) the uniform magnetic field in the volume $L \times L \times L$ where $L = 2\ m_a$ is the length of the antenna core. Creating of large enough calibrating solenoids is an obvious problem, especially when we take into account the necessity of positioning the magnetic antennas at the center of this device and in parallel to its field. Therefore, one usually uses the Helmholz rings for calibration procedures at the magnetic observatory, since the access into the working zone is quite comfortable in these devices.

Physical idea of the Helmholz rings is rather simple: one can treat them as the ends of a solenoid wiring with missing inner portion (see Fig. 3.4). A couple of the wide coaxial (this is important!) rings generates a rather uniform field distribution in an area around the central point of the ring's common axis. The major advantage of such construction is a free access to the working area where the field

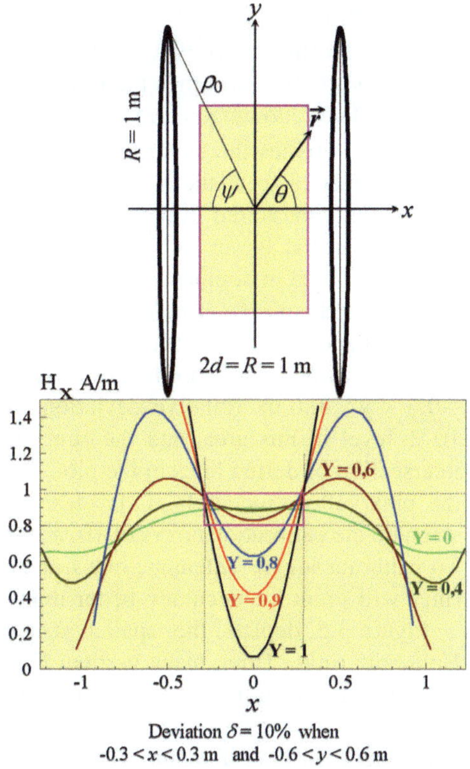

Fig. 3.4 Helmholz rings (*upper diagram*) and the spatial dependence of axial field (*lower plot*). The *pink* rectangles denote a zone where the relative deviations of the field amplitude do not exceed 10 %

sensor is positioned. The field components (in A/m) produced by the Helmholz rings are found from Yanovsky (1953):

$$H_x = \frac{Iw}{\rho_0}\sin^2\psi \sum_{n=1}\left(\frac{r}{\rho_0}\right)^{2(n-1)} \Pi_n^x \tag{3.6}$$

$$H_y = \frac{Iwy}{\rho_0^2}\sin^2\psi \sum_{n=1}\left(\frac{r}{\rho_0}\right)^{2n-1} \Pi_n^y \tag{3.7}$$

Here $\Pi_n^x = P_{2n-2}(\cos\theta)P_{2n+1}^1(\cos\psi)$, $\Pi_n^y = P_{2n+1}(\cos\psi)[P_{2n+1}(\cos\theta) - \frac{\cos\theta}{2n+1}P_{2n+1}^1\cos(\theta)]$ ($P_n(x)$ is Legendre function, and $P_n^1(x)$ is associated Legendre function), I is the current in the ring, and w is the number of turns in the winding.

We use the non-system A/m units, which are connected with the magnetic field measured in Tesla by the relation: 1 μA/m = 0.4 π pT. The A/m units are convenient in technical applications. For example, the ratio of vertical electric field measured in V/m to the horizontal magnetic field in A/m provides the well-known impedanceof the free space of 120 π Ohms.

The optimal form of the rings corresponds to the condition $R = 2d$. In this case, the second term $(r/\rho_0)^2 P_3^1(\cos\psi)$ in the sum of Eq. (3.6) turns into zero, and the spatial distribution becomes the most uniform one. We show this configuration in Fig. 3.4. The following parameters were specified: $R = 2d = 1$ m, and the current amplitude satisfies the condition $(Iw/\rho_0)\sin^2\psi = 1$A/m. The abscissa x is measured in meters from the system center along the major axis of symmetry. The ordinate y is measured in meters along the perpendicular to this axis.

The upper diagram in Fig. 3.4 illustrates the system configuration and the values specified in computations. The plots in the lower frame of this figure depict the 'longitudinal' $H_X(x)$ field component (in A/m) computed for the fixed transverse distances Y. These plots show that the H_X field amplitude becomes almost invariable when the coordinates satisfy the condition $-0.3 < x < 0.3$ m and $-0.6 < y < 0.6$ m. Relative deviations of the field amplitude do not exceed the 10 % level in this area, and the zone itself is outlined by the pink rectangle at intersection of dotted lines in the plot. This zone is shown in the upper diagram by the pink/yellow rectangle. It is clear now that particular Helmholz rings substantiate the calibration errors of 10 %, provided that the field sensor is 0.6 m long. To calibrate longer antennas, one has to construct greater rings. These enlarged rings will grant the accuracy better than 10 % for small antennas.

Figure 3.5 depicts the spatial structure of magnetic field formed by the Helmholz rings. Thick black vertical lines show position of the rings with 2 m in diameter and separated by the distance 1 m. The coordinates along the axes x and y are given in meters. The maps show the field distribution for the H_X and H_Y components in the XY plane. The centers of calibrating rings are found at points $y = 0$ and $x = \pm 0.5$ m. The upper map shows spatial distribution of the 'longitudinal' field component H_X, and the lower one corresponds to the 'transverse' H_Y component. One may notice from the plots of Fig. 3.5 that the magnetic field of Helmholz rings is mainly directed along the axis of calibrating system. Amplitude of the H_X field component remains approximately constant within the cross-like zone around the center of the system. The field sensor we calibrate should be placed right there.

The rings are convenient in use and provide an appropriate accuracy of calibration. For example, the field deviations are less than 1 % within the area $-0.15 < x < 0.15$ and $-0.2 < y < 0.2$. It is easy to see that the Helmholz rings of 8 m diameter separated by 2 m and being ideally parallel and coaxial guarantee a calibration error smaller that 1 % for the 'log' magnetic antenna 1 m long. We must stress that high accuracy of the calibration might be reached when the rings are absolutely parallel and coaxial. If this is not so, the angular deviations $\psi = \psi + \delta\psi$ appear in Eqs. (3.6) and (3.7). One might obtain the relations for evaluating the field disturbance by differentiating these equations with respect to ψ. By doing so and after performing some re-arrangements, we obtain that the particular design with the ring separation equal to their radius results in the normalized deviation $dH_X/H_X = 3.5\delta\psi$. One can readily calculate that the angular error of one degree (about 1/60 of radian) results in the amplitude error of ~6 %.

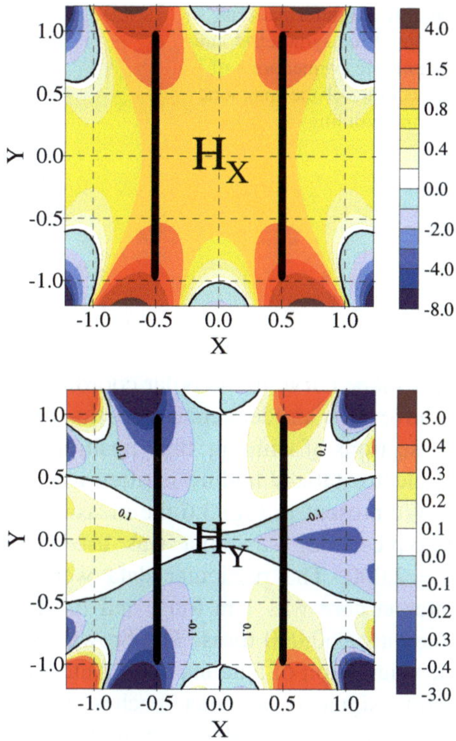

Fig. 3.5 Lines of equal field amplitude around the Helmholz rings. The *thick black lines* show the positions of ring winding

This estimate demonstrates that a calibration facility must be of large size and be accurately positioned. Performing such an installation in the field measurements becomes an unattainable task. Therefore, a simplified calibration scheme is used for the magnetic antennas in practice. Each antenna, when manufactured, is supplied by a pair of its own small calibrating rings, which are positioned at the edges of antenna winding. The antenna winding is assembled at the ferromagnetic core about 1 m long, and it is calibrated by a manufacturer who has the stationary, accurate Helmholz rings of great size, say, of 10 m diameter and 5 m separation. Such a facility enables us to calibrate the 'log' antenna with the error smaller than 1 %. The built-in portable calibrating coils are also tested in this stationary facility concurrently with the magnetic antenna. As a result, the consumer knows that the built-in rings provide the signal of, say, 1 mA/m amplitude when the applied calibrating current is equal to,say, 10 mA. From now on, both the sensitivity and the frequency characteristics of a particular magnetic antenna might be checked directly at the observatory. Thus, the periodic routine calibration of magnetic antennas becomes possible during the course of the field measurements.

Finally, we must comment on the magnetic antenna directivity and its angular pattern. Sensitivity of an ideal magnetic loop is the function of wave arriving angle A_Z. It varies as the $\sin(A_Z)$ function, and we say that the antenna angular pattern is

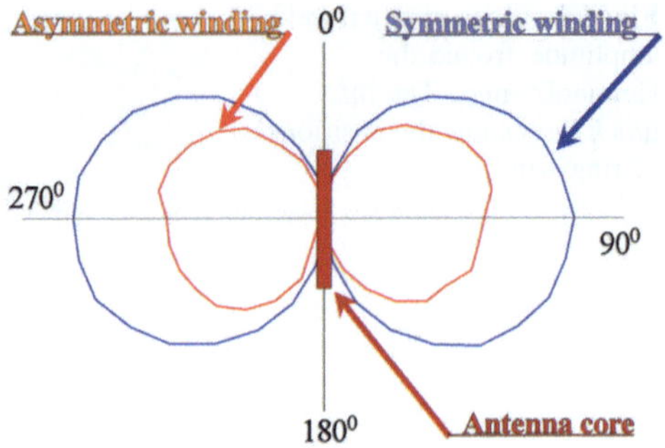

Fig. 3.6 Angular patterns measured for symmetric and asymmetric connections of the sections of the magnetic antenna winding

an 8-shape curve in the polar coordinates. The antenna must be exactly symmetric to have such an ideal characteristic. This concerns with all the elements: the core, the sections of winding, their positions, etc. In addition to symmetric construction, an electrostatic shielding is often used (Burrows 1978).

We present in Fig. 3.6 the measurement results of the antenna angular pattern. A typical 'log' antenna was used with the ferrite core of the permeability $\mu = 2{,}000$. The particular data were collected in the field measurement. The antenna core was 1 m long, and its winding contained 16 sections and the total amount of 60,000 turns. The brown rectangle in the middle of the figure denotes the orientation of antenna core. Antenna angular patterns of Fig. 3.6 show that the output signal of antenna reaches its maximum when the electromagnetic wave arrives from the right or from the left, the arrival azimuths are 90 and 270° correspondingly. In this case, the magnetic field vector is parallel to the antenna core, and the induced voltage reaches its maximum.

Figure 3.6 demonstrates deformation of the antenna angular pattern due to the violation of symmetry in the connection of separate sections of antenna winding. The distorted pattern tends to acquire the 'butterfly' outline. Fortunately, the self-capacitance of the sections was not great in the particular winding, so that deformation was not crucial.

Antenna directivity might be distorted by the influence of local singularities of the observatory, and we do not speak about an impact of the second magnetic antenna positioned nearby. These could be the natural geological non-uniformities in the crust or the local magnetic anomalies. It might be the steel oil-pipe or water-pipe lines at some distance from the observatory, etc. At any rate, the experimentalist must know the real angular patterns of the magnetic field sensors. This allows in principle for compensating the angular deviations in the data collecting system. Otherwise, the systematic errors are possible in the source bearings.

A simple gadget was used for tests of the angular pattern at the field site. A small induction coil antenna was used, which contained about 300 turns winded at a PVC pipe of 2 cm diameter and ~30 cm long. This antenna was connected to the standard LF signal generator via a ballast resistor. The current in the coil is the voltage applied divided by the ballast resistance. A fishing-line was tied to one end of the PVC pipe (a radiating coil antenna), and the other end of this line was fixed at the pole stuck over the center of the buried receiving antenna. The antenna-inductor was moved around the receiving antenna along a circle of ~ 10 m radius, with the fishing line stretched out. The inductor was placed on the ground at the points of the circle separated by 10°, and measurements were made of the signal detected by the receiver. Thus, the angular pattern was obtained rotated by 90° with respect to the core orientation, which is insignificant when one tests the symmetry of angular dependence.

References

Burrows ML (1978) ELF communications antennas. Peter Peregrinus, London

Fuks BA, Shabat BV (1964) Functions of complex variable and some applications, Nauka Publ, Moscow, p 387 (in Russian)

Huang E, Williams E, Boldi R, Heckman S, Lyons W, Taylor M, Nelson T, Wong C (1999) Criteria for sprites and elves based on Schumann resonance observations, J Geophys Res, 104(16)943–16,964

Morse PH, Feschbach H (1953) Methods of theoretical physics. McGraw-Hill, New York, Toronto, London

Nickolaenko AP, Hayakawa M (2002) Resonance in the Earth-ionosphere cavity. Kluwer Academic Publishers, Dordrecht, Boston, London, p 380

Yanovsky VM (1953) Terrestrial magnetism, 2nd edn. State Technical-Theoretical Pubs, Moscow (in Russian)

Chapter 4
Spectra of Continuous SR Background

There are three types of natural ELF radio signals: the continuous SR, ELF–flashes and ELF transients (Q-bursts). This chapter deals with the continuous background signal. It is formed by a superposition of pulses arriving from ordinary lightning strokes distributed all over the globe, and those pulses arrive at an average rate of 50–100 events per second. We address the properties of natural SR background. It is impossible to mention every details of enormous information on the ELF background, so that we describe the most common characteristics. An example is presented of the most complete traditional data set at a sub-polar observatory. It contains the power spectra of three orthogonal field components, the complex cross-spectra including the Poynting vector, the time domain histograms of the wave arrival angle, and individual records of ELF transients. Illustrated samples are given of the data accumulated during a day. Daily motion is presented of the global thunderstorm activity around the observatory and the diurnal variations are estimated of its level. Thus we illustrate a standard signal processing of the resonance signal and the typical information on the Earth–ionosphere cavity that was acquired by using the ordinary SR data.

4.1 SR Background Signal

The continuous SR signal is used in the majority of studies. In the simplest approach, the continuous digitized ELF radio signal is processed and accumulated in the computer memory. The processing is performed in the laboratory with some delay with respect to the measurement date. An advantage of such an approach is a principal possibility to return to a particular archive record and to process it in adequate manner. However, accumulation of original data demands a huge operation resource of the system and an extended memory. Indeed as we already found, the 24-bit ADC (analogue–digital converter) must be preferred in the data acquisition system. Otherwise, it is impossible to let the industrial interference 'pass through' with minor impact on the natural radio signal. The sampling frequency must be high enough for enabling the future processing of the Q-bursts

A. Nickolaenko and M. Hayakawa, *Schumann Resonance for Tyros*,
Springer Geophysics, DOI: 10.1007/978-4-431-54358-9_4,

from the record, and it is chosen to be ~200 Hz. Obviously, a considerable data amount appears containing the time records of three field components. This information must be stored in the computer memory, re-written to a portable carrier and translated to the laboratory. Data processing and obtaining of geophysical information are postponed in this case till a more appropriate time.

The technical means appeared recently that enable us to continuously accumulate the detailed time domain data. Prior to this, a preliminary spectral and correlation processing was performed at the observatory, and the results of this processing were accumulated in the computer memory. The amount of archived information is substantially reduced in this case, and concurrently, the important 'operational' information is obtained about the Earth–ionosphere cavity and global lightning activity directly in the course of observations. A combined approach was also used when the averaged SR spectra were monitored simultaneously with the separate records of the waveforms of intense ELF transients.

Monitoring of the SR background suggests application of the FFT algorithm, which provides the power spectra of vertical electric and two orthogonal horizontal magnetic field components. A field record is sub-divided into relatively short fragments, usually ~10 s long, so that the spectral resolution is about 0.1 Hz. Individual power spectra are averaged over the ensemble of realizations accumulated during 5–10 min interval. These averaged, statistically significant, stable spectra (power spectra and the cross-spectra) are stored in the computer memory. The three field components of Q-bursts were recorded into separate files directly in the time domain, and this was done in the real time mode including a short pre-history of the event. Sometimes, an option is added in the software that allows an operator to visually inspect the recent waveform and to decide whether it should be saved or not.

We show in Fig. 4.1 a typical set of SR spectra accumulated during a day of observations. Data were recorded at the Lehta observatory (Karelia, Russia, 64.427°N and 33.974°E) on July 25, 1998 (Belyaev et al. 1999). Three field components were recorded: vertical electric E and two orthogonal horizontal components H_{WE}, H_{SN}. The ELF signal was stored on a PC. The special software enabled us to perform the preliminary signal processing in the real time. The spectral estimates were accumulated at the hard disk of PC. The sampling frequency was equal to 204.8 Hz, and the receiver bandwidth covered the interval from 4 to 40 Hz. Preliminary experience showed that the sampling frequency should exceed the receiver upper cut-off frequency by a factor of 5–10 for obtaining the detailed waveforms of ELF transients. This is why the sampling frequency was f_s ~205 Hz. Spectral processing was based on FFT procedure applied to the 2,048 point segments. Therefore, the spectral resolution was 0.1 Hz. The power and the cross-spectra including the Poynting vector (being in fact the second statistical moments of random variables) of all three field components were averaged over 10 min intervals and stored by the computer. The software worked in the real time and saved the following functions:

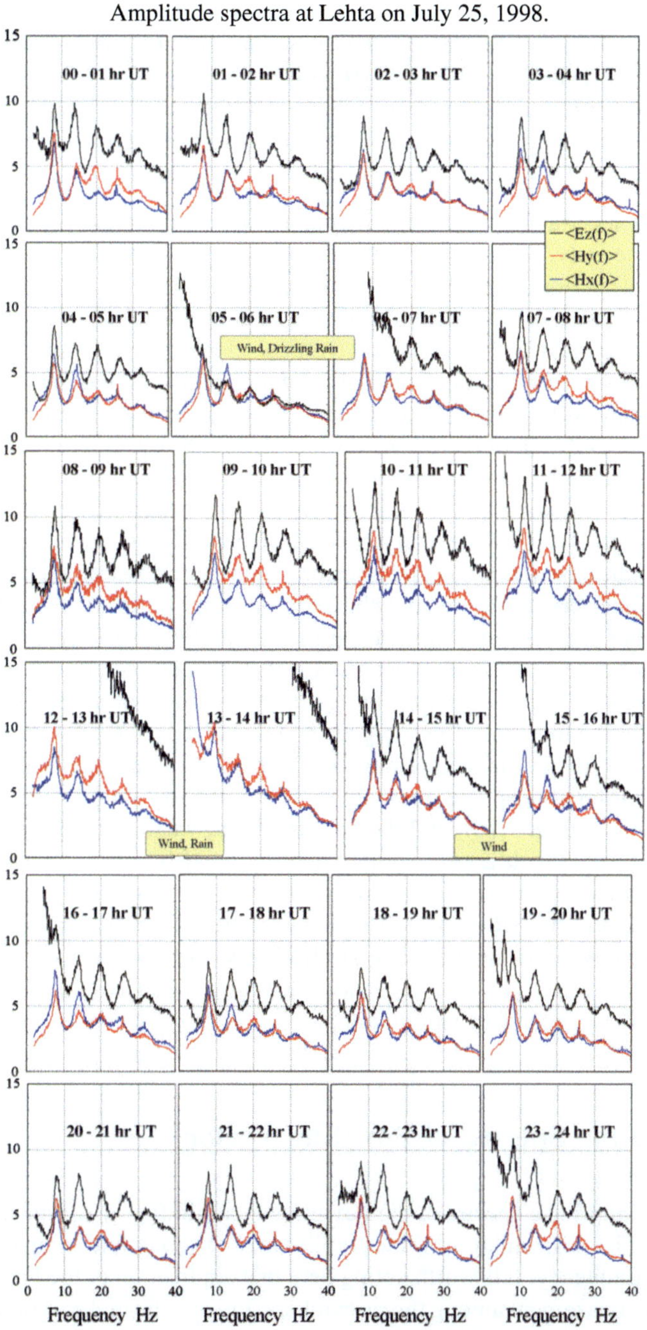

Fig. 4.1 Survey of SR spectra observed in the vertical electric and horizontal magnetic fields at Lehta observatory on July 25, 1998. Amplitude spectra at Lehta on July 25, 1998

- Average power spectra of the field components $\langle |E(f)|^2\rangle$, $\langle |H_X(f)|^2\rangle$, and $\langle |H_Y(f)|^2\rangle$
- Average complex spectra of the Poynting vector components $P_X(f) = -\langle E(f)H_Y^*(f)\rangle$ and $P_Y(f) = \langle E(f)H_X^*(f)\rangle$
- Average complex cross-spectrum $\langle H_X(f)\, H_Y^*(f)\rangle$
- Histograms of the wave arrival angles and the time domain Poynting vector components $P_X(t) = -\langle E(t)H_Y(t)\rangle$ and $P_Y(t) = \langle E(t)H_X(f)\rangle$.

The cross spectra allow for the unambiguous direction finding (Poynting vector) and simultaneously, for establishing the magnetic field polarization. This latter is used in the detection of SR line splitting (see Chap. 8).

Figure 4.1 surveys the amplitude spectra of SR additionally averaged over 1 h intervals. Data include the spectra of vertical electric field component E (black line) and the horizontal magnetic field components west-east $H_{WE} = H_X$ (blue line) and south-north $H_{SN} = H_Y$ (red line). Three spectra are shown in individual panels for the 24 h UT observations on July 25, 1998. Up to five resonance peaks (modes) are recognized in the amplitude spectra, while the fine structure addressed in Chap. 2 did not completely vanish even in the averaging over 1 h.

Plots in Fig. 4.1 indicate that the electric antenna is sensitive to the local weather conditions: the wind and drizzling rain (the observation interval 5–7 h UT) cause interference by raising the spectral amplitudes at lower frequencies. With a further increase of wind and rain (interval 12–13 h UT), the interference grows to such an extent that the resonance pattern is completely lost in the electric field component. After cease of the rain, but with continuing wind (interval 15–17 h UT), the local interference goes down, and still, the SR spectrum remains exaggerated in the first mode vicinity. Since the horizontal magnetic antennas are isolated from the atmosphere, they are buried in the dry sand inside the hermetic wooden box (see Fig. 2.10), and the impact of bad weather is less pronounced in the magnetic records. It appears only when the wind becomes rather strong.

There is a characteristic narrow maximum around 24–25 Hz frequency, which is often present in the SR records. This is undoubtedly the man-made interference, which is detected from time to time at all observatories. However, its source has not been established yet, and it is not clear how it appears and what process governs its time dependent amplitude.

4.2 SR in the Spectra of Poynting Vector

We plot the amplitude spectra of the Poynting vector in Fig. 4.2. Data are presented similarly to Fig. 4.1, which show the amplitude of the complete complex vector as a function of frequency. It is worth noting here that application of the Poynting vector might cause amazement among the researchers. Indeed, the real part of this complex vector is the power flux of the radio wave at a given frequency. However, only the standing waves are observed in the cavity resonator formed by the perfectly conducting walls. The energy flux is absent there, and the Poynting vector becomes a

Fig. 4.2 Survey of the amplitude spectra of Poynting vector recorded at Lehta observatory on July 25, 1998. Amplitude spectra of Poynting vector at Lehta on July 25, 1998

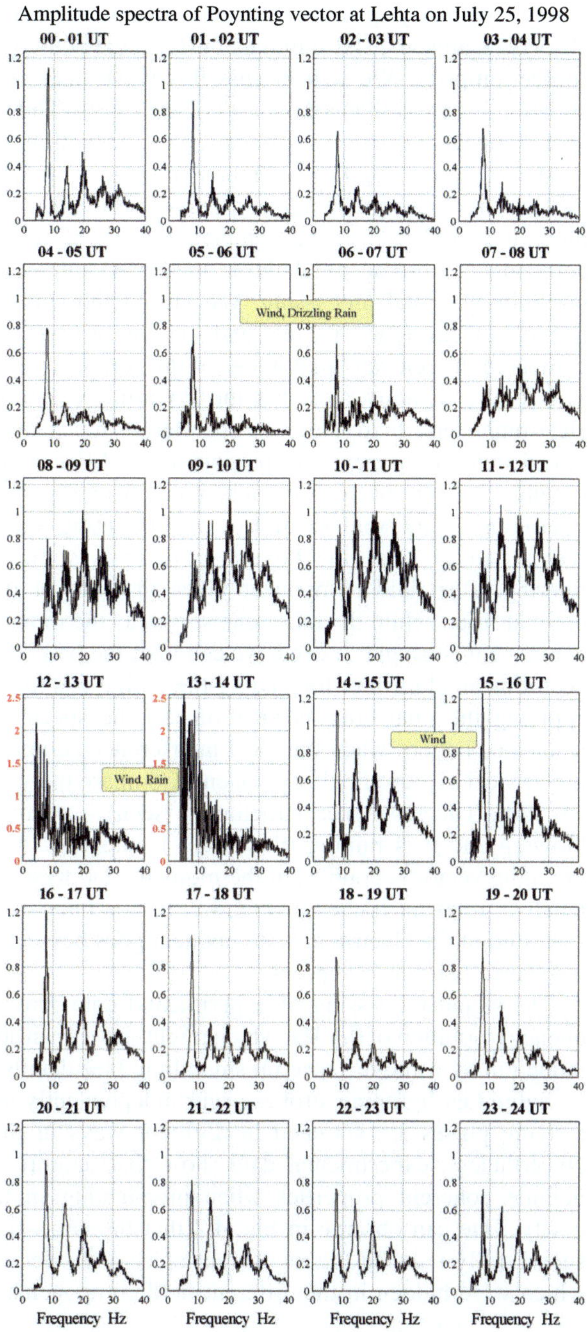

purely imaginary quantity. When the walls acquire a finite conductivity, wave absorption appears, and simultaneously, a real non-compensated wave travels to the source antipode. We will discuss this wave later, and here we note that such a situation is connected to the losses of the cavity. Thus, the real part of the Poynting vector appears in a cavity with the wave attenuation.

We show in Fig. 4.2 the hourly averaged amplitude spectra of the Poynting vector (the real and imaginary parts included), which correspond to the data presented in Fig. 4.1. The complex spectra of two orthogonal Poynting vector components were accumulated at the observatory every 10 min in the course of preliminary signal processing. We picked the complex power flux at every frequency and averaged its real and imaginary parts over one hour. After this we computed the relevant amplitudes of the complex vector and plot them in Fig. 4.2. This figure demonstrates that the Poynting vector is a robust characteristic of electromagnetic field, since its spectra are much more stable than those of individual field components. In particular, the SR peaks might be recognized in the Poynting vector spectra for every interval of Fig. 4.2. Only exceptionally strong interferences are able to destroy the Poynting vector spectrum. Such stability seems natural since the Poynting vector is similar, in a way, to the optimal filtering. One field component plays the role of the response function of the optimal filter, while the other one serves as the input signal in which the coherent component is sought. Indeed, we can treat the procedure of obtaining the Poynting vector as an algorithm for finding the global electromagnetic resonance signal by applying the dual cross-correlation. So the spectra of power flux preserve the resonance pattern when the local interference completely destroyed the spectra of the vertical electric field component: Compare the spectra in the intervals 5–7 and 15–17 h. The resonance structure is seen to disappear when a strong wind is blowing and it is raining (the 12–13 h interval). Although the amplitude of the Poynting vector is doubled at the moment, and to stress this, we apply the red axis labels in relevant frames of Fig. 4.2. In this interval (see Fig. 4.1), the amplitude spectrum of the electric field component goes beyond the limits of the plot, and the resonance pattern disappears.

The signal processing of experimental data provides a slightly excessive frequency resolution of 0.1 Hz, which might serve as an additional resource in the noise reduction by using the spectral smoothing [see details in Belyaev et al. (1999)].

Individual lightning strokes occur independently over the whole globe, and the arriving pulses are summed in different ways in the various field components. Nevertheless, experimental data show that natural radio noise activity acquires definite coherent properties when propagating in the closed Earth–ionosphere cavity. One can observe in Fig. 4.2 that the cross-spectra of the Poynting vector successfully reduced the incoherent noise and selected the “interrelated” signals, which were hidden in the different field components of natural ELF field.

Plots of Figs. 4.1 and 4.2 demonstrate the resonance nature of terrestrial ELF radio noise and its temporal variations. Surveying the data in form of separate spectra of the field components is cumbersome and sometimes inconvenient, e.g., spectra obtained at different times occupy different panels, and their comparison is hampered. Dynamic spectra or digital sonograms are a much more convenient form of data presentation over the time–frequency plane as seen in Fig. 4.3, where we show the dynamic spectra of all three field components for the same date of July 25, 1998. The spectral intensity is shown in arbitrary units by the color inking in the plot.

Figure 4.3 contains dynamic spectra of every field component shown as the color map. The Universal Time (UT) in hours is shown on the abscissa of every map, and the frequency is plotted along the ordinate in Hz. One may see that a dynamic spectrum presents the results of measurements in a very compact form. One may observe the clear horizontal traces of SR peaks in the vicinity of 8, 14, 20 etc. Hz. The dynamic spectra visibly outline the intervals of the high interference caused by the local weather conditions (wind and rain). The intervals are seen as vertical strips of brown color. Dynamic spectra demonstrate obviously that the magnetic field sensors are less sensitive to the weather. However, the severe rain and wind cause the vertical strip around 14 h. Since the wind coming at the magnetic antenna shelter causes some trembling, the magnetic spectra at the lowest frequencies are enhanced, in such a way and that they 'pull' or 'elevate' the whole resonance spectrum. The peaks are still resolved, but their amplitude is modified by the interference.

The spectral fine structure that remained visible in Fig. 4.1 after 1 h integration is also present in the dynamic spectra. We observe it in the form of 'spots' or in the form of structured, 'trembling' lines of constant amplitude in the maps. The sonograms or dynamic spectra are both convenient and informative, which will be often used in the following sections.

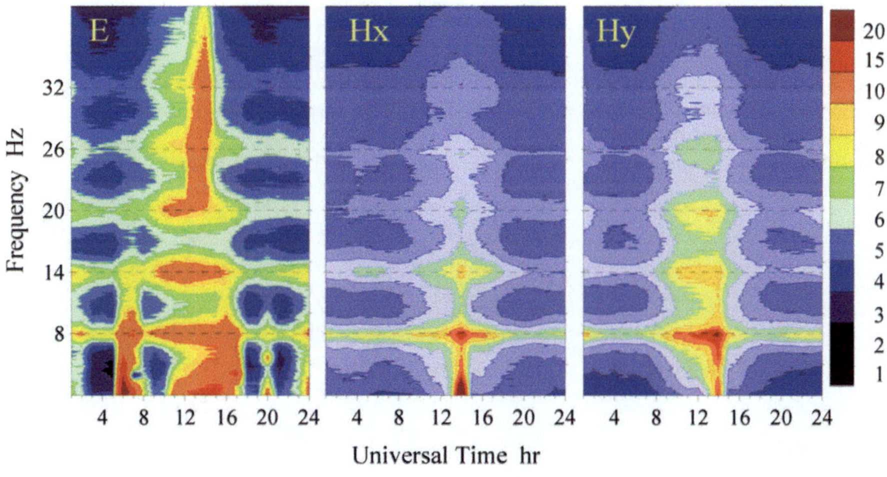

Fig. 4.3 Dynamic spectra (sonograms) of SR records of different field components recorded on July 25, 1998

4.3 Source Motion as seen in the Poynting Vector

One of the major advantages of the Poynting vector is a possibility to unambiguously find the direction of the wave arrival. Amplitude of the Poynting vector presents the median power flux, and its orientation shows the direction from the source. Thus, one can simultaneously monitor the current thunderstorm activity and its median position with respect to the observatory. The above information included the Poynting vector components computed directly in the time domain for elementary realizations of 10 s duration. This allowed us to obtain and store the arrival angle histograms for the 10 min intervals with high temporal resolution (Belyaev et al. 1999). As a result, one obtains the statistical ensemble of wave arrival angle every 10 min, its average value, and the standard deviation.

Poynting vector computations in the time domain became possible after matching the amplitude-phase characteristics of the electric and magnetic channels. For this purpose, an additional phase rotating circuit was introduced into the electric antenna channel. This element equalized the phase characteristics in the lower part of SR band. When such a circuit is not available, the Poynting vector losses its genuine linear polarization and acquires a fallacious elliptical polarization (Belyaev et al. 1999).

We show in Fig. 4.4 typical diurnal variations of the source bearing and the amplitude of the Poynting vector. These reflect the daily drift and intensity of the global thunderstorm activity. To show the true direction to the source instead of the wave arrival, we changed the sign of the Poynting vector. Data presented in this figure were obtained in the following way. The preliminary processing at Lehta yielded the components of Poynting vector in the following way:

$$P_X = -\sum_{k=1}^{2048} E(k) H_Y(k) \tag{4.1}$$

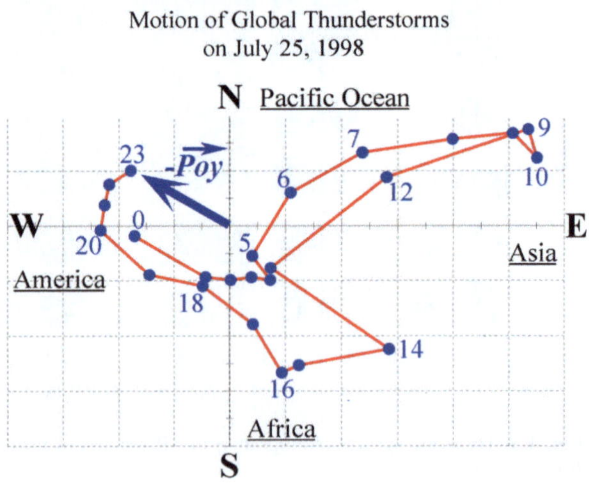

Fig. 4.4 Lehta observations of the diurnal motion of global thunderstorms on July 25, 1998

$$P_Y = \sum_{k=1}^{2048} E(k) H_X(k) \tag{4.2}$$

Here $E(k)$, $H_X(k)$, and $H_Y(k)$ are the samples of measured three field components and k is the number of a sample in the particular data segment.

Thus, the Poynting vector components (and hence amplitude and orientation) were recorded every 10 s concurrently with the relevant spectra in the frequency domain. The $-P_Y$ versus $-P_X$ data are shown in Fig. 4.4 by blue dots. Varying time (UT) is the parameter, and relevant numbers are printed beside the points. The dots are connected by the red line, and the cardinal directions are indicated in the plot. These latter correspond to the directions from the Lehta observatory toward the major global thunderstorm centers in Asia, Africa, and America. Quite similar diurnal variations are obtained from the integrated spectra of the Poynting vector, but we do not show them.

Every dot in Fig. 4.4 demonstrates the hourly averaged position of the tip of the negative Poynting vector, i.e., the intensity (the distance from the origin of the coordinate system) and the direction toward the source. As one may observe, the global thunderstorms circle the globe during the day. Activity rapidly increases in Asia from 6 to $\sim$10 h UT, and the direction to the source remains practically invariable. Asian thunderstorms were the most intense on the globe during that particular day. In early afternoon hours, activity moves to Africa and Madagascar. American lightning strokes dominate in the evening from 18 to 23 h UT. During the UT nighttime, the global thunderstorms become inactive, and the amplitude of Poynting vector is small, while its position becomes unstable. This is a commonly accepted behavior of terrestrial thunderstorms, and the plot in Fig. 4.4 is in accord with this conception except one detail. The nighttime thunderstorm activity of the particular day of July 25 was definitely concentrated in Africa, and it was about 1/3 of the day time level. Thus SR indicated the nocturnal thunderstorms in Africa.

Diurnal variations shown in Fig. 4.4 are not reproduced day after day and from a season to season (compare Figs. 4.4 and 5.18). Nevertheless, the general behavior remains stable. In the UT morning, the natural ELF radiation arrives predominantly from the east when the East-Asian thunderstorms 'wake up'. Closer to afternoon time, the wave comes form Africa, and the American thunderstorms dominate in the UT evening time. Approximately 1/4 of the diurnal records corresponds to a noticeable night thunderstorm activity in Africa. Still, the SR intensity comes down by a factor of 2–4 during the UT night, and the source bearing often strongly fluctuates.

4.4 Source Parameters Deduced from Continuous SR Data

SR spectra are rarely used as a whole while these are obtained and recorded at the majority of working SR facilities. Usually, the peak amplitude, peak frequencies, and sometimes the quality factors are used in the interpretations of experimental data (see e.g. Sentman 1987; Boccippio et al. 1995; Füllekrug et al. 1995, 1999; Williams et al. 1996; Füllekrug and Fraser-Smith 1997; Neska and Sátori 2006; Sátori et al. 2006, 2008, 2009). These resonance parameters are often found by fitting the observed resonance curves with the Lorentzian curve. The particular outline $S(f) = \frac{S_n}{1+[2Q_n(f-f_n)/f_n]^2}$ is often exploited, and we compare this with Eq. (1.2). In a sense, such a processing returns us to the early studies by Madden and Thompson (1965) who used the same curve and by Nelson (1967) who suggested the SR mode tracker. The distinction is in a huge amount of information processed nowadays with the help of computers. We must note in this context the original algorithm of obtaining the resonance parameters by Sátori et al. (1996a), and Sátori and Zieger (1996b). We note that storing of only resonance parameters of discrete modes results in the information loss on the real outline of SR line. For example, an impact of the poor local weather (wind and rain, see Fig. 4.1) might pass unnoticed in the data and even cause errors. This point is relevant to practically every unusual SR spectra, so it is desirable to store the power and cross-spectra as a whole since a modern computer memory performs such a task easily.

The typical monthly averaged SR data are given in Fig. 4.5. Upper frames show the dynamic power spectra (sonograms) of the two perpendicular horizontal magnetic field components $H_X = H_{WE}$ and $H_Y = H_{SN}$. The data were collected at Lehta observatory in August 1999 (Nickolaenko et al. 2001). Prior to the final processing of accumulated data, the selection of spectra was performed. The final ensemble included the 10 min power spectra without local interference. After this, the spectra of a given hour were averaged over the days of observations, and the average spectra were stored. Such spectra are rather stable, and the fine structure of resonance is substantially reduced.

Averaging the power spectra over the month period allows us to eliminate or at least reduce the impact of local interference that leads to the removal of some records, for example, the power spectra in the time intervals 5–7, 12–14, and 14–16 h of Fig. 4.1. Involvement of spectra of different days in a month, but pertinent to a given hour allows for compensating the gaps in the series. This is important for establishing parameters of the global thunderstorm activity from the SR data or in comparing the diurnal–seasonal patterns at different observatories.

The other approach is possible when the data succession is chosen for the interval containing no gaps. However, the difficulties arise as a rule when comparing records at different field-sites since the poor weather conditions or the other local interference take place at different observatories on different days. It often turns out that a series of continuous records at two distant sites do not practically overlap. This is why we use below the diurnal variations averaged over the period of a month.

Sonograms present the time–frequency data in a very compact form. Separate resonance peaks are noticed as red-brown tracks separated by blue valleys of spectral minima. Dynamic spectra clearly show that temporal variations simultaneously occur at all frequencies. This means that the field intensity varies, as a whole, hence, alterations are caused by changes in the source intensity.

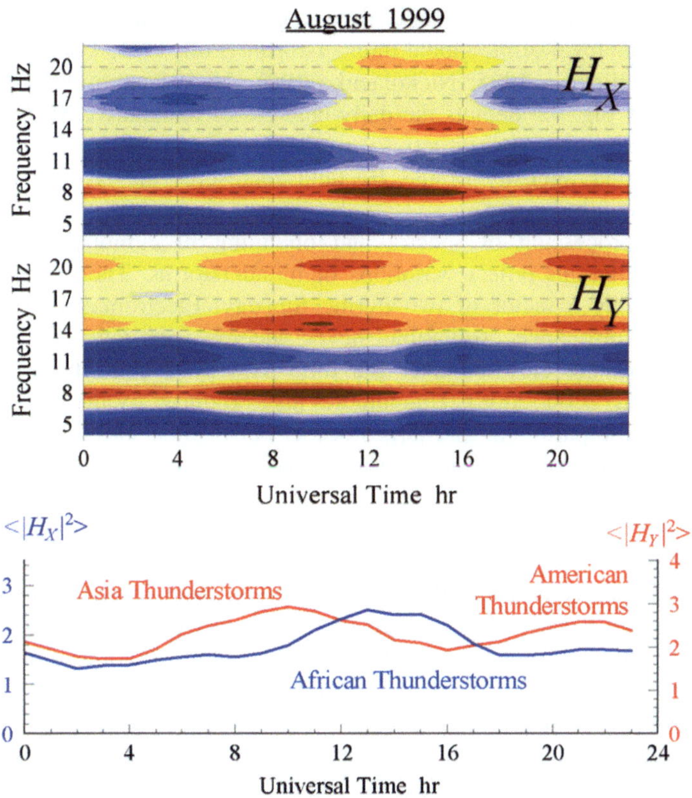

Fig. 4.5 Global thunderstorm activity in the SR records, The upper maps demonstrate dynamic spectra of two orthogonal magnetic field components averaged in August 1999. The lower plots show diurnal variations of the integral intensity of SR oscillations conditioned by Asian and American thunderstorms (*red curve*—$H_Y = H_{SN}$ field component) and by the Africa source (*blue curve*—$H_X = H_{WE}$ field component)

For instance, the field intensity H_X simultaneously increases at three SR modes around 13–14 ht UT. We know that African thunderstorms prevail on the planet during this time (WMO 1956; Christian et al. 2003). The core orientation of the particular magnetic antenna makes it sensitive to the sources in this particular sector. The spectral intensity of the orthogonal H_Y field component is high in the intervals from 10 to 12 and 20 to 22 UT thus reflecting an increase in Asian or in American activity. Unfortunately, plots in Fig. 4.5 do not show the wave arrival angle. Nevertheless, the sonograms clearly indicate that temporal variations are conditioned first of all by alterations in the level of the global thunderstorm activity. Variations of the distance from the observer to the sources plays the secondary role. The simultaneous and consistent variations of the oscillation intensity indicate a variations of the source intensity, while the motion of thunderstorms will modify the intensity of separate modes in different way. Thus, the cumulative ELF field intensity (intensity integrated over the frequencies) reflects variations of the level of the global thunderstorms (Polk 1969, 1982; Fraser-Smith et al. 1991; Füllekrug 1995; Nickolaenko 1997; Nickolaenko et al. 1998). The lower frame in Fig. 4.5 depicts the daily changes of cumulative SR intensity. As one might expect, the African activity is clearly seen in the H_X field component (blue curve), and intensity of the H_Y field represents the Asian and America lightning strokes (red curve).

Statistical processing of spectra and obtaining of cumulative intensity of resonance oscillations reflect variations of thunderstorms. Their intensity varies approximately by a factor of two during the day. Electromagnetic monitoring indicates that resonance oscillations never become very low. In other words, diurnal patterns are elevated over the time axis by a characteristic 'podium' (Yatsevich et al. 2006, 2008). This latter is also produced by the global lightning strokes, but these particular discharges occur 'out of time' (see below).

By similar processing of the long-term SR records covering a few years and performed at different observatories, one may evaluate the seasonal, annual and inter-annual trends in the global thunderstorm activity and its redistribution between the separate global thunderstorm centers. Since the electric activity is caused by the air convection, thunderstorms must carry information on the soil temperature (see below). We will separate the universal time (intensity) and the local time (motion) components in the diurnal intensity variations. Finally, the SR background will be used as a source of information for constructing the formally rigorous solution of inverse electromagnetic problem, which resolves the position and the intensity of relevant lightning centers (see Chap. 10).

References

Balser M, Wagner CA (1960) Observation of Earth-ionosphere cavity resonances. Nature 188:638–641

Barr R, Jones DL, Rodger CJ (2000) ELF and VLF radio waves, J Atmos Solar-Terr Phys, 62(1):689–1718

Belyaev GG, Schekotov AYu, Shvets AV, Nickolaenko AP (1999) Schumann resonances observed using Poynting vector spectra. J Atmos Solar-Terr Phys 61:751–763

Boccippio DJ, Williams ER, Heckman SJ, Lyons WA, Baker IT, Boldi R (1995) Sprites, ELF transients and positive ground strokes, Science, 269, pp 1088–1091

Christian HJ, Blakeslee RJ, Boccippio DJ, Boeck WL, Buechler DE, Driscoll KT, Goodman SJ, Hall JM, Koshak WJ, Mach DM, Stewart MF (2003) Global frequency and distribution of lightning as observed from space by the optical transient detector, J Geophys Res, 108(D1): 4005, doi:10.1029/2002JD002347

Fraser-Smith AC (1982) ULF/lower ELF electromagnetic field measurements in the polar caps. Rev Geophys 20:3. doi:10.1029/RG020i003p00497

Fraser-Smith AC, McGill PR, Bernardi A, Helliwell RA, Ladd ME (1991) Global measurements of the low frequency radio noise, in "environmental and space electromagnetics". In: Kikuchi H (ed), Springer, Tokyo, 191–200

Füllekrug M (1995) Schumann resonances in magnetic field components, J Atmos Terr Phys, 57(7):479–7484

Füllekrug M, Fraser-Smith AC (1997) Global lightning and climate variability inferred from ELF magnetic field observations, Geophys Res Lett, 24, pp 2411–2414

Füllekrug M, Fraser-Smith AC, Bering EA, Few AA (1999) On the hourly contribution of global cloud-to-ground lightning activity to the atmospheric electric field in the Antarctic during December 1992. J Atmos Solar-Terr Phys 61:745–750

Jones D Ll (1974) ELF propagation theory. In: Holtet JA (ed) ELF-VLF radio wave propagation. D. Reidel Publishing Company, Dordrecht-Holland, pp 207–223

Jones D Ll (1999) ELF sferics and lightning effects on the middle and upper atmosphere. Modern Radio Sci. In: Stuchly MA (ed), Published for URSI by Oxford University Press, pp 171–191

Madden T, Thompson W (1965) Low frequency electromagnetic oscillations of the earth-ionosphere cavity. Rev Geophys 3:211–254

Nelson PH (1967) Ionospheric perturbations and Schumann resonance data, Ph.D. Thesis, MIT, Cambridge Mass

Neska M, Sátori G (2006) Schumann resonance observation at Polish polar station at Spitzbergen and geophysical observatory in Belsk. Przeglad Geofizyczny 51(3–4):189–198 (in Polish)

Nickolaenko AP (1997) Modern aspects of the Schumann resonance studies. J Atmos Solar-Terr Phys 59(7):805–816

Nickolaenko AP, Sátori G, Zieger B, Rabinowicz LM, Kudintseva IG (1998) Parameters of global thunderstorm activity deduced from long term Schumann resonance records. J Atmos Solar-Terr Phys 60:387–399

Nickolaenko AP, Shvets AV, Yatsevich EI (2001) Deducing variations of the global thunderstorm activity from the Schumann resonance data. Radiophys Electron 6(1):71–77

Nickolaenko AP, Hayakawa M (2002) Resonance in the Earth-ionosphere Cavity. Kluwer Academic Publishers, Dordrecht, Boston, London, p 380

Ogawa T, Tanaka Y, Miura T, Yasuhara M (1966) Observations of natural ELF and VLF electromagnetic noises by using ball antennas. J Geomagn Geoelectr 18:443–454

Polk C (1969) Relation of ELF noise and Schumann resonances to thunderstorm activity, in Planetary Electrodynamics. In: Coronoti SC, Hughes J (eds), 2, Gordon and Breach, New York, pp 55–83

Polk C (1982) Schumann resonances. In: Volland H (ed), Handbook of Atmospherics, vol.1, CRC Press, Bocca Raton, Florida

Price C, Asfur M, Lyons W, Nelson T (2002) An improved ELF/VLF method for globally geolocating sprite-produced lightning, Geophys Res Lett, 29(3), doi: 10.1029/2001GL013519

Price C, Greenberg E, Yair Y, Sátori G, Bor J, Fukunishi H, Sato M, Israelevich P, Moalem M, Devir A, Levin Z, Joseph JH, Mayo I, Ziv B, Sternlieb A (2004) Ground-based detection of TLE-producing intense lightning during the MEIDEX mission on board the space shuttle Columbia. Geophys Res Lett 31(20):L20107. doi:10.1029/2004GL020711

Price C, Melnikov A (2004) Diurnal, seasonal and inter-annual variations in the Schumann resonance parameters, J Atmos Solar-Terr Phys, 66(1):179–1185, doi:10.1016/j.jastp.2004.05.004

Price C, Asfur M (2006) Can lightning observations be used as an indicator of upper-tropospheric water-vapor variability. Bull Amer Meteorol Soc 87(3):291–298. doi:10.1175/BAMS-87-3-291

Price C, Pechony O, Greenberg E (2007) Schumann resonance in lightning research. J Light Res 1:1–15

Sátori G, Szendröi J, Verö J (1996) Monitoring Schumann resonances–I. Methodology. J Atmos Terr Phys, 58(1): 475–1482

Sátori G, Zieger B (1996) Spectral characteristics of Schumann resonances observed in Central Europe. J Geophys Res, 101(29):663–29,669

Sátori G, Neska M, Williams E, Szendröi J (2006) Signatures of the day-night asymmetry of the Earth-ionosphere cavity in high time resolution Schumann resonance records. Radio Sci, 42, RS2S10, doi:10.1029/2006RS003483

Sátori G, Williams E, Lemperger I (2008) Variability of global lightning activity on the ENSO time scale. Atmos Res 91(2–4):500–507. doi:10.1016/j.atmosres.2008.06.014

Sátori G, Mushtak V, Williams E (2009) Schumann resonance signatures of global lightning activity, in "lightning: principles, instruments and applications".In: Betz HD, Schumann U, Laroche P (eds), Springer, Dordrecht, pp 347–386

Sentman DD (1987) Magnetic elliptical polarization of Schumann resonance. Radio Sci 22:595–606

Sentman DD (1995) Schumann resonances. In: Volland H (ed) Handbook of atmospheric electrodynamics, 1, pp 267–298, CRC Press, Boca Raton, London, Tokyo

Williams E, Wong C, Boldi R, Lyons W (1996) Dual Schumann resonance methods for monitoring global lightning activity. Proceedings of 10th international conference on atmospheric electricity, Osaka, Japan, 704–707 (June 10–14)

WMO (1956) World distribution of thunderstorm days, part 2: tangles of marine data and world maps, OMM-No. 21, TP. 21

Yatsevich EI, Nickolaenko AP, Shvets AV, Rabinowicz LM (2006) Two component source model of Schumann resonance signal. J Atmos Electr 26(1):1–10

Yatsevich EI, Nickolaenko AP, Pechony OB (2008) Diurnal and seasonal variations in the intensities and peak frequencies of the first three schumann-resonance modes. Radiophys Quantum Electron 51(7):528–535

Chapter 5
Regular SR Parameters

This chapter presents the classification and samples of all three types of ELF radio signals. The continuous signal is a composition of individual pulses from the global lightning activity. Signals of the second kind are the ELF-flashes: the intense pulses from the nearby thunderstorms occurring within 1,000–2,000 km distance from the observatory. An ELF-flash overloads the input circuits, and is attributed usually to natural interference. The third type is the ELF transients or Q-bursts. These pulses arrive from the distant powerful strokes, while their amplitude surpasses the continuous background by a factor of 3–10. Q-bursts are recorded at a rate of one pulse in a minute. Afterwards, we concentrate on the 'geophysical' data obtained from the continuous SR records. We show in particular how to estimate the area occupied by the global thunderstorms from the diurnal/seasonal variations of the observed peak frequency in the resonance spectra. Resonance data were verified by orbital optical observations of midnight lightning flashes. Diurnal motions and seasonal drifts of global thunderstorms are estimated by using the electric and magnetic field records, as well as seasonal variations of the effective area occupied by the lightning strokes. By using the singular spectral analysis (SSA), the annual and semi-annual variations were obtained. These are present in the long-term records of the SR intensity (amplitude) and of the peak frequency of the first mode. The presence of inter-annual variations is obvious. An exceptional similarity is demonstrated between the semi-annual component in the first mode frequency and analemma (median solar time). By using the three-source model, a possible impact was computed of the El Nino Southern Oscillation (ENSO) on the observed SR frequency. Another possible explanation is discussed that exploits the global inter-annual variations of the ionosphere height. By using the so-called 'terminator effect', we demonstrate difficulties in correct interpretation of resonance records.

A. Nickolaenko and M. Hayakawa, *Schumann Resonance for Tyros*,
Springer Geophysics, DOI: 10.1007/978-4-431-54358-9_5,

5.1 Classification of Natural ELF Radio Signals

Experimental studies of SR started in 60 s. It was demonstrated (Balser and Wagner 1960a, b) that the power spectra of the natural ELF radio noise contain peaks at frequencies 8, 14, 20, and 26 Hz when accumulated during a time substantially greater than the period of the basic resonance frequency (intervals a few minutes long). The classification of natural ELF radio signals was introduced by Ogawa et al. (1966), and it includes three classes:

1. The ELF background or the continuous noise produced by radio pulses arriving from the customary terrestrial lightning activity. The background intensity remains stable at times from ten minutes to an hour.
2. ELF flashes, pulses of great amplitude surpassing the amplitude of the SR signals by a few orders of magnitude. These pulses arrive from the "nearby" lightning strokes (the distance up to 2,000 km), and they overload the receiver, so that the recorded waveforms are usually clipped.
3. Q-bursts (the term "ELF transients" is also used) are the discrete pulses of about 1/3 s duration exceeding the background level by a factor reaching 10.

Figure 5.1 depicts typical waveforms of all three types. SR background is always present in the record in a form of continuous noise composed of pulses coming from global thunderstorms at a rate of 50–100 events per second. ELF flashes are observed when thunderstorms take place around the site at distances of hundreds of kilometers. The overloading of circuits and relevant transient processes might distort the waveforms of ELF flashes and the subsequent records at an interval of about a few seconds. The Q-bursts (the quiet bursts according to Ogawa et al. (1966)) arrive from powerful strokes remote to the distance up to 20 Mm (1 Mm = 1,000 km). These signals are also called ELF transients (Jones and Kemp 1971), their amplitude exceeds the background noise by the factor of 10, and they arrive at a rate of about one pulse in a minute.

Special attention is directed to the Q bursts in the literature. Initially, this interest was explained by the possibility of resolving the inverse electromagnetic problem and establishing both the coordinates of the parent discharge and its spectrum. We discuss this inverse problem in Chap. 9. Here, we remark that the technique was elaborated and applied by Jones (1970a, b, 1999), and Jones and Kemp (1970, 1971). It was demonstrated by Burke and Jones (1992, 1995, 1996) that spectral amplitude of the parent strokes decreases with frequency, and the term "red spectra" was introduced for such powerful strokes (see also Boccippio et al. 1995; Williams et al. 2007).

Investigations showed that ELF transients or Q-bursts are often associated with the so-called "red sprites", which are luminous effects observed in the mesosphere (middle atmosphere) over the clouds after powerful lightning strokes. This is why the Q-bursts were used for the global electromagnetic sensing of the red sprites (see e.g. Boccippio et al. 1995; Füllekrug et al. 1998; Füllekrug and Reising 1998; Huang et al. 1999; Lyons et al. 2000; Füllekrug and Constable 2000; Hobara et al.

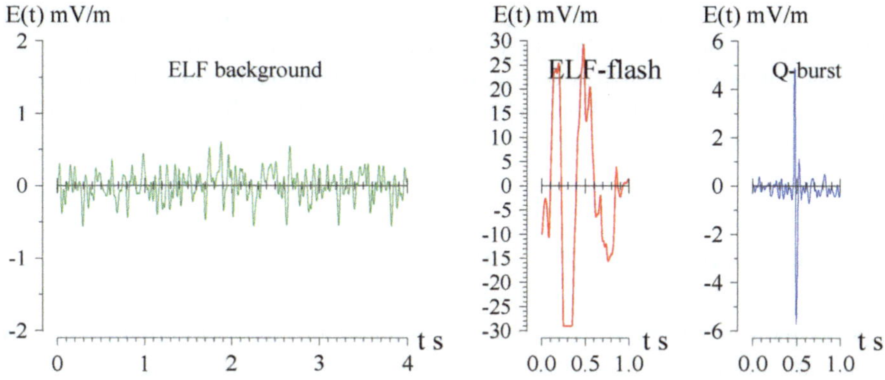

Fig. 5.1 Typical waveforms of natural ELF radio signals: ELF flash, Q-burst and continuous background signal

2001). Red sprites are discussed in many publications; we mention only the recent monographs and special issues (Füllekrug et al.2006, Special Issue: Journal of Atmospheric and Solar-Terrestrial Physics (2003), Planetary Atmospheric Electricity 2008). The 'red' spectrum of radiation from the sprite parent lightning stroke is explained by the discharges with the long and multiply bent current channels having the extraordinarily durable continuing current. At any rate, experimental data and interpretation indicate that Q-bursts might be used for the remote sensing of sprites all over the globe (Whitney et al. 2011).

The pioneering experiments in SR were expensive and laborious since the data acquisition, accumulation, and processing were done by the analog equipment. As a rule, a special device was applied in the spectral processing called a sonograph, and sometimes the computers were used for the purpose. Signal records were made by the mirror galvanometers, by the high-speed chart recorders, or by filming or by the tape recorders. The latter fixed the frequency modulated signal, as the frequency band of SR is too low for performing its direct records on magnetic tapes. The sonographs were widely applied in the measurements in the western countries. This device initially recorded the realization of ELF radio noise of 10 min duration on the magnetic drum. Afterwards, the record was replayed with the speed 10 times higher, and this raised the frequencies by a factor of ten. The signal 'compressed' in time was fed to the bank of filters in the audio frequency range (or to a single tunable filter whose frequency is switched with every turn of the drum). Spectral data were presented in two modes. One of them showed the averaged power spectrum of the whole record. While, the other mode provided the sonogram or the dynamic spectrum over the time–frequency axis. The latter was obtained as a result of the exposition of a cylinder of the photo paper.

The less expensive and more convenient alternative techniques were permanently sought in the ELF signal accumulating and processing. For example, the Ukrainian group elaborated a special analog spectrum-analyzer for the dual coherent records of SR (Bliokh et al. 1971). Japanese groups applied in addition to sonograms, the digital processing of continuous data series (Ogawa et al. 1968, 1979; Ogawa and Murakami 1973; Ogawa and Otsuka 1973). A cardinal solution for the SR monitoring was suggested and realized in the work by Nelson (1967). He made and exploited, during a year, the SR mode tracker: an analog device that continuously recorded the amplitude, peak frequency, and Q-factors of the first SR mode.

5.2 Monitoring of SR Parameters

The SR mode tracker was an analog device automatically performing the current processing of natural radio noise. The results were recorded by the chart recorder. It was elaborated by Nelson (1967) when the first operational amplifiers appeared together with the metal–oxide–semiconductor (MOS) transistors. The circuit was called "gyrator", which transforms the complex impedance of a capacitor into the equivalent inductance impedance. The effect is achieved by placing the capacitor into the negative feedback of the operational amplifier. The gyrator was included into the parallel resonance circuit tuned to the 8 Hz frequency and having the quality factor (the Q-factor) of about 10. By using the special loop of negative feedback, the resonance circuit follows the peak frequency of the first SR in the input noise. Thus, the first resonance mode tracker appears that records the output data on the paper. Data included the peak amplitude, frequency, and Q-factor. The first device worked during a year (Nelson 1967) and provided such interesting results that similar devices were manufactured for other modes and exploited for several years by many research groups: T. Ogawa and K. Sao in Japan and C. Polk in the USA. The records, covering almost six years, were made by C. Polk at the West Greenwich observatory, Rhode Island, USA. These first SR investigations were supported by the US Navy, with the aim on the elaboration of the ELF radio link with the submarines ("Sanguine-Seafarer" Project). The natural radio signal acts as interference for such a communication channel line (special issue of IEEE Trans 1974, Com-22, No. 4).

An apparent advantage of the mode tracker is the automated signal processing performed in the course of measurements. An experimentalist obtains the 'ready' data in the form of temporal variations of amplitude, peak frequency and the Q-factor (or the width of resonance line). The information extent and its compact form remain beneficial even now. For instance, the continuous registration is carried out of the SR parameters at the Hungarian Nagycenk observatory since May 1993 (Sátori 1996; Sátori and Zieger 1996; Sátori et al. 1996). However, the data is obtained nowadays by using the computers instead of analog devices.

We know from the early monitoring of the global electromagnetic resonance that there is a tendency of replication of particular patterns day after day. The curves might deviate, of course, owing to random nature of the field sources. The daily configurations become rather stable after averaging over the interval of a month or so. This is why one finds predominantly the diurnal variations averaged over the month in the literature.

5.3 Peak Frequencies and Effective Zone Occupied by Global Thunderstorms

We show in this section how the information on the global thunderstorm activity is extracted from the classical SR records. We pick the monthly averaged records of the diurnal frequency variations of the first SR mode on the yearly span (Ogawa et al. 1968). The vertical electric field component was recorded at the Kyoto observatory (35° N, 137° E). The first peak frequency reaches its maximum $F_1^{MAX}(UT)$ and minimum $F_1^{MIN}(UT)$ value during the day, the UT is the Universal (Greenwich) time. By following Nickolaenko and Rabinowicz (1995), we introduce the frequency range $dF_1 = F_1^{MAX}(UT) - F_1^{MIN}(UT)$ and present the relevant data in Table 5.1. The record by Ogawa et al. (1968) started on February 1967 and continued until January 1968. The first and the second columns in this table show the month and the year of observations correspondingly. The third column contains the average frequency range for the given month.

SR data will be compared with the orbital observations of the first optical transient detector (OTD) (Orville 1981; Orville and Henderson 1986). This satellite counted the number of the midnight optical flashes. The fifth and sixth columns of Table 5.1 contain the number of optical flashes and the year of observations (the relevant months are listed in the first column).

Experimental data of Table 5.1 are presented in Fig. 5.2 together with supplementary curves. The abscissa of the major (lower) frame shows the month, and two functions are plotted along the ordinate on the logarithmic scales. The blue line with dots corresponds to the left ordinate, and it shows the annual variations of the diameter of the effective circular area occupied by the global lightning activity. The red curve with squares is plotted on the right ordinate, which shows the number of midnight optical flashes detected by the satellite OTD on the orbit.

A short explanation is necessary. Diurnal variations of the peak frequency of the SR are due to the motion of the source with respect to the observer, and the effect appears due to small Q-factors of the resonance oscillations (e.g. Nickolaenko and Hayakawa 2002). Diurnal variations of the peak frequency are the most pronounced when the resonance is excited by a point source (Jones 1969; Galejs 1970; Ogawa and Otsuka 1973), i.e., when all planetary strokes are concentrated at one point. Relevant model variations of the first peak frequency are shown in the left upper inset of Fig. 5.2 for three different source widths $D_1 < D_2 < D_3$. In the model computations demonstrated in the inset, the random independent strokes were uniformly distributed within a circle of diameter D positioned at the equatorial point where the local time was 17 h. The observer was found at the Kyoto latitude and the zero longitude. During the day, the center of the source moves around the globe, and the curves were obtained showing the relevant diurnal variations $F_1(UT)$ in the left inset. The minimum source–observer distance corresponds to 17 h UT and the distance becomes the largest one at 05 h UT. At 11 and 23 h, the center of the source crosses the nodal line of the first SR mode, and the frequency F_1 exhibits an abrupt change. A detailed description

Table 5.1 Frequency range (Hz) and the effective source diameter (h) found from the SR data compared with the number of midnight flashes detected by the OTD orbiter

Month	Year of Schumann resonance observations	Monthly averaged frequency range dF_1 Hz	Effective source diameter h	Number of midnight flashes	Year of OTD observations
January	1968	0.15	5.5	2,440	1978
February	1967	0.2	4.5	1,890	1978
March	1967	0.07	9	2,113	1978
April	1967	0.08	8	3,947	1978
May	1967	0.07	9	3,433	1978
June	1967	0.1	7	3,369	1978
July	1967	0.12	6.2	3,481	1978
August	1967	0.4	2.8	2,517	1978
September	1967	0.2	4.5	1,826	1977
October	1967	0.12	6.2	2,181	1977
November	1967	0.12	6.2	2,174	1977
December	1967	0.15	5.5	2,891	1977

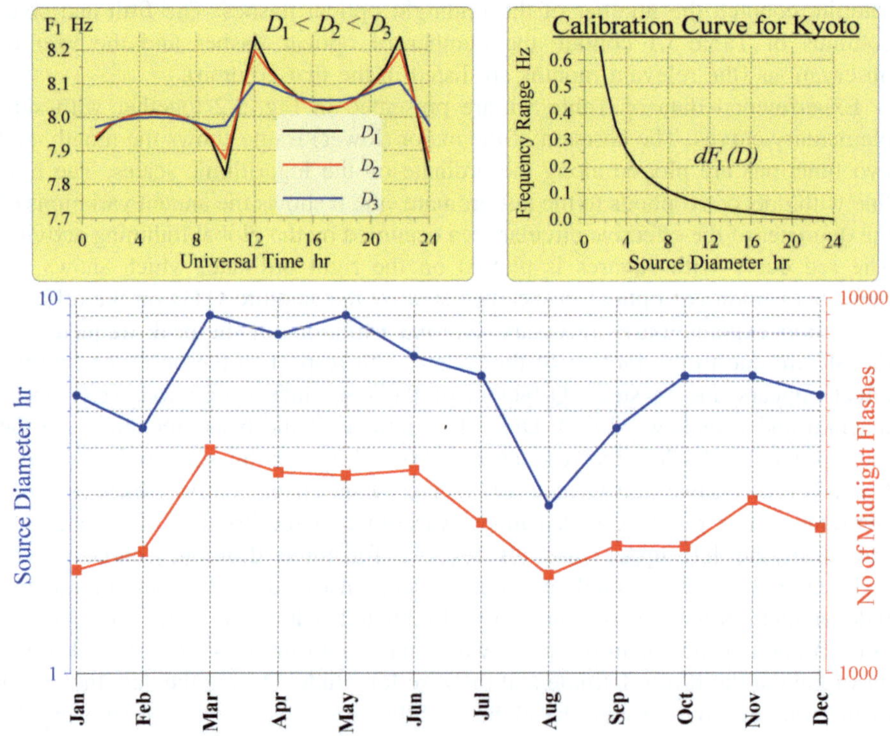

Fig. 5.2 Comparison of optical orbital data (*red line* with *squares*) with the effective diameter of the source (*blue line* with *dots*) found from the range of the first SR peak frequency. The calibration curve for the Kyoto observatory relates the range dF_1 with the source effective diameter D

might be found in Nickolaenko and Hayakawa (2002). Here, we demonstrate that widening of the area, covered by lightning strokes, tends to reduce the range of frequency variations dF_1.

By using the daily patterns, the frequency range dF_1 is found as a function of the source width *D*. Thus the calibration curve is obtained shown in the right upper inset of Fig. 5.2 (Nickolaenko and Rabinowicz 1995). One can transform the observed ranges $dF_1(UT)$ into the effective source widths *D* with the help of this curve. The related quantities are collected in the third and the fourth columns of Table 5.1, and the *D*(*UT*) function is plotted by the blue line with dots in Fig. 5.1.

Thus, two curves of Fig. 5.2 were obtained: the red line with dots presents orbital observations, and the blue one depicts the effective source area inferred from the SR records. To improve correspondence of these two curves, the blue line (SR data) was shifted forward by one month. Such a shift might be justified by relatively long (11 years) mutual delay in observational data. Statistical processing (regression analysis) of two series revealed the following link between the number of midnight flashes *N* and the frequency range dF_1:

$$N = (dF_1)^{-0.44} \times 10^3 \tag{5.1}$$

By applying the same regression analysis, one may connect the number of midnight flashes *N* with the effective area *S* (in km^2) of the circular zone where the lightning strokes are concentrated. This relation has the following form:

$$N = 4.5\,(S)^{1/3} \tag{5.2}$$

and the relevant cross-correlation coefficient is very high: $R = 0.9$.

The regression coefficient $A = 4.5$ in Eq. (5.2) corresponds to the number of midnight flashes in the unit (1 km^2) area of the ELF source. The index $B = 1/3$ characterizes the growth in the flash rate with the effective area of the global lightning activity.

The correlation found in such dissimilar data is easily explained. Daily variations of the peak frequency arise from the motion of thunderstorms around the globe because the varying source–observer distance modulates the spectral pattern and the peak frequency in the cavity (Nickolaenko and Hayakawa 2002). Model computations (see insets in Fig. 5.2) showed that the diurnal frequency range reduces with an increase of zone occupied by lightning strokes (Jones 1969; Galejs 1970, 1972; Nickolaenko and Rabinowicz 1995; Nickolaenko et al. 1998). Such a decrease is quite natural: the diurnal variations vanish when the lightning sources are uniformly distributed over the globe.

Obviously, the model is very simple of a single area occupied by thunderstorms, and we discuss its applicability below. Popularity of such a model is explained by the fact that it has only a couple of free parameters: the source size and position of its center. Global thunderstorms take place predominantly in tropical regions over the continents where the Asian, African, and American centers are found. The lightning activity tends to 'step over the oceans' in its diurnal motion from one center to another. Nevertheless, the model of a single moving source is efficient when explaining the SR observations, especially at the first resonance mode. The success is explained by the wavelength equal to the Earth's circumference. These long radio waves do not resolve such insignificant details as continents or the oceans. A more sophisticated source distribution is necessary for explaining experimental data at higher modes.

Despite the high cross-correlation of data relevant to Eq. (5.2), we must be cautious and consider these relations as provisional. The reason is that only the midnight flashes were taken into account, and the resonance data covered just the one-year span. From the physical point of view, a positive correlation seems natural between the number of optical lightning flashes and the width of zone covered by the global thunderstorm activity. Therefore, further investigations would be desirable. Similar processing was performed of the longer series of SR data collected at the Hungarian Nagycenk observatory (Nickolaenko et al. 1998). The effective source width reached its minimum in the winter and summer, and it was maximal in the intermediate seasons. Such behavior was attributed to an impact of the monsoon periods on the global thunderstorms.

Detailed long-term observations of the diurnal pattern of the SR frequencies in the vertical electric field allowed for deducing the duration of global seasons (Sátori 2003; Sátori et al. 2003, 2009). In the boreal simmer the global thunderstorms occupy the northern latitudes, and they reside at the southern latitudes in the winter and rapidly move from one hemisphere to another in the spring and autumn. Such a motion is clearly seen in the Nagycenk observatory records as the turning upside down of the peak frequency daily patterns.

Three types of diurnal variations were recoded during the year. The first type, "January" or winter type is observed from November to April for about 165 days per year. Variations of the summer or "July" type are observed from the late May to the late August during ~98 days. The autumn or "October" variations occupy the September–October period and last for ~60 days. Other variations cannot be attributed to any of these types, and they are observed during the remaining 42 days of the year in spring. During this time interval, the area occupied by thunderstorms substantially spreads reducing the frequency range of diurnal frequency variations. Changes in the shape of diurnal frequency variations are attributed to the meridional shift of the global lightning activity in different seasons. Durations of the seasons are uneven in the sense of the global thunderstorms. Thunderstorms stay in the Southern hemisphere during 165 days versus 98 day or their "Northern stay". In spring and fall, thunderstorms quickly transit across the equator and simultaneously spread over the planet. Despite the thunderstorm concentration over the continents, their northward or southward drift depends on the temperature of the tropical Pacific Ocean. A comparison with the ocean surface temperature revealed that the northern drift of thunderstorms (in the end of boreal winter) is associated with an increase of the middle Pacific surface temperature by a few centigrade.

We want to stress that obtaining the comprehensive geophysical information from the SR data is not the goal of this book. Our purpose is demonstration of some characteristic links between the geophysical information and ELF radio

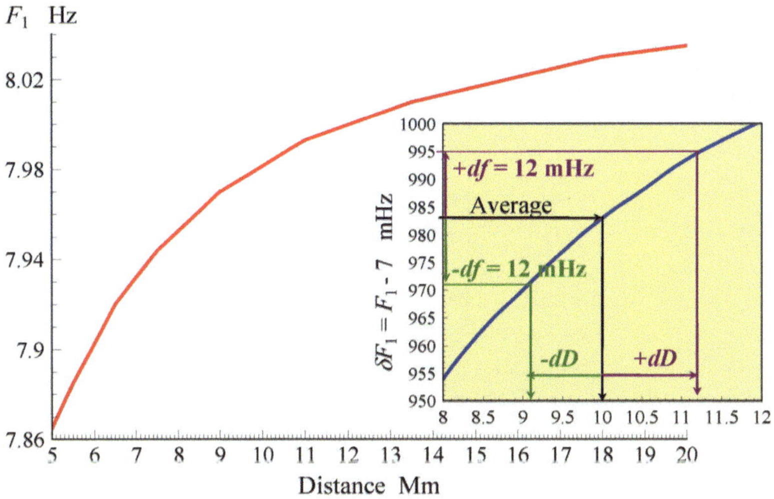

Fig. 5.3 Daily range of the first mode frequency in the horizontal magnetic field against the median distance from the observatory to the compact source of 5 Mm width

waves. We want to show on a few particular examples in what way such data were obtained from the SR monitoring. Our goal would be achieved when a reader is able to similarly process his own data or combinations of these with the records of other observatories.

5.4 Peak Frequencies of Magnetic Field Component

In the previous section we used the vertical electric field component whose daily range of peak frequency depends on the source size. In contrast to the electric field, the first SR peak frequency of the horizontal magnetic field is proportional to the contemporary source–observer distance (Nickolaenko and Hayakawa 2002). To demonstrate this property, we apply the model of a single compact source with the center positioned at different distances from the observer. The source is characterized by a parabolic distribution of individual stroke distances within the interval of 5 Mm. The first peak frequency of the horizontal magnetic field component increases as the distance expands between the observatory and the source center. Computations data are shown by a red curve in Fig. 5.3. The abscissa depicts the distance in Mm from the observer to the center of zone occupied by thunderstorms. The first mode frequency of the magnetic field is shown on the ordinate in Hz. The red curve demonstrates the distance dependence of the first mode frequency pertinent to the source of 5 Mm width. The inset in Fig. 5.3 illustrates the same dependence in more detail, and the $\delta F_1 = (F_1 - 7)$ value is shown in millihertz along the ordinate.

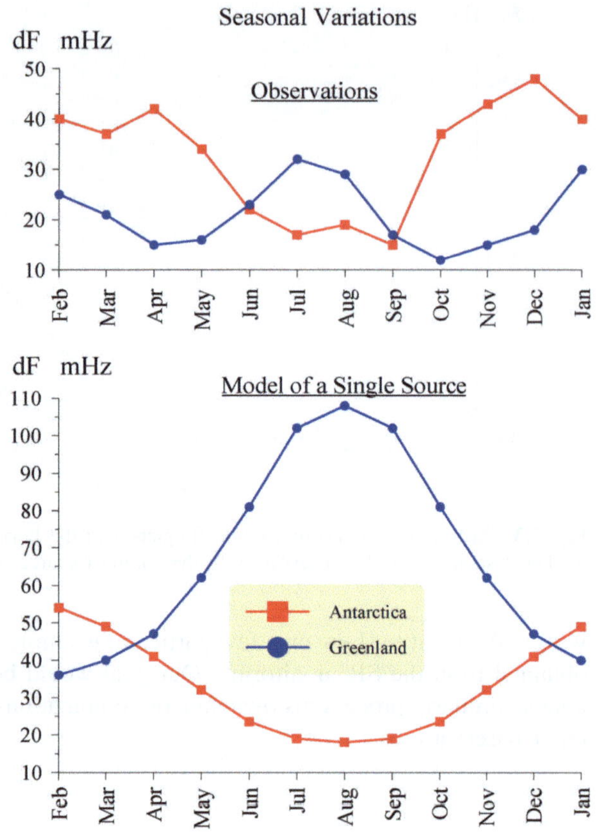

Fig. 5.4 Experimental data and interpretation in the model of a single thunderstorm center of finite size

The red curve in Fig. 5.3 is in fact a calibration curve that allows us to estimate the source distance by the contemporary peak frequency of magnetic field. Simultaneously, this curve connects the daily variations of the first mode frequency in the H–component with changes in the source-observer distance. Indeed, let the daily averaged source–observer distance be 10 Mm. In this case, the median daily value of the first peak frequency is equal to 7.983 Hz, as black arrows demonstrate in the inset of Fig. 5.3. Let us suppose that the frequency varies by ± 12 mHz during the day. The violet and green arrows explain obtaining the minimal source—observer separation of 9.1 Mm and the maximum distance of 11.2 Mm. We apply this property to interpret experimental data shown in Fig. 5.4.

The measured data were presented in the paper by Schlegel and Füllekrug (2000). These ELF records were performed at the observatories Arrival Heights, Antarctica (78° S and 167° E) and Sønderstrømfjord, Greenland (67° N and 51° E) collected by the system of global ELF/VLF monitoring installed by Stanford University (Fraser-Smith et al. 1991). The upper frame in Fig. 5.4 reproduces experimental data by Schlegel and Füllekrug (2000). The months are plotted on the abscissa and the average deviations of the first peak frequency from its median value are shown along the ordinate. The horizontal magnetic field component was recorded, and unfortunately no data was given concerning the median peak frequency. Plots in the upper frame of Fig. 5.4 depict deviations of the first resonance frequency measured simultaneously in the Northern hemisphere (blue line with dots) and in the Southern hemisphere (red line with squares). By using the calibration curve of Fig. 5.3 and the experimental records of frequency deviations, we estimated the range of annual meridian drift of the global thunderstorms. Then we "reconstructed" the frequency deviations and showed these in the lower frame of Fig. 5.4 in the same format.

Experimental data has a rare advantage of simultaneous measurements on the Northern (Greenland) and the Southern (Antarctica) hemispheres. The global thunderstorms move northward and southward with the season. Since the first peak frequency of horizontal magnetic field is proportional to the source–observer distance, the northern and southern observatories must register the opposite annual variations in the daily mean values of the first SR frequency. The idea is confirmed by the upper frame in Fig. 5.4, which shows an increase in the frequency deviation at Greenland combined with the decrease at Antarctica during the boreal summer. There are pronounced semi-annual variations, but we ignore them now. The annual thunderstorm drift must be seen in the daily median frequency, but this data is absent. We will have to model somehow the annual motion of thunderstorms. The "fast" diurnal variations are superimposed on the slow seasonal trends. These are caused by the daily circling of the globe by thunderstorms and depend on the daily minimal and maximal source–observer distances.

Let us accept that the lightning source circles the globe along the fixed parallel. For the observer at latitude υ, the daily range of the source–observer distance (in Mm) is $\delta D = 2\theta_0/9$ where $\theta_0 = (90° - \upsilon)$. This relation is readily obtained by considering two limiting cases. One is the minimum distance when the source and the observer occupy the same meridian: $D_{MIN} = \theta_S - \theta_0$. The second is the maximum distance when the source and the observer occupy the "opposite" meridians: $D_{MAX} = \theta_S + \theta_0$. By using particular coordinates, we estimate the distance ranges of $\delta D = 2.66$ Mm for the Antarctic observatory and $\delta D = 5.11$ Mm for the Greenland site.

When obtaining the lower plots in Fig. 5.4, we must introduce the median source–observer distance. By assuming that thunderstorms are positioned on the equator (averaged over the year), we obtain that for the observatories at the 67° and 78° latitudes the median distance to equator (annually averaged source) is 8 Mm. In the same fashion as we used before, we find that for the 14 mHz downward variation at Antarctica, the minimum distance was 7.3 Mm. The 12 mHz upward variation for Greenland correspond to the 8.7 Mm distance. Obviously, the real annual drifts are underestimated in our approach. Still we accept that the median distance for observatories to thunderstorms is 8 Mm and its annual variations is 0.7 Mm. In this case the source latitude as a function of month M is $\theta_S = 6° + 0.7 \cdot 9° \cdot \cos(M\pi/6)$, where the latitude is measured in degrees. It is easy now to compute the diurnal variations at the Antarctica and the

Greenland sites for every month of the year and calculate the relevant daily deviations shown in the lower frame of Fig. 5.4. The red model curve with squares correspond to Antarctica, and the blue line with dots depicts the Greenland model data.

Figure 5.4 indicates that simultaneous ELF records in the Northern and Southern hemispheres reveal the seasonal North–South motion of the global thunderstorm activity. There is a quantitative similarity between the model and observational data. However, the model was oversimplified, and it does not explain all details of these variations. More realistic models must include the longitudinal dependence of thunderstorm distribution, say, to include three major thunderstorm centers or to apply the LIS–OTD orbital data. The experimental data on the median frequencies would be helpful also. Still, the concept of a compact source successfully explains the impact of the seasonal North–South drift of the activity.

We demonstrated that the model of a single moving source agrees fairly well with the SR observations, being rather efficient. By applying this model to the experimental data on the SR frequencies, one can obtain the effective diameter of zone occupied by the lightning strokes (vertical electric field component) and the amount of the annual North–South drift of the center of such a zone (horizontal magnetic field).

5.5 Long-Term Variations of Peak Frequencies

The previous section demonstrated the advantage of positioning of the SR observatories at high latitudes. First of all, the source–observer distance only slightly varies during the day. In addition, records of the Northern and Southern hemisphere complement each other in the sense that thunderstorms come closer to one observatory and simultaneously retreat from the other due to the seasonal meridional drift. The cumulative resonance intensity of the records performed at such observatories presents the level of the global thunderstorm activity and it is practically independent of the diurnal and annual motion. Unfortunately, measurements at the sub-polar observatories are hard to do, so that there are rare relevant publications.

SR monitoring at the middle latitude sites allowed for obtaining many interesting features of global thunderstorms. For instance, the periodic variations were found ranging from the annual and semi-annual to a few days scales (Füllekrug and Fraser-Smith 1996, 1997; Sátori and Zieger 1996, 1999; Zieger and Sátori 1999; Sátori et al. 1999). We demonstrate in Fig. 5.5 the effective source width found from the first SR records at the Nagycenk observatory (47.6° N, 16.7° E). Data cover the interval from January 1995 to June 1998, and the month is shown on the abscissa. The upper plot demonstrates the collection of daily variations of the first mode frequency observed in the vertical electric field.

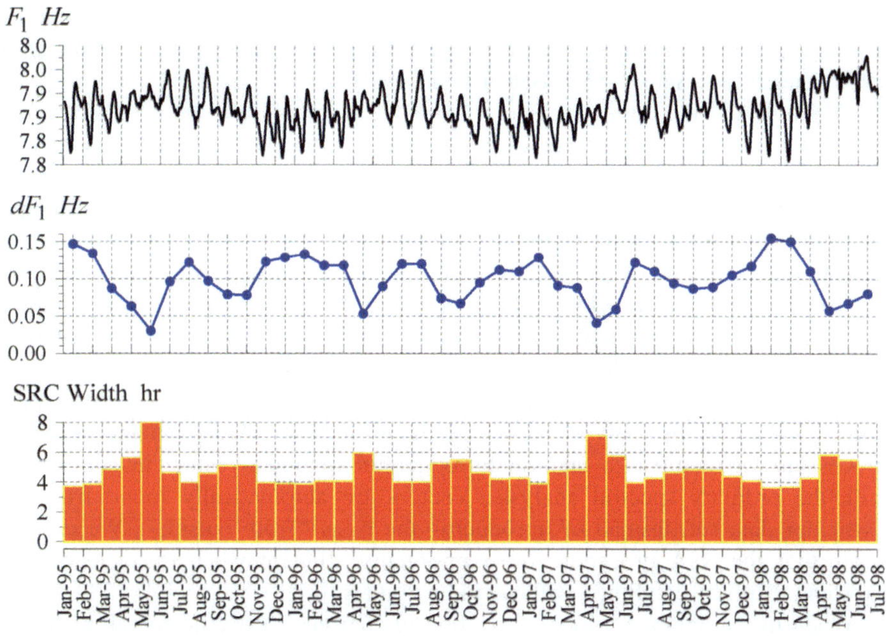

Fig. 5.5 Survey of diurnal/seasonal changes of the first SR mode frequency and seasonal variations of the effective source width

A few words should be said about the data presentation because it will be used in many cases. The abscissa of the upper plot is divided into several strips each corresponding to the particular month of the year. Every strip contains the 24 h (UT) daily variations. Thus, the upper graph combines 42 strips of individual diurnal variations $F_1(UT)$ corresponding to every month. The ordinate shows the frequency in Hz, and the monthly averaged data are attached one after another. Such a form is rather informative, compact, and simple, because it allows observing the diurnal patterns and their monthly/annual changes concurrently.

The middle plot in Fig. 5.5 shows monthly variations of the frequency range $dF_1 = F_1^{MAX}(UT) - F_1^{MIN}(UT)$ where $F_1^{MAX}(UT)$ and$F_1^{MIN}(UT)$ denote the daily maximum and minimum values of the peak frequency. Thus, one obtains a single point at the middle plot from the 24-point pattern of the upper plot. Bars in the lower diagram show the seasonal variations of the effective source width that we discussed in Sect. 5.3. The particular diameter of zone occupied by the global thunderstorms was derived from the frequency range dF_1 by using the calibration curve computed for the Nagycenk observatory (Nickolaenko et al. 1998).

One can note the periodic components present in seasonal variations of the frequency range and the effective source width. The annual period is clearly seen in the minima of frequency range occurring in May each year (it corresponds to the thunderstorm spreading over the globe conditioned by the monsoon period). The semi-annual variation is conditioned by less pronounced dF_1 minima in the

September–October interval. While, annual and semi-annual variations were noted also in the diurnal/seasonal patterns of the SR modal intensity (Nickolaenko et al. 1998; Zieger and Sátori 1999; Sátori et al. 1999).

To single out the annual and semi-annual variations from the data series, one can apply the "Caterpillar" algorithm (Danilov 1996; Danilov and Zhiglyavsky 1997; Golyandina et al. 2001). This algorithm is a realization of the Singular Spectral Analysis (SSA) procedure, which is regarded also as Method of Principal Components (MPC). From the formal point of view, the "Caterpillar" algorithm is an automated version of the MUSIC (multiple signal classification procedure, see e.g. Marple (1987)). The idea of this processing is based on forming a rectangular matrix from the observational data series and finding the eigen-functions (vectors) and the eigen-values of this matrix. The eigen-vectors form the basis, or the set of characteristic functions pertinent to the particular time series. The eigen-values correspond to the intensity of particular eigen-functions. An advantage of such a processing technique is a 'natural introduction' of eigen-vectors in comparison with the ordinary spectral analysis. Vectors are automatically found from the particular data set, instead of introducing the basis "from outside". We demonstrate that the SSA procedure is exceptionally efficient when the data contain substantial trends combined with periodic signals.

When initial data are a long realization of stationary periodic signal, a 'square wave', for instance, the SSA processing finds out the sum of sinusoids of appropriate frequencies and amplitudes, so that its output asymptotically coincides with the series of ordinary Fourier transform. Application of the 'Caterpillar' software provides the results similar to the ordinary filtering when the bank of filters is tuned to the annual and sub-annual components (Sátori and Zieger 1996). The important distinction is that the SSA processing finds automatically the preliminary unknown periodic components. The sinusoidal waveform of the components extracted is not obligatory, but they are definitely orthogonal.

Figure 5.6 demonstrates the properties of SSA procedure when applied to a specially designed testing signal. The upper plot in Fig. 5.6 shows the conventional data on passengers 'transported' by a prosperous company during 12 years. We see that the number of passengers grows in time, and it undergoes seasonal changes.

Initial one-dimensional series of data $x_k = x(t_k)$ was transformed into a matrix having L elements in the row. This means that signal samples with index $1 \le k \le L$ are placed in the first row of the matrix. A parameter L is called the Caterpillar length and is equal to 12 months in this particular sample. The second row of data matrix is formed by elements with $2 \le k \le L + 1$. The third row includes the elements with $3 \le k \le L + 2$, etc. The process continues until the rectangular matrix is filled. Then, the eigen-vectors and eigen-values are found. The most intense principal component (PC) in the test data was the inter-annual trend (PC1), which is shown by the violet line in the second plot of Fig. 5.6. The annual (PC2 and PC3) and semi-annual (PC4 and PC5) variations have smaller intensity, which are shown by red and blue quasi periodic curves. Their combination is depicted by the lower green line in Fig. 5.6. Since the annual term and its

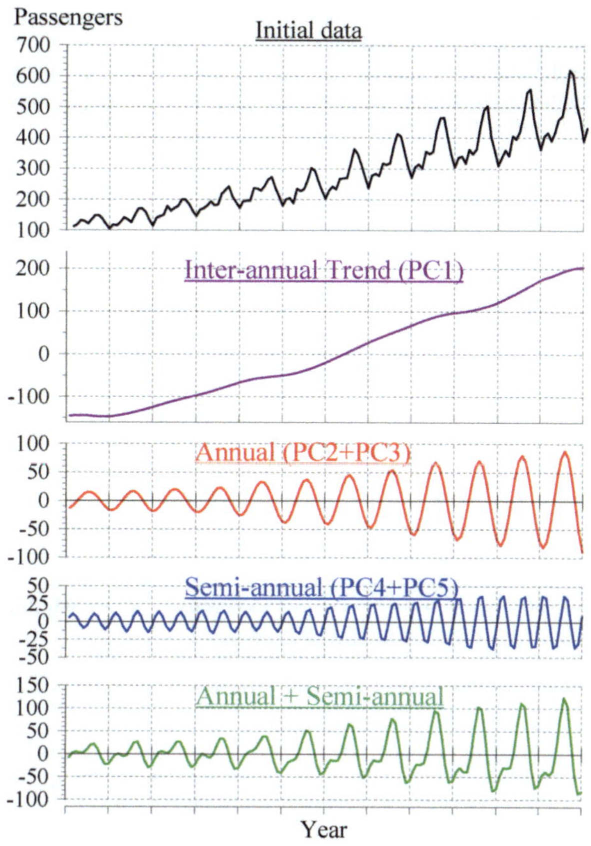

Fig. 5.6 Processing of the test data by the "Caterpillar" algorithm

'second harmonic', the semi-annual variation, have a definite phase shift, their sum found by the 'Caterpillar' procedure has a rather complicated 'beating' outline. It is worth noting here that quasi-periodic components present in the signal are grouped in pairs similarly to the customary Fourier-series expansion comprising the sine and cosine components. The SSA algorithm has a large dynamic range. Its major advantage is an unfailing separation of the trends from the periodic series, as this could not always be done by the conventional processing.

We show the monthly variations in Fig. 5.7 of the effective source width measured in hours (1 h = 15° = π /12). The abscissa shows the months from May 1993 to July 1998. Four upper plots are relevant to the source width *D* in hour. The upper black line depicts the source width deduced from the range of the first SR frequency at the Nagycenk observatory. This curve presents the initial data involved in the SSA processing. It is natural to choose the "Caterpillar" length $L = 12$ months, because this ensures separation of the annual and semi-annual components from the inter-annual trends and the unavoidable random deviations. The second red line in Fig. 5.7 shows the major principal components PC1 and PC2 corresponding to the annual variation which has the period of ~12 months

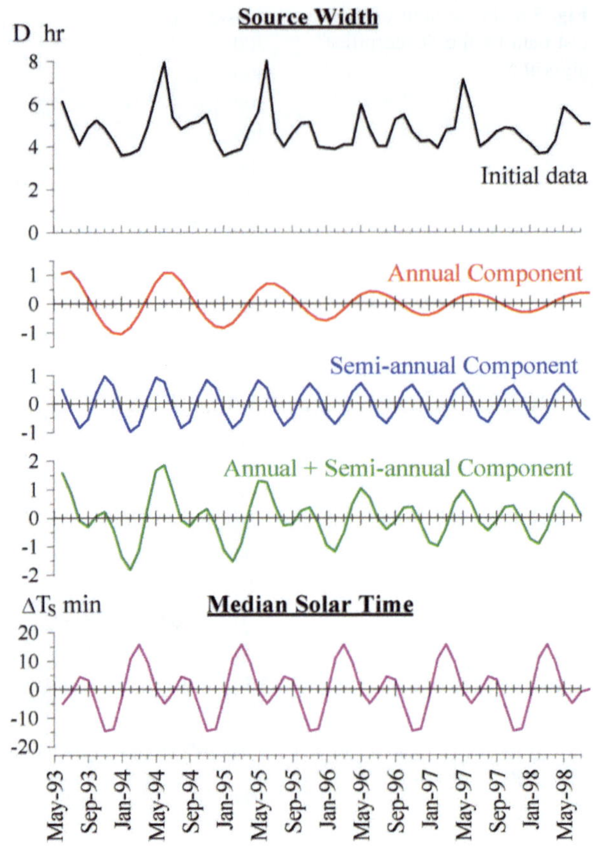

Fig. 5.7 Seasonal variations and the principal components of the effective effective: source width compared with the median solar time

and gradually decreasing amplitude. The third plot presents the semi-annual period corresponding to principal components PC3 and PC4. The fourth green curve is the sum of annual and semi-annul variations. Amplitudes of annual and semiannual components are about 1.5 and 1 h correspondingly. Their sum vary from −2 to +2 h around the median 3 h value. Thus, the effective source diameter may vary during the year by 30° or by 3.333 Mm around the average size of 45° or 5 Mm.

The characteristic annual and semi-annual variations of the source width are combined in the curve of a 'beating mode' with decreasing amplitude. Such a waveform reminds us of variations related to a physical process with no relevance to the global thunderstorms at the first glance. We speak about annual variations of the median solar time being the astronomical data on the seasonal variations of the noon moment. This variation is called analemma, and we show it in minutes by the lower curve in Fig. 5.7. Annual deviations of the median solar time (position of the center of the Solar disk in the sky) from the astronomical UT time arise form the small eccentricity of the Earth's orbit. Owing to the orbital eccentricity, the angular velocity varies periodically of the Earth's motion around the Sun, and the visible Sun crosses the local meridian ahead or behind the astronomical local noontime. Deviations reach $\sim$20 min. The analemma ΔT_S we show in the lower plot of Fig. 5.7 was calculated by using the following formula (Nickolaenko and Hayakawa 2002):

$$\Delta T_S = 0.128 \cdot \sin A_N + 0.167 \cdot \sin 2A_N \tag{5.3}$$

Here $A_N = 2 \cdot d_N\, \pi/365$ is the normalized orbital angle of the Earth. It depends on the day in the particular month. It is convenient to use the recast day number

$$d_N = [10 + (M - 1) \times 365/12 - d_D] \tag{5.4}$$

where M denotes the month and d_D is the day number in the month.

Of course it is not easy to find the reason explaining the similarity of the green and magenta curves in Fig. 5.7, as well as an obvious three month delay. Different data—astronomical and electromagnetic observations—hint that there exists an intricate connection between the global thunderstorms and the median solar time. We think that this is not a coincidence, but simultaneously we do not know the exact nature of such a relationship. Most probably, the connection is as follows. The effective source width regularly reaches its higher value when the monsoon period starts in May. This effect is caused by the annual North–South drift of thunderstorms: these follow the annual motion of the Sun in the sky. The process is driven by the tilt of the Earth's rotational axis to the plane of its orbit. We expect the dry and rainy seasons in the tropics forming the two-harmonic curve of the annual variation of the source size. Obviously, the tilt of the terrestrial poles causing the seasonal variations is combined with the non-even motion of the planet around the Sun. Thus, the two processes appear each having two harmonics and being synchronized in time by the orbital motion: one is the median solar time (the analemma) and the second is the tilt of rotational axis (seasonal drift of thunderstorms and the monsoon periods). At any rate, the similarity is intriguing.

Summarizing above, we note that monitoring of the SR (we omit higher modes as these data might be found in the literature) indicates the annual variations associated with the North–South drift of thunderstorms and the semi-annual changes caused by the spreading of the global activity twice per year. Annual variations are attributed to the non-symmetric distribution of continents around the equator. The continents occupy the northern hemisphere, and since the thunderstorms occur over the ground, their general amount increases during the boreal summer. This is the reason why the SR intensity increases in summer. On the other hand, the summer and the winter thunderstorms are concentrated within the compact areas, and thus there arise the semi-annual variations. SR data reflect both the annual North–South drift and the intermediate spreading of activity.

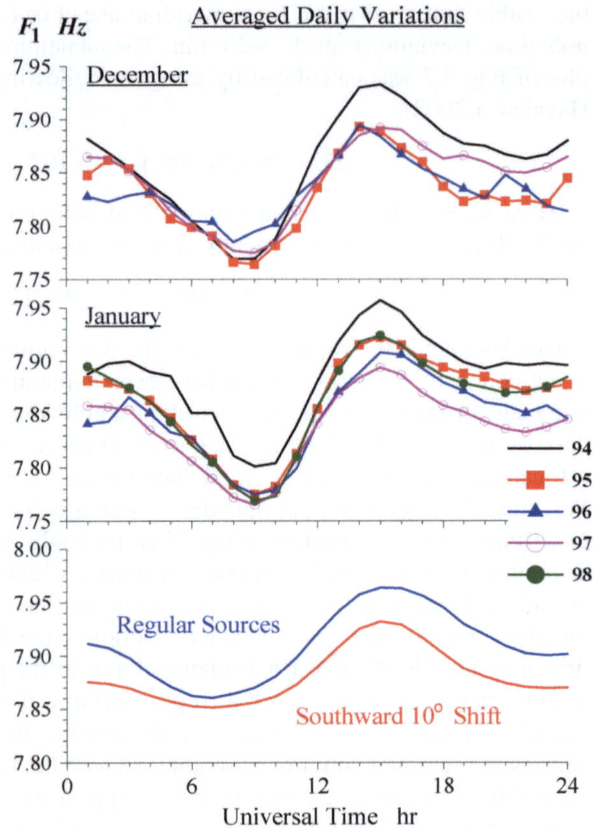

Fig. 5.8 Diurnal variations of the first peak frequency observed at Nagycenk in December–January of the period of 1994–1998. The lower plot presents the model variations for the regular position of the global thunderstorm centers and for African and American storms shifted southward by 10°

5.6 Source Position or Ionosphere Modification?

The upper plot in Fig. 5.5 demonstrates that monthly averaged daily patterns of the first SR modal frequency tend to reproduce year after year. Some inter-annual deviations are also present, which can lead to interesting conclusions. Figure 5.8 presents characteristic daily variations for December and January of a few successive years when the thunderstorms are concentrated in relatively small areas. The upper frame contains daily patterns of December 1994–1997. The downward shift of the whole curve is obvious during the years of 1995 and 1996, and the curve of 1997 occupies an intermediate position. The middle plot in Fig. 5.8 shows similar variations observed in January 1994–1998. The lowest curve here corresponds to the year 1997. The greatest systematic frequency shift reaches −0.05 Hz, and it was observed during the winter of 1996–1997 (Sátori and Zieger 1999).

There are two different explanations of the shift in the resonance frequency pattern. One of them is the meridional drift of the global lightning activity (Sátori and Zieger 1996, 1999; Nickolaenko and Rabinowicz 1998), and the alternative interpretation exploits a global alteration in the ionosphere height (Nickolaenko and Rabinowicz 1998). The meridional shift was suggested and discussed by Sátori and Zieger (1996, 1999), but with no supporting model computations. We model and discuss both the explanations (Nickolaenko and Hayakawa 2002). The Nagycenk observatory has a favorable position with respect to the African and the South American thunderstorms: the median source–observer distance is close to the nodal line of the third SR mode in the vertical electric field component. This means that $P_3(\cos\alpha) \approx 0$ when α is equal to the median angular distance from the observer to the African or American thunderstorms. Hence, a slight northward or southward drift of activity changes the third mode records in the cardinal way. Experimental data were interpreted by the meridional drift of African thunderstorms by 4°–8°, which was conditioned by El Nino Southern Oscillation (ENSO) episode. We present in Fig. 5.8 the computational data supporting this qualitative conclusion (the lower plot), but using only the first mode frequency.

The model of three major global centers of lightning activity were used in particular computations. The centers are found in the South-East Asia, Africa, and America and have the activity varying in time, as seen in Table 5.2 (WMO 1956). The first column contains the UT in hours. Columns 2–4 describe relative variations of the activity in Africa, America, and Asia (in arbitrary units). The fifth column shows the effective diameter of the area occupied by thunderstorms as the function of the month number, which we evaluated from the diurnal frequency range of SR oscillation in years 1995 and 1995 (Nickolaenko et al. 1998).

Every center contributes to the SR intensity in correspondence with the data of Table 5.2. The circular areas are centered at given coordinates with the thunderstorms uniformly distributed within the circles. All centers have equal diameters, which we pick for a given month from the fifth column of Table 5.2. For instance, we use the lines 12 and 24 when computing the 'regular' patterns of December shown in the lower plot of Fig. 5.8. The model incorporates the seasonal changes of the coordinates of the centers that were estimated from the monitoring of SR frequencies and amplitude (Nickolaenko et al. 1998):

$$\theta_S = \theta_1 + \theta_2 \cos\left[\frac{(M+3)\pi}{6}\right] + \theta_3 \cos\left[\frac{(M+3)\pi}{3}\right] \tag{5.5}$$

and

$$\varphi_S = \varphi_1 + \varphi_2 \cos\left[\frac{(M+3)\pi}{6}\right] + \varphi_3 \cos\left[\frac{(M+3)\pi}{3}\right], \tag{5.6}$$

here M is the month number and θ is the co-latitude, so that $\theta = \pi/2 - \upsilon$ where υ is the geographic latitude and φ is the longitude, all angles being in radians. Parameters relevant to the yearly motion of the centers are listed in Table 5.3.

The blue smooth line in the lower plot of Fig. 5.8 was computed for the regular position of thunderstorm centers in December ($M = 12$). The SR intensity was computed in the model of three centers of global thunderstorsm acitivty being the

Table 5.2 Intensity and size of three major thunderstorm centers

UT h	Intensity a.u.			Effective diameter h
	Africa	America	Asia	
1	0.51	1.38	0.12	3.8
2	0.43	1.17	0.15	3.7
3	0.32	0.91	0.22	4.3
4	0.24	0.78	0.38	5.6
5	0.18	0.65	0.6	6.7
6	0.13	0.52	0.88	5.3
7	0.14	0.43	1.24	4.9
8	0.22	0.4	1.46	5.0
9	0.35	0.35	1.65	5.0
10	0.55	0.28	1.58	4.9
11	0.88	0.22	1.37	4.2
12	1.38	0.18	1.14	3.4
13	2	0.17	0.95	3.6
14	2.4	0.18	0.7	3.7
15	2.7	0.28	0.6	4.6
16	2.47	0.47	0.5	5.3
17	2.03	0.87	0.4	5.7
18	1.63	1.42	0.3	4.8
19	1.3	1.9	0.3	4.3
20	1.02	2.24	0.2	4.8
21	0.8	2.45	0.15	4.9
22	0.6	2.3	0.1	4.7
23	0.51	2.08	0.08	4.1
24	0.45	1.75	0.08	3.9

Table 5.3 Parameters of regular seasonal drift of three global thunderstorm centers

	$\theta_1°$	$\theta_2°$	$\theta_3°$	$\varphi_1°$	$\varphi_2°$	$\varphi_3°$
Africa	100	5	5	28	8	8
America	104	30	−15	−66	12	−6
Asia	87.5	10	0	120	20	0

sum of individual intensities from every center at a given hour UT (Nickolaenko et al. 1998; Nickolaenko and Hayakawa 2002). This is due to the Poisson nature of the pulse flux formed by radiation of independent lightning strokes. The inter-pulse interference in such a signal becomes compensated, and the cumulative intensity is the sum of intensities of separate pulses. Similarly, the intensities of thunderstorm centers are also summed. The model suggests that the strokes are uniformly distributed in the circular area of the diameter specified in Table 5.2. The random amplitudes of the currents in the strokes were normally distributed. The power spectra for every center were computed by incorporating positions of the centers relative to the observer. The resulting spectrum was the sum of weighted contributions from the three global thunderstorm centers, and the relevant coefficients were picked from the columns 2–5 of Table 5.2 for the particular UT time. Since the thunderstorm activity moves around the globe during the day, the weighting coefficients depend on time. The particular propagation constant was used in computations $\nu(f) = (f - 1.5)/6.3 - if/100$. Such a dispersion corresponds to the SR peak frequencies of 7.8, 14.1 and 20.4 Hz.

The smooth red line in the lower plot of Fig. 5.8 presents the model diurnal variations of the first peak frequency when the African and American centers are shifted southward by 10°. The position of the Asian center remained unchanged. As one may see, the frequency shift for this particular geometry provides a desirable displacement of the daily pattern. Thus, computational data support the conclusion that a relatively small southward transfer of the global thunderstorms might cause an inter-annual variation noted experimentally in the first peak frequency. The conclusion by Sátori and Zieger (1996, 1999) has got a quantitative support: the global thunderstorms have 'dived' southward in the winter of 1996–1997, and then they returned to the usual position. Such behavior seems reasonable in the context of the concurrent El Nino episode.

At the same time, we remark that regular decrease of the first SR frequency might be explained in terms of a completely different viewpoint, especially, when all resonance peaks simultaneously shift by approximately the same amount. Unfortunately, such information was not available in the published data; still we describe an alternative explanation. Analogous simultaneous variations of resonance parameters are considered later when treating the abrupt ionosphere disturbances caused by the Solar Proton Events (SPEs), cosmic gamma flares, or the seismic activity.

Monitoring and the campaign measurements of SR in 60-s has revealed definite sensitivity of the resonance parameters to abrupt changes in the lower ionosphere caused, for example, by the high altitude nuclear explosions. The effect of the "Starfish" explosion was noted as an abrupt simultaneous reduction of all resonance frequencies by $\sim$0.5 Hz (Gendrin and Stefant 1962; Madden and Thompson 1965; Nelson 1967). Our goal is to evaluate the role of possible natural variations in ionosphere height.

Let the global thunderstorms remain at their regular positions, while the modification occurs in the height profile of atmosphere conductivity. We estimate the impact of such a modification by excluding any influence of the source–observer distance. For this purpose, we turn to the uniform distribution of lightning discharges over the planet and compute the power spectrum of the vertical electric field. One has to connect the propagation constant ν and the ionosphere parameters for describing the spectral modifications related to the ionospheric disturbance. The modified Greifinger and Greifinger (1978) model was used for the purpose having the upper reference height (Nickolaenko and Hayakawa 2002). Much more complicated models could be used, but we apply this simplest approach, as the sophisticated models provide similar results and do not introduce a novel physical sense.

The model by Greifinger and Greifinger (1978) exploits the exponential vertical profile of air conductivity and introduces two characteristic heights. The lower height is found as an altitude where the conductivity currents become equal to the displacement currents at a given frequency. This is the height to which the electric field penetrates of an incident ELF radio wave, so that it is regarded as the electric height h_E. The upper height is the level to which the magnetic field reaches of the ELF radio wave, and it is called the magnetic height h_M (see Chap. 14). Both the heights are the logarithmic functions of frequency for the exponential conductivity profile of the lower ionosphere (Nickolaenko and Rabinowicz 1982, 1987). We apply here the model with the upper reference height (Nickolaenko and Hayakawa 2002). In particular, we introduce the electric reference height $G_h = 89$ km where the conductivity currents become equal to the displacement currents at the $F_g = 10$ kHz frequency. Obviously, the electric height will go down with the decrease of frequency.

The electric and magnetic height are found from Nickolaenko and Rabinowicz (1982, 1987), and Nickolaenko and Hayakawa (2002):

$$h_E = G_h + \zeta \cdot \ln(f/F_g) \tag{5.7}$$

$$h_M = h_E - 2\zeta \cdot \ln(2\zeta k) \tag{5.8}$$

The height scale ζ of the conductivity profile lies in a range of 3 and 5 km for the ambient day and ambient night conditions. After computing the characteristic heights for the given frequency, one can calculate the relevant propagation constant by using the relations:

$$\frac{\nu(\nu+1)}{(ka)^2} = \frac{h_M + i\zeta_M \frac{\pi}{2}}{h_E - i\zeta_E \frac{\pi}{2}}, \tag{5.9}$$

$$\nu(f) = \sqrt{\frac{1}{4} + \frac{h_M(f) + i\zeta_M \frac{\pi}{2}}{h_E(f) - i\zeta_E \frac{\pi}{2}}(ka)^2} - \frac{1}{2} \tag{5.10}$$

Formally, ζ_E is the scale height at the altitude h_E and ζ_M is that at the altitude h_M. For simplicity, a single scale profile is used $\zeta_M = \zeta_E = \zeta$, k is the wavenumber in free space, and a is Earth's radius. The SR power spectrum might be computed at any given frequency when we know the $\nu(f)$ value. By varying the frequency and the scale height, we obtained a set of power spectra of vertical electric field component. These were combined into the 2D maps of power spectra presented in Fig. 5.9.

The color map in Fig. 5.9 depicts the power spectra of electric field in the frequency range 6–22 Hz for the height scales varying from 3 to 5 km. Three SR modes are seen in Fig. 5.9 as 'ridges' extended upward and to the left. Since the conductivity profile is attached to the upper reference frequency $G_h = 89$ km, an increase of the height scale lowers the electric height of ionosphere at a given frequency. The losses increase in this case and the peak frequency decreases. This

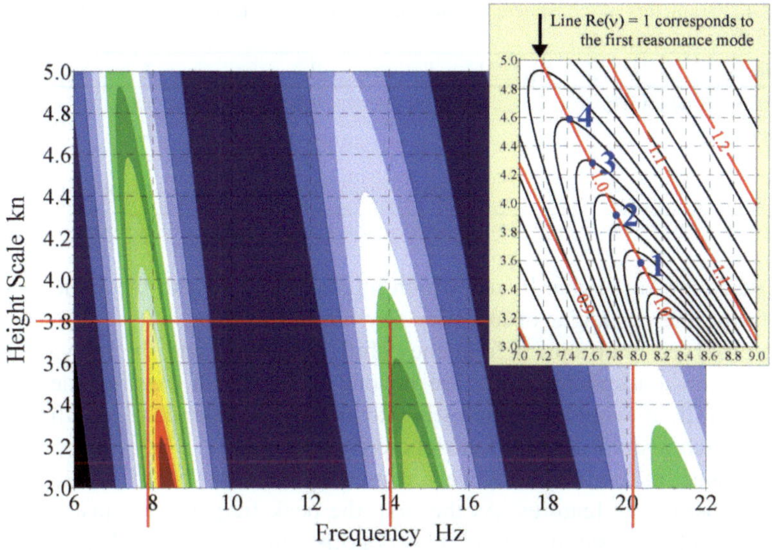

Fig. 5.9 SR peaks over the frequency—height scale plane. The inset shows the contour map in the vicinity of the first mode. The *red lines* here plot a few fixed values of $\mathrm{Re}\{\nu\} = 0.8, 0.9, 1.0, 1.1, 1.2$, and 1.3

is why the resonance peaks shift to the lower frequencies and become wider when the height scale increases.

The inset in Fig. 5.9 shows the detailed picture in the close vicinity of the first SR mode. It combines two contour maps over the plane frequency–height scale. The black and white map depicts the 'elliptical' lines of constant resonance intensity that are extended to the upper left. The first mode peak is seen in this map as a 'ridge' connecting the tips of resonance curves. The second map is superimposed on the first one, and it shows to the profiles $\mathrm{Re}\{\nu(f)\} = \mathrm{const}$. The red lines show the constant $\mathrm{Re}\{\nu\}$ levels ranging from 0.8 to 1.3. The inset clearly demonstrates that the first resonance frequency corresponds to the condition $\mathrm{Re}\{\nu(f_1)\} = 1$ for an arbitrary height scale. The numbered blue points mark the peak frequencies of 8, 7.8, 7.6, and 7.4 Hz, so that one may observe that their values correspond to the scale heights equal to 3.6, 3.95, 4.3, and 4.6 km.

By using plots of Fig. 5.9, one can evaluate the height scale of the conductivity profile and the characteristic ionosphere heights from the observed peak frequencies. Relevant data are collected in Table 5.4. We see that a decrease in the first SR frequency from 8 to 7.6 Hz corresponds to the global reduction of the h_E by 5 km. The inter-annual of shift −0.05 Hz was present in the experimental data of Fig. 5.8, so that it might be caused by the global reduction in the ionosphere height by approximately 0.8 km. We must remark here that a particular ionosphere model is a mere illustration, and the 'integer' values we use just explain the idea instead of deducing the exact ionosphere model. The realistic height h_E and the

Table 5.4 Parameters of ionosphere profiles corresponding to the particular first mode frequency in Fig. 5.9

Point No.	1	2	3	4
Resonance frequency Hz	8	7.8	7.6	7.4
Scale height ζ	3.6	3.95	4.3	4.6
h_E km	63.3	60.87	58.1	55.8
h_M km	111.7	113.3	114.8	116.1

height scale are somewhat lower, which might be found with the same technique when fitting the particular observations by the model data.

The observed inter-annual frequency shift was small, and the relevant modifications in the ionosphere height are beyond the reach of existing measurement facilities. Still, it would be interesting to obtain 'independent' confirmation or denial of such a small regular ionosphere change.

These spectral modifications were caused by the ionosphere profile, and they have two distinctive features. As the first, the peak frequencies must vary simultaneously and consistently at all observatories in the world. This property is exploited for detection of the ionosphere modifications caused by an intrusion of the cosmic gamma rays or the energetic solar particles (protons or electrons). The second distinctive property is that these globally observed modifications must occur at all SR modes simultaneously: the resonance frequencies increase or decrease all together. Detection of such changes is facilitated when they start abruptly. In short, the distinctive features are: the alterations caused by the ionosphere modification occur simultaneously all over the globe and are present at all SR modes (see also Sect. 9.3).

In our opinion the most probable cause of the observed inter-annual trends in the first SR frequency was a dislocation of the field sources rather than changes in the ionosphere height.

To conclude the section, we describe in what a way one can estimate the parameters of ionosphere profile from the SR data. First of all, a particular model should be chosen. We described the model with the upper reference height $G_h = 89$ km and the relevant upper $F_g = 10$ kHz frequency. The models with the lower reference heights are also used in the literature, say, with $G_h = 55$ km and the lower $F_g = 10$ Hz. We will apply such a model in the subsequent sections, but here let us return to Fig. 5.9. By fixing a particular height scale (the red horizontal line in Fig. 5.9), we can deduce the relevant peak frequencies of different resonance modes (the vertical red lines). By varying the scale, we obtain different sets of peak frequencies listed in Table 5.5.

By picking the typical SR frequencies (8, 14, 20 Hz) and interpolating the data in Table 5.5, we estimate the effective planetary height scale of the ionosphere profile. The result is $\zeta \approx 3.8$ km. This estimate agrees with the scales used in the literature (Nickolaenko and Rabinowicz 1982, 1987; Sentman 1990, 1996; Füllekrug 1995, 2000; Fraser-Smith and Bannister 1998; Pechony and Price 2004).

Table 5.5 The first three peak frequencies for different scale heights of the ionosphere profile

ζ km	f_1 Hz	f_2 Hz	f_3 Hz
3.5	8.05	14.29	20.15
3.8	7.88	14.02	20.06
4.0	7.74	13.82	19.87

5.7 SR Amplitude and Global Thunderstorm Activity

Monitoring of modal amplitudes of the SR oscillations provides information on the global lightning activity. Amplitude data is easy to interpret in comparison with the frequency variations, since the signal amplitude depends first of all on the current level of the global thunderstorm activity and on the source—observer distance. Data on SR amplitude were widely published in the literature (Polk 1969; Sentman and Fraser 1991; Williams 1992; Füllekrug 1995; Sentman 1995; Chrissan and Fraser-Smith 1996; Nickolaenko et al. 1996, 1998, 2011; Nickolaenko 1997; Füllekrug and Fraser-Smith 1997; Füllekrug et al. 1999; Belyaev et al. 1999; Price and Melnikov 2004; Price et al. 1999; Sátori et al. 1999; Zieger and Sátori 1999; Hobara et al. 2000a, b, 2001; Schlegel and Füllekrug 1999, 2000; Sekiguchi et al. 2004, 2006; Pechony 2007; Yatsevich et al. 2006, 2008; Yatsevich 2009; Hayakawa et al. 2011), and the major results are summarized as follows:

1. SR intensity reflects variations in the global thunderstorm activity.
2. Resonance amplitudes behave similarly even when recorded at the observatories remote from each other.
3. There is a positive correlation between the Schumann resonance intensity and the fair weather field, which indicates the same source of these phenomena—the planetary thunderstorms.

We show in Fig. 5.10 the characteristic results of SR monitoring: the seasonal-diurnal variations of the modal intensity of the vertical electric field component (Nickolaenko and Hayakawa 2002). Three contour maps are shown in this figure over the month (the abscissa)—UT (the ordinate) planes. The upper map shows the daily–seasonal dynamics of the first SR mode. The second and the third maps represent similar data for the second and third SRs. Records cover the period from January 1994 to July 1998 (Sátori and Zieger 1999). This figure demonstrates that general dynamics of resonance intensity is similar at each mode in respect to the time of the day and to the season. The general behavior and characteristic details are simultaneously present in all three maps. Such a resemblance proves that the intensity of natural ELF radio signal reflects the level of thunderstorm activity, while the motion of lightning activity that varies the source–observer geometry plays a less important role. The spatial distribution of each mode of oscillations is indeed described by a particular Legendre polynomial. Therefore, alterations in the source–observer distance must provide different effects in the three maps of Fig. 5.10, but such distinctions are insignificant. One may conclude that the

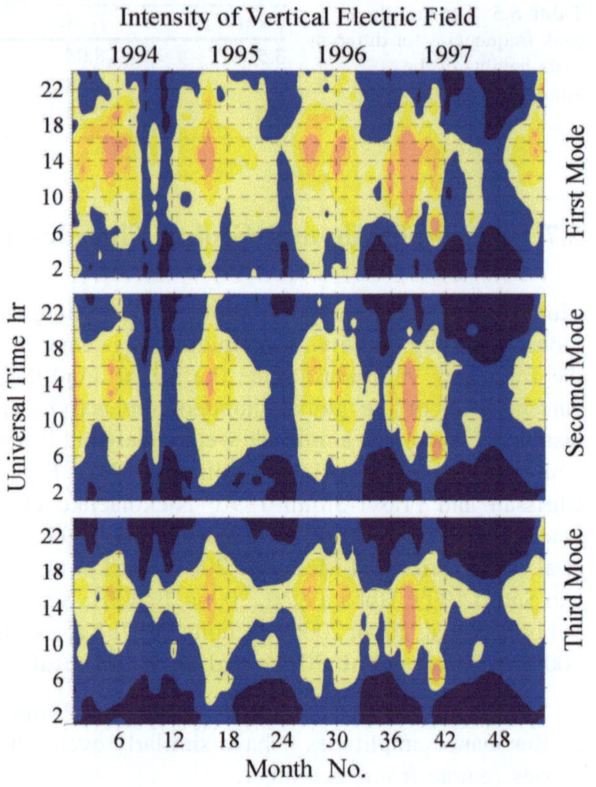

Fig. 5.10 Diurnal–seasonal variations in the intensity of three SR modes monitored at the Nagycenk observatory from January 1994 to July 1998

diurnal and seasonal drift of the global thunderstorms is reflected by the daily patterns of peak frequencies, while the amplitude portrays the source intensity.

Observational data and the model results indicate that the intensity of separate resonance modes fairly well represents the current level of the global thunderstorm activity. However, this is done in a much better way by the cumulative SR intensity, which is the sum of intensities of individual modes. Such a conclusion was achieved in many works discussed in Chap. 4 of the book by Nickolaenko and Hayakawa (2002). The diurnal—seasonal variations of the cumulative SR intensity (the first three modes) is depicted in Fig. 5.11. The dominance of African thunderstorms is obvious, which is seen as a maximum in the vicinity of 14–15 h UT. Contributions are also seen from the Asian (9–10 h) and American (20–21 h) thunderstorms. Activity of these latter centers is highly variable during the year. SR data perceive a noticeable increase in the global thunderstorm activity in the Asian and African centers in the spring of 1997.

We must emphasize that all data used in this section for estimating the global thunderstorm activity was recorded at a single ground-based observatory positioned in the middle Europe, i.e., far away from the thunderstorm centers. We did not use any records of meteorological networks or of scanning the Earth from the

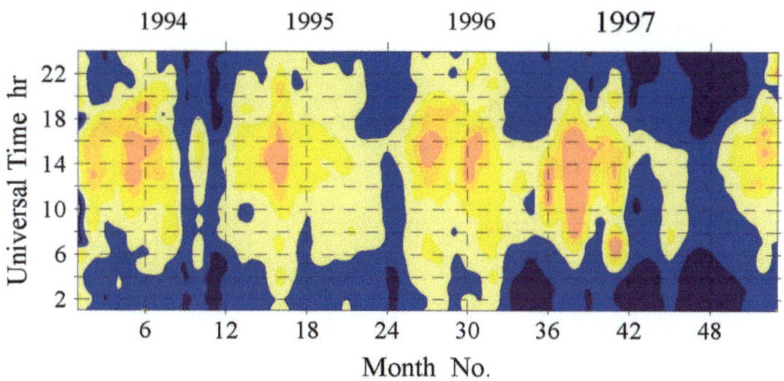

Fig. 5.11 Cumulative SR intensity over the month—UT plane

space, or did not undertake round-the-world expeditions. Everything was obtained from the SR records.

We applied the singular spectral analysis (SSA) to the cumulative intensity as we have done to the records of resonance frequencies. We derived the monthly averaged diurnal variations of the cumulative intensity, which present the thunderstorm activity. Then we integrated the intensity over the 24 h of the day and obtained the 'initial series' shown in the upper plot of Fig. 5.12. The lower plots in this figure exhibit the principal components derived by the SSA procedure. The abscissa presents the months from the August 1993. The cumulative SR intensity (an estimate of the global lightning activity) is plotted on the ordinate in arbitrary units. The upper black line shows the 'initial data' of the cumulative SR intensity. The second plot (the red line) is the annual variation present in the initial data and having the period close to 12 months. Its amplitude increases in time. The third blue line demonstrates the semi-annual variation, which is not very regular owing to small semi-annual variations combined with the slower trends. The lower green line depicts the sun of annual and semi-annual variations. One may note that in distinction from the source width shown in Fig. 5.7, the source intensity does not show any variations reminding us of analemma. This means that the seasonal spreading of the global thunderstorm activity over the planet has a minor impact on their cumulative intensity.

As we already mentioned, the similar results might be obtained by using a more traditional processing: by applying the digital band-pass filters (Zieger and Sátori 1999; Sátori et al. 1999). Such a processing was applied to the original day after day records and revealed the following periods in the SR intensity: 185, 152, 108, 73, 55, 47, 27, 20, 13.5, 10.4, and 8.6 days. The same procedure was applied to the concurrent records of the solar wind velocity, and the periods found were compared with the above series. The following SR periods were absent in the solar wind: 108, 73, 47, 20, and 10 days. Therefore we can attribute the processes with these particular periods to purely atmospheric phenomena. Of course, it is clear that periods 35, 27, 13.5, and 8.6 days are associated with variations of the solar wind, so one may say that these have a cosmic origin. It is worth noting that

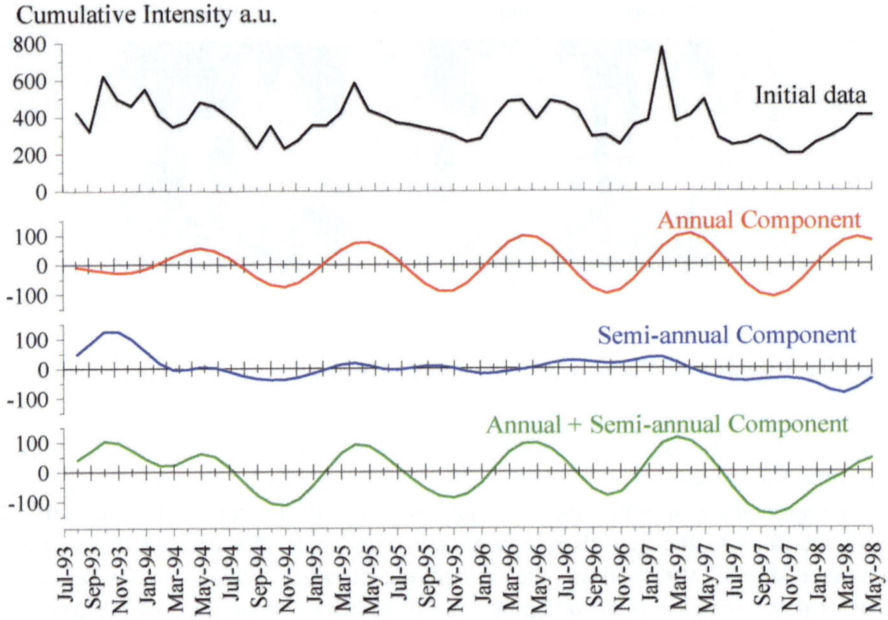

Fig. 5.12 Principal components in seasonal variations of the daily averaged integral SR intensity being the estimate of the global thunderstorm activity

amplitude of the basic 27-day period was smaller than its second harmonic of 13.5 days (Zieger and Sátori 1999).

We must remark that periods of 20, 10, and 5 days were found not only in the natural radio noise, but also in amplitudes of the man-made radio signals (e.g. Shimakura et al. 1991; Füllekrug and Fraser-Smith 1997; Hayakawa 2001). These periods on radio signals are often explained by the influence of planetary Rossby waves (Ahlquist 1982; Harth 1982; Hayakawa 2001) or by the atmospheric gravity waves propagating in the atmosphere and modulating the thunderstorm activity and the lower ionosphere (Molchanov et al. 2001).

A special investigation was performed (Sátori et al. 1999) of the periodic components present in the data collected at two remote observatories: Nagycenk (Hungary) and West Greenwich (USA). The records of SR parameters were processed by the digital band-pass filters tuned to annual and semi-annual frequencies. Measurements showed that annual variations occur in-phase at both the observatories, and this proves that the component has the same origin: seasonal variations in the global thunderstorm activity.

Another special technique was suggested for selecting the semi-annul variations of African and South American thunderstorm activity (Füllekrug and Fraser-Smith 1997). They analyzed the data recorded in different intervals of the day. The semi-annual component was revealed in the most obvious way in the records of the interval 12–16 h UT. We know that African thunderstorms dominate in this interval having the clear semi-annual component of two rainy and two dry seasons per year in the equatorial Africa. The semi-annual component was absent in the ELF data from the interval 16–22 h UT when the American activity prevailed. Such behavior is explained by the seasonal variations of the land temperature (see below). Here we must remark that association between the annual and semi-annual components and particular time intervals could be noticed in the maps of Figs. 5.10 and 5.11. Indeed, monthly variations in the central part of maps (the activity of Africa) contain two distinct maxima in a year period. The upper part of maps around 20 h (American thunderstorms) contains a single maximum. After finding this distinction in the ELF electromagnetic data, one can readily explain it by departures in the geographical configuration of Africa and America.

First of all, we must recall that SR oscillations are driven by the global thunderstorm activity. The lightning discharges are concentrated in the tropical area, i.e. close to the equator. The equatorial Africa occupies a substantial part of the continent, and two seasons follow each other here. One is the dry season, and thc other is the rainy season. The equatorial America is a narrow neck of land separating two oceans. Proximity of the Pacific and Atlantic oceans removes the possibility of dry seasons. This is the reason why the semi-annual variation is irrelevant to American thunderstorms.

This peculiarity of amplitude variations seems odd in the light of semi-annual component clearly seen in the peak frequencies of electric and magnetic field components. In fact, there is no contradiction, since the observed peak frequency depends on the whole ensemble of sources. It is impossible to 'cut off' or to 'add' the contribution to the peak frequency from a particular thunderstorm center. Similarly, the frequency variations cannot be attributed to a particular center. Diurnal variations of SR frequency are governed by the global positioning of sources and contain a noticeable semi-annual component. The frequency range depends predominantly on the source width or on the effective area covered by the electric activity. This latter definitely varies with the semi-annual period, while the cumulative intensity of thunderstorms might remain invariant.

We showed that modern monitoring of the SR has revealed the semi-annual variation in both the intensity and the peak frequency. It is interesting whether these two components were also present in the SR records performed in 60–70 s. For this purpose we turned to the analog record made by Sao group at the Tottori observatory (35.5° N and 134.4° E) from August 1967 until November 1970 (Sao et al. 1971, 1973). The record was performed with the help of a SR mode tracker, which monitored three resonance modes. The data were recorded on paper charts, and this archive was digitized and processed. The level of global thunderstorm activity was evaluated by using the cumulative intensity of the first three SR modes. A distinction of the data was that a single horizontal magnetic field component H_{SN} was recorded, so that the antenna angular pattern had the maxima oriented to the East and to the West. As a result, a characteristic twin-peak was present in the amplitude record conditioned by crossing the node of antenna pattern by the moving activity in the Asian center.

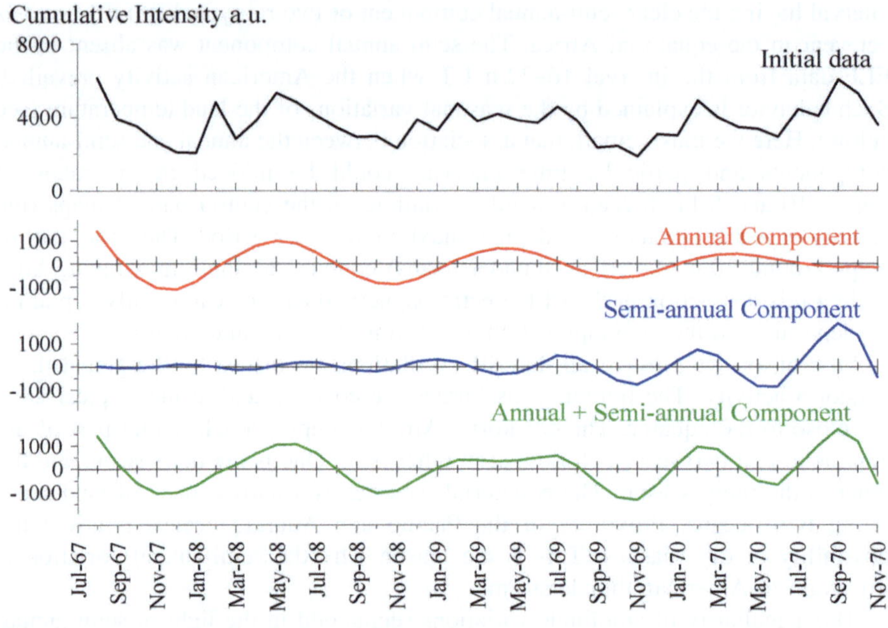

Fig. 5.13 Global thunderstorm activity estimated from the records of Tottori station in 60–70's. Annual and semi-annual harmonics are present in the SR record

The sum of three amplitudes squared provided the integral resonance intensity that served as a proxy for the level of global thunderstorm activity. A substantial increase in the activity was noted during boreal summer. Besides, the winter lightning activity was clearly seen in February conditioned by thunderstorms in the Sea of Japan. The final results of the signal processing are collected in Fig. 5.13. The abscissa shows the months from July 1967 to November 1970. The cumulative intensity and its principal components are plotted along the ordinates in arbitrary units. The upper plot illustrates the initial variations in the cumulative resonance intensity. One can immediately note the presence of annual and semi-annual components in this record. We extracted these components by applying the SSA (the Caterpillar procedure). The annual component is shown by the red curve in the second plot. It carries 44 % of the original intensity. The second harmonic or the semi-annual variation (blue line) contains 22 % of signal intensity. The processing suggested that amplitude of the annual component decreased in time, and that of semi-annul component increased. The sum of these two terms is shown by green line, which was responsible for $\sim$80 % of seasonal amplitude variations.

Measurements at Nagycenk in 90s showed a moderate increase of the annual component as seen in Fig. 5.12, while similar data from the Tottori observatory show a decrease in this component. Deviation might be caused by an unidentified inter-annual trend in the global thunderstorms similar to the solar cycles. At any rate, we must note a remarkable correspondence of the curves. Their periods and patterns agree in spite of completely different receiving equipments applied, and the field-sites were separated by the distance by ~1/3 of the Earth's circumference and in time by approximately 30 years.

Processing of the archive data on the SR amplitude recorded in Japan from August 1967 to November 1970 confirmed the existence of annual and semi-annual components. A comparison of these data with the Hungarian records of 90s demonstrates their high reciprocity, which possibly suggests some physical mechanism driving the global thunderstorm activity, causing the inter-annual changes.

The semi-annual variations are usually explained in terms of the Earth illumination by the Sun (e.g. Williams 1994). There are several climate zones on the Earth, and one of them is the tropical belt. The Sun is positioned right above the equator twice in a year, and there are two seasons—the dry and the rainy periods. Thus, the semi-annual variations are associated with these two seasons and are linked to the equinox times. There are four seasons at the middle latitudes, and alterations are synchronized with the summer and winter solstice times.

Monitoring of SR peak frequencies and the range of their daily variations implies that the area covered by the global thunderstorm activity expands twice in a year, and the maximum spreading is synchronized with the monsoon periods. Seasonal changes in the daily frequency range are found to correlate with the orbital optical observations, with the number of flashes from the midnight thunderstorms. Monitoring of ELF natural radio noises has revealed the semi-annual variations in the lightning activity of Asia and Africa. Their absence in the thunderstorms of America might be explained by differences in the geographical configuration. The tropical lands of Indonesia and Africa are extended along the parallel thus suggesting an opportunity and space for the thunderstorm spreading around the equinox periods. Concerning America, there is only a narrow strip of tropical land that connects the North and the South Americas, while both the continents predominantly occupy the middle latitudes. This neck of land does not allow for the thunderstorm spreading.

If we compare the annual and semi-annual variations concurrently present in the records of the peak frequencies and peak amplitudes (Figs. 5.6 and 5.12), we would observe the closeness of their phases. The yearly maximum of activity is observed in May and the minimum is found in the October–November period. The relative modulation of the activity reaches ± 25 % against its median level. The area covered by thunderstorms tends to increase concurrently with the growth of their cumulative intensity. The normalized variations in the effective diameter of zone occupied by the activity is also about ± 25 % with respect to the median size being ~5 h.

Such a connection is not a surprise. It is clear from physical considerations that a rise in the thunderstorm activity must be associated with the increase of the area. Indeed, a convective thunderstorm cell has some characteristic size and a definite productivity as a machine separating electric charges. The breakdown of the air slab takes place when the cloud voltage exceeds a relevant "standard value". It is clear then that a thunderstorm cell of a median productivity provides the typical

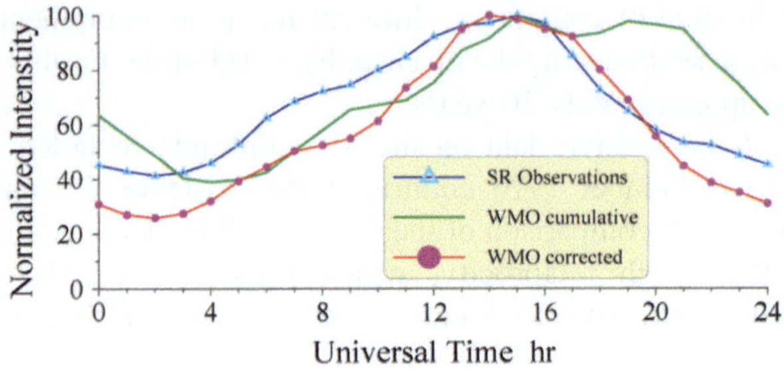

Fig. 5.14 Comparison of median diurnal variations in the global thunderstorm activity according to the WMO (1956) data and the cumulative intensity of SR on the 1993–1998 year span. Time correction −2 h for Asia and Africa, −4 h for America

lightning discharges with a 'standard' frequency. The value is well known, and it corresponds to one stroke in 20–40 s (Uman 1987; McGorman and Rust 1998; Rakov and Uman 2003). The global thunderstorm activity must be increased when one wants to raise the level of natural ELF radio noise, and this might be achieved only by escalating the rate of lightning strokes. This means that the number of thunderstorm cells should be increased, as these have the 'nominal' electric productivity. A greater number of cells means the corresponding enlargement of the area covered by the activity, and so the global electric activity of atmosphere must be proportional to the area covered by thunderstorms. Correspondingly, the growth of SR intensity means an inevitable decline in the diurnal range of the peak frequencies.

To conclude this section, we estimate the global thunderstorm activity by using the integral intensity of SR oscillations. For this purpose, we integrate the cumulate intensity shown in Fig. 5.11 form month to month for the fixed time of day, i.e., along the horizontal lines in the plot. We obtain, as a result, the long-term average diurnal pattern of normalized SR intensity. The data are shown in Fig. 5.14 by the blue line with triangles. The estimate of the global thunderstorm activity based on the SR monitoring reaches its maximum (100 %) at 15 h UT and the minimum (41.4 %) at 2 h UT. Thus, the cumulative intensity of the global electromagnetic resonance varies by a factor of 2.5 during the day.

We show in the same frame the classical curve (green line) of diurnal variations of the global thunderstorm activity obtained by the World Meteorological Organization (WMO 1956). This curve sums the intensity of the three global thunderstorm centers in Asia, Africa, and America (see Table 5.2). One may observe a good correspondence of these two curves in Fig. 5.14. Simultaneously, the patterns do not coincide; they have different maxima and minima, while the range of variations is similar. We know the history of obtaining the meteorological data based on registrations of the thunder sound, and we have in mind a great time delay of the records, so that we will try to fit the observational SR data by the patterns of WMO. We maintain the form of variations unchanged and modify the position of the curves for Asia, Africa, and America over the time axis. For

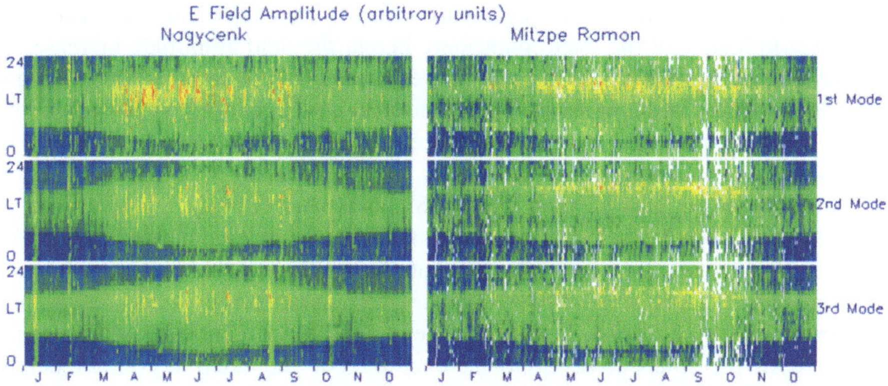

Fig. 5.15 "Terminator effect" in the SR records of Nagycenk and Mitspe Ramon

obtaining the best fit (red curve with dots), we must move ahead all individual WMO curves: those for Asia and Africa—by 2 h, and for America—by 4 h. The resulting red line has the peak close to the SR observations. An important distinction is that the range of diurnal variations in electromagnetic data is smaller than that of the climatological data. We will return to this issue later when discussing the presence of a 'podium' in the SR diurnal variations. The podium arises from the ordinary thunderstorms occurring on the 'wrong' time, in particular, from the nocturnal activity.

5.8 "Terminator Effect" in SR Records

The designation "terminator effect" appeared in the SR studies after the long-term records of signal amplitude were arranged over the plane month–local time (Melnikov et al. 2004, see also Sátori et al. 2006). The plots are given in Fig. 5.15, in which the data were picked from the records of Nagycenk (47.6° N; 16.7°) and Mitspe Ramon (31° N and 35° E) observatories. The left frames depict the seasonal–daily variations of the three SR modes recorded in the vertical electric field component at Nagycenk and averaged over eight years of monitoring (1994–2001). The right frames depict similar data from Mitspe Ramon averaged over four years (1999–2002). The dates are shown along the abscissa ranging from January 01 to December 31. The local time is plotted on the ordinate. Relative variations of the field amplitude are illustrated in color: the lower levels are blue and the higher amplitudes are red. Data are given for three SR modes.

Characteristic diurnal–seasonal variations with lens-like forms were found in the records of many observatories, which were immediately interpreted as an impact of the day–night asymmetry of the ionosphere. The boundary between the sunlit and the dark sides of the globe is called terminator; therefore the phenomenon got the name of "terminator effect". The statement was made that under the influence of the local ionosphere height the field amplitude increases at the dayside of the globe, particularly at the sunrise, and it decreases at the night side, at the sunset at the observatory. Unfortunately, this concept is not supported by the model data. Our computations (see Fig. 5.21) showed that the effect of ionosphere non-uniformity cannot exceed the 5–10 % level at SR frequencies (Nickolaenko 1986; Rabinowicz 1986, 1988; Bannister 1999).

The lower ionosphere boundary has a smaller height at the dayside of the globe in comparison with the night side, so that the Earth–ionosphere cavity is non-uniform (e.g. Bliokh et al. 1968). The height non-uniformity must result in a redistribution of the resonance fields in the cavity, and similar effects are widely exploited in the microwave cavity resonators. For the relatively small height deviation which is the case, the redistribution of the electric and magnetic fields obeys the following conservation law: the complete electromagnetic intensity remains a constant in the unit volume of the cavity $|\varepsilon_0 E|^2 + |\mu_0 H|^2 = \text{const}$. This means that an increase in electric field at a point is compensated by the decrease in the magnetic field at this point. In other words, since the electric field concentrates at the dayside with the lower height, the magnetic field is smaller here. At the same time, the magnetic field relocates to the wider side where the electric field is smaller. Therefore, the lens-like structures observed in the electric and in the magnetic fields must have the opposite sign (increase versus decrease): to be of different color in the maps similar to those of Fig. 5.14. The observational data by Melnikov et al. (2004) show the simultaneous increase in the both field components. Nevertheless, it was interpreted as an impact of the ionosphere height.

There are two more aspects making the day-nigh interpretation flawed from the physical point of view, and we mention them below.

The first problem follows from the principles of diffraction. The impact of ionosphere asymmetry might be perceived when comparing the two distinct propagation geometries: when the propagation path is positioned in the ambient day or in the ambient night conditions. The propagation path is the great circle arc connecting the points of source and the receiver. One can speak about propagation in the day (night) conditions when the path and the first Fresnel zone are found within the day (night) hemisphere. Unfortunately, the ambient day as well as ambient night conditions are not realized at the first SR mode: the wavelength is equal to the Earth's circumference, so that the whole globe is found within the first Fresnel zone. The radio waves of such a wavelength cannot 'resolve' the details like different hemispheres, and the cavity must behave as an average uniform, neither day nor night, resonator. Besides, the position of the most intense sources of natural ELF radiation is linked to the evening terminator line, so the purely day or nigh conditions are never realized at higher modes also.

The second peculiarity of observational data in Fig. 5.15 is in their outstanding similarity while the relevant frequencies (and hence, the wavelength) depart by the factor of 2.5–8, 14, and 20 Hz. The diffraction patterns for the waves of such different wavelengths must substantially deviate, but this is not the case as seen from the figure. Remarkable similarity of experimental variations at different modes indicates that the driving mechanism was something else, but not the wave diffraction at the day–night interface.

The problems in interpreting the experimental data disappear when one turns to the variable source intensity. Such an interpretation explains all properties of observational data. Moreover, the lens-like patterns are readily obtained in the uniform cavity model with the OTD source distribution. We demonstrate below that the 'terminator' effect reflects the everyday development/cease in the global thunderstorm activity seen at an observatory occupying specific longitude. The daily onset of the global lightning activity at the South-East Asia center coincides with the sunrise at the European observatories, while its ending in America is close to the passage of evening terminator over Europe. Thus, the misinterpretation has arised.

Let us demonstrate the seasonal–diurnal changes similar to the observations that we obtained in the model of a uniform cavity. We used the sources distributed in accordance to the data of OTD pertinent to a given month. The diurnal motion of thunderstorms was modeled with the help of special masks. In fact, one can derive the monthly averaged diurnal variations of the source distribution by using the original OTD data from the site www.ghrc.msfc.nasa.gov (Pechony 2007; Nickolaenko et al. 2006a). The OTD measurements and the results are presented in the comprehensive paper by Christian et al. (2003). The optical flash rate is given in the flashes per square kilometer per year for the global map with the $2.5 \times 2.5°$ resolution. We transformed these data into the number of optical flashes observed at a given cell of map during the year and constructed the map shown in the upper left frame of Fig. 5.16. Thunderstorms are concentrated over the continents (Christian et al. 2003; WMO 1956), and one can recognize the outlines of Africa, America, Europe, Australia, etc. in the frame. The lower distinct boundary is always seen in the map around the 40° S latitude. The upper edge of activity depends on the month, while the West–East extension also varies of the zone covered by lightning strokes. We will use the yearly averaged thunderstorm distribution for simplicity.

We use the masks for modeling the daily and seasonal changes in the thunderstorm distribution. The simplest mask is the 'Day' model, which is shown in the upper left frame of Fig. 5.16. This mask opens (i.e., activates) the strokes in the sunlit hemisphere of the planet, and at its center the sun is in the zenith point for the given month and given time of day. In the Day Model (DM), thunderstorm activity is ceased at night, so that the solar terminator line defines the unit step modulation function being equal to unity on the dayside and to zero on the night side of the globe. The right upper frame in Fig. 5.16 shows the Day model at the winter solstice of December 22 on the UT noon. Thunderstorms of Africa and South America are active at the moment.

Since the continents are distributed non-uniformly along the West–East and the South–North lines, the diurnal rotation of the planet is associated with changes in the number of lightning strokes (Christian et al. 2003). The South–North asymmetry of the continental distribution is responsible for the annual variations: due to

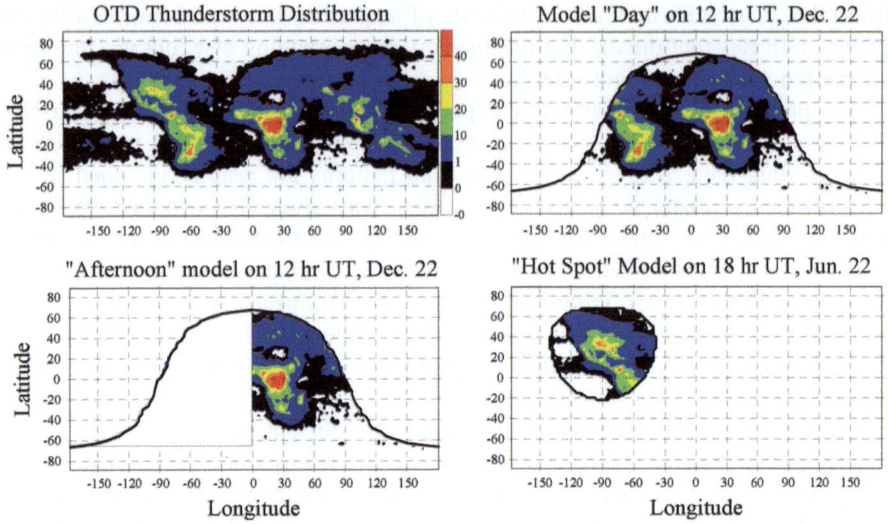

Fig. 5.16 Initial global map of thunderstorm distribution according to the OTD observations and effect of imposing different masks causing the daily and seasonal variations: the "Day", "Afternoon" and "Hot Spot" masks. The latter has the 3 h radius

the tilt of Earth rotation axis to the orbital plane, the drift of thunderstorms with season to the North or to the South from the equator. The West–East asymmetry is responsible for the daily variations. The concept is not new, however, it was not used in modeling of the global electromagnetic resonance until recently (Hayakawa et al. 2005; Nickolaenko et al. 2006a, b; Pechony 2007; Pechony et al. 2007).

Three types of masks are shown in Fig. 5.16. The left upper map is the initial distribution. The DM is the mask coincident with the terminator line, it activates the dayside thunderstorms. A more realistic model accounts for the development of air convection, emergence of clouds, and their electrification. This is the reason why the center of the DM is usually shifted by a few hours along the parallel to the evening time. We use the 3 h delay in computations.

The second, "Afternoon" mask (AM) model activates the thunderstorms found in the afternoon sector of the dayside. This model is shown in the lower left map of Fig. 5.16 for 12 h UT on the winter solstice (December 22). The African thunderstorms are active at the moment when this model is applied. The afternoon sector moves westward in time, and the activity shifts from Africa to America. The source–observer geometry varies correspondingly, and we also expect the variatim of the number of lightning strokes.

The Hot Spot (HS) model activates thunderstorms within the circle of predetermined radius; the center of the circle is coincident with the center of the day hemisphere. This model turns into the DM model when the circle radius becomes equal to 6 h. The right lower map in Fig. 5.16 demonstrates the active zone of 3 h radius for 18 h UT on the summer solstice day (June 22) when the thunderstorms are active in Northern America. The HS moves around the planet to switch on and off the thunderstorms in particular areas. The latitude of its center varies with the season by the ±23.5° being the tilt if the Earth's rotational axis. The latitude of the center is 23.5° N in the summer solstice, and it is 23.5° S in the winter solstice. The longitude of the local noon point also varies with month in accordance with the solar time equation as seen by Eq. (5.3). All these factors are included into computations. To account for the retardation between the maximum of solar illumination and the local thunderstorm activity, we shift the position of the HS along the parallel eastward by 3 h (45°). This shift is not shown in Fig. 5.16.

Computations imply the coordinates of the center of the dayside of the globe. The solar terminator line (the DM model) is positioned at the 90° angular distance from the center. At this point, the center of solar disk is placed at the horizon in the absence of atmospheric refraction. One may introduce the solar terminator line as the geometric locus from which the upper edge of the Sun's disk is seen at the horizon in the presence of refraction, as this is done when computing the sunrise and the sunset times (Nickolaenko and Hayakawa 2002). In this case, the angular distance from the center of the dayside increases to 90.85° or to 10 094.4 km. We will ignore such insignificant corrections. In the HS model the boundary is found similarly, and only the distance is reduced to 3 (45°) or to 4 h (60°) (compare this size with the data in Fig. 5.5 and Table 5.2). After forming the particular mask and finding its center, one can calculate the current number of active strokes or the level of thunderstorm activity pertinent to the given season and time of day (see Fig. 5.17). The current intensity of electromagnetic fields is computed by summing the weighted contributions of the particular cells 'opened' by the mask.

The models of the global thunderstorm activity are shown in Fig. 5.17. The time UT is plotted on the abscissa in hour. The cumulative flash rate is shown along the ordinate. The frames from left to right correspond to the Day, Afternoon, and HS models. The color lines show the characteristic daily variations for the middle months of winder, spring, summer, and autumn. The black lines with dots illustrate the averaged daily variations. All three models have much in common. The summer lightning activity is higher than during the winter. The winter thunderstorms start at later UT hours than the summer activity. This result agrees with many electromagnetic observations (Nickolaenko and Hayakawa 2002; Sátori 1996; Sátori et al. 1999; Heckman et al. 1998; Price and Melnikov 2004; Melnikov et al. 2004; Yatsevich et al. 2011). Models predict a large contribution from the American thunderstorms in the global activity in summer, which is also in accord with observations. The details in the diurnal patterns depend on the particular model. The narrow masks increase the fine structure of the model curves. Contributions are clearly seen in all the models from the three major centers of global thunderstorm activity in Africa, America and South-East Asia.

The level of thunderstorm activity varies during the day by a factor of 2.5 in the Day model, of 5 in the Afternoon model and of 7 in the Hot Spot model. These relative diurnal variations noticeably surpass the observed ranges in the amplitude of the fair weather field and in the intensity of the global electromagnetic

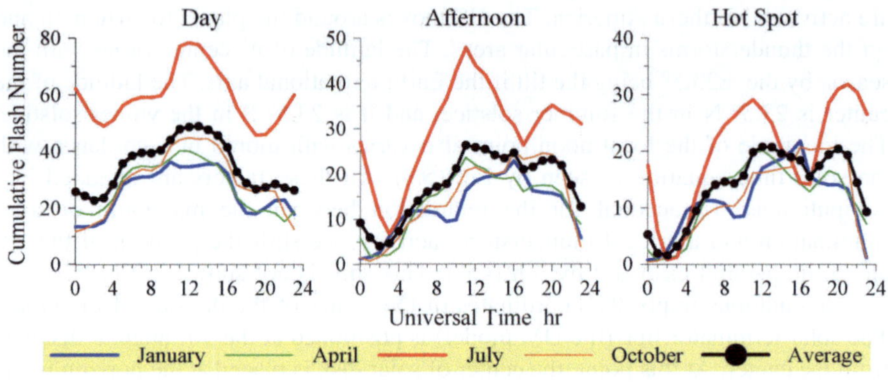

Fig. 5.17 Diurnal variations of the cumulative OTD flashes for different months

resonance. Deviations in the diurnal range possibly arise from the "podium" lightning activity always present in the experiment, but left disregarded in the modeling.

All three models suggest that electromagnetic intensity is directly proportional to the stroke rate at the current moment of time, which means in fact that the median radiation from a lightning stroke does not vary with the time of day and is independent of the coordinates of particular stroke. Such an assumption seems sound since the lightning discharge is a breakdown of the insulating air slab, and the atmosphere composition and the mechanism of the breakdown remain invariant over the planet.

A model with the globally distributed lightning activity was applied in the pioneering work by Ogawa and Murakami (1973) who have used the world distribution of the thunderstorm activity published by WMO (1956), whose global maps were obtained by recording the thunder sound from the network of meteorological observatories all around the world. We use the averaged map of the optical orbital observations by OTD. Similar results are obtained when the source distribution is constructed directly from the OTD data base (e.g. Hayakawa et al. 2005; Nickolaenko et al. 2006a; Sekiguchi et al. 2006; Pechony et al. 2007; Hobara et al. 2011), which allow for obtaining the 24 maps corresponding to every hour UT averaged over the month of observations. These 288 maps might be used in SR computations instead of the mask models.

To check the adequacy of the mask models, the data were compared with observations, and we demonstrate an example of such comparisons: the model and observed Poynting vectors. We suggested in the model that SR background is the Poisson flux of independent pulses with the rate of 100 events per second. The sources occupy independent random positions. The Poynting vector components are the second order mixed statistical moments of the field components E, H_X, and H_Y. Similarly to the power spectra, one can compute the cross-spectra of the vertical electric field and horizontal magnetic field components, which corresponds to a particular 2.5° × 2.5° cell of the global map describing the source distribution for particular month and time of day. The center of the k-th cell is remote from the observer by the θ_k distance with the bearing B_k. The cells are activated by the mask, and we use the HS mask for demonstrations. This means that the spectra computed for the geometry of the given active cell are multiplied by the number of strokes N_k recorded at this cell by OTD. Contributions of all active cells are summed thus providing the second statistical moments at the given frequency ω. Formally, the spectrum of the Poynting vector components is described by the following equation:

$$P_{Y,X}(\omega) = \frac{\nu(\nu+1)}{\omega\varepsilon_0(4ha)^2|\sin\pi\nu|^2}\sum_{k=1}^{K} N_k P_\nu[\cos(\pi-\theta_k)]\{P_\nu^1[\cos(\pi-\theta_k)]\}^* \begin{Bmatrix} \sin(B_k) \\ \cos(B_k) \end{Bmatrix} \tag{5.11}$$

Here N_k is the number of strokes at the particular cell, the θ_k and B_k are the angular distance from the observer to the cell and the cell bearing correspondingly, $P_\nu[\cos(\pi-\theta_k)]$ and $P_\nu^1[\cos(\pi-\theta_k)]$ are the Legendre and associated Legendre functions, ν (ω) is the propagation constant, h is the effective height of the Earth–ionosphere cavity, a is the Earth's radius, and ε_0 is the dielectric constant of vacuum. In computations of Legendre functions, we use the zonal series representations with accelerated convergence (see Chap. 14).

The real part of the complex Poynting vector is the power flux in the cavity at the given frequency. By integrating the Re$\{P_X\}$ and Re$\{P_Y\}$ over the frequencies covering the three SR modes, we obtain the averaged components of the power flux and its median direction. These data were computed for the characteristic months of four seasons in the HS model of the 3 h radius (the left plots in Fig. 5.18). The right plot in this figure shows the experimental data measured at the Lehta observatory (64° N and 34° E) on 14–15 August 1998 (Belyaev et al. 1999). We remind that the median direction toward the source is obtained when we pick the negative sign of the both Poynting vector components. The plots in Fig. 5.18 have the form of loops for which the time is a parameter. The distance of particular point (the numbers indicate the UT hour) from the origin indecates the intensity of the power flux, and the arrow shows the instant direction to the source. We observe that thunderstorm activity circles the globe clockwise during the day, and the model phasor of the Poynting vector is almost as good as the observed one. The other type of diurnal variations with the nocturnal African thunderstorm, activity was shown in Fig. 4.4 of Chap. 4.

Figure 5.18 suggests that model computations based on the OTD data provide the reasonable data, which agree with observations qualitatively at least. The axis scales are the same in the model plots. This allows for visual estimating of seasonal–daily variations of the thunderstorm activity. It is clear that thunderstorms are more intense in summer, which is conditioned by both increases in the stroke rate and the North drift, i.e., a closer positioning with respect to the observatory. We illustrated that computations based on the OTD data agree well with the SR monitoring.

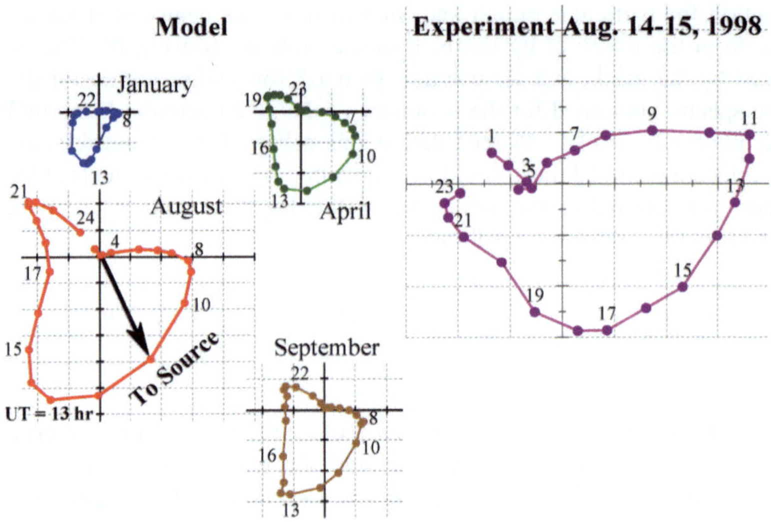

Fig. 5.18 Diurnal motion and intensity on global thunderstorm activity as seen from the SR records at the Lehta observatory. The *left four plots* show the Hot Spot (*HS*) 3 h model data for four seasons, while the *right plot* depicts the experimental record

Figure 5.18 also contains a characteristic feature that was observed in the SR records: the power flux from the thunderstorms in the South–East Asia often is equal or exceeds that from the America. The experimental plot clearly demonstrated this property that was confirmed by the records at the Israeli observatory Mitspe Ramon.

After verifying the HS model data with the help of observations, we can proceed with the modeling of the annual—daily SR variations at a few observatories, especially, those sited at Europe.

Since the SR is a global phenomenon, its intensity depends first of all on the level of the global thunderstorm activity, and therefore the daily patterns at different observatories must have the universal character. Now, we have a numerical tool for checking the soundness of such expectations. The integral intensity is shown in Fig. 5.19 of the vertical electric field component of SR oscillations. The 2D color maps were constructed over the month–UT plane computed for a set of observatories. The observatory name and the coordinates are printed at every map. The month is shown on the abscissa, and the UT in hour is shown on the ordinate. The range of UT variations is extended to 27 h to clarify the upper edge (the evening time) of the lens-like outline. The same color scale is used for inking all the maps.

The left upper map in Fig. 5.18 shows the yearly dynamics of the field source intensity: the level of the global thunderstorm activity pertinent to the HS model of the 3 h radius. The rest of maps correspond to observatories positioned in the Northern and Southern hemispheres and in the Eastern and Western hemispheres,

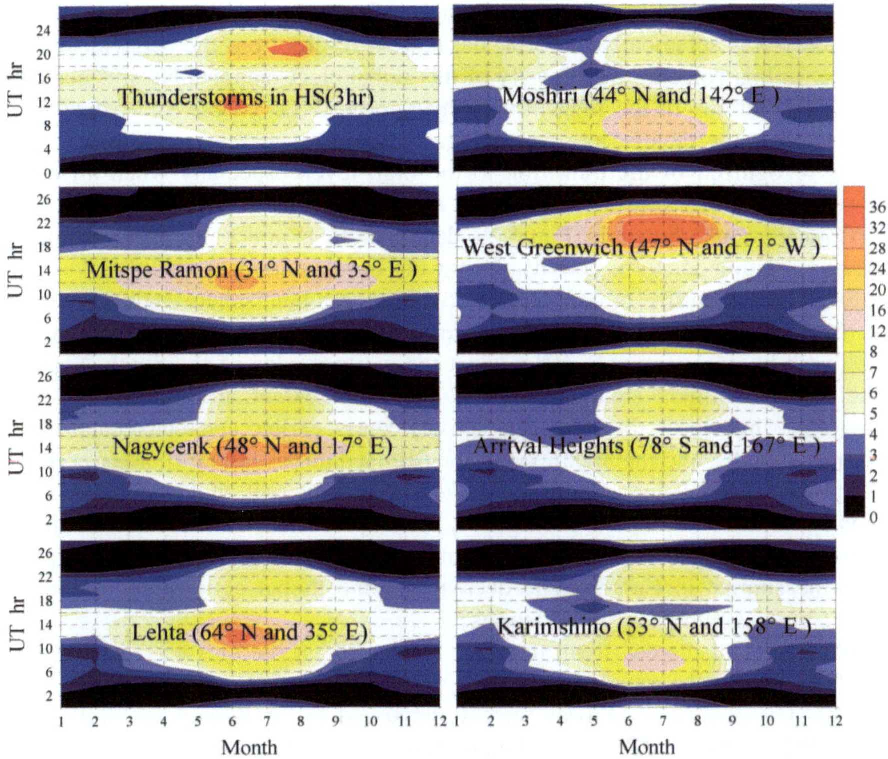

Fig. 5.19 Model lens-like seasonal–diurnal patterns computed for a set of observatories in the uniform Earth–ionosphere cavity with the Hot Spot source of 3 ht radius

where the long-term SR records were performed. Obtaining the similar experimental maps from accumulated data would not be a serious problem. The set of observation sites is quite sufficient for discussing the property of SR records, which reflect changes in the global thunderstorm activity rather than the local ionosphere conditions. Of course, there are definite deviations in exact times of transfer from the small (dark in the maps) levels of signal to the high intensity (bright fields in the maps). However, these deviations cannot destroy generally the same shape of all the maps: transition from low to higher levels occurs in a similar way everywhere on the globe. It is important to stress that the dark-bright outline of all the maps is governed by the source activity, and simultaneously, this outline reminds us of dynamics observed in SR records.

Computational results for the European observatories Nagycenk (48° N and 17° E), Lehta (64° N and 34° E), and Mitspe Ramon (31° N and 35° E) shown in Fig. 5.19 leave an impression that variation of the SR amplitude is linked to the motion of the day-night interface of the observatories. Such a conclusion was made in the Melnikov et al. (2004) paper. In particular as it is seen from Fig. 5.19, this feature is stressed by the closeness of the local and the universal times at these sites. We must remind that no day–night interface was included in computations: the cavity was spherically uniform: the pattern results from the behavior of field source. In this context, the intensity of the global electromagnetic resonance (SR) reflects variations in the planetary electric activity. Of course, individual maps deviate in many details, as seen in Fig. 5.19. However, the common outline is easily recognized of the transition to the high level, especially when all of the maps are constructed in the UT instead of the LT.

Computations shown in Fig. 5.19 demonstrate a noticeable increase in the duration of thunderstorm activity during the boreal summer. This detail is conditioned by the outline of continents in the Northern hemisphere. Indeed, with the increase of northern latitude the western shore of the North America moves westward by approximately 60° against the western edge of the South America. The Asian continent similarly, but to the minor extent, expands to the East. Therefore the world thunderstorm activity starts earlier and ends later in the summer months, and the high level of the SR oscillations lasts for longer period at all observatories. Thus, seasonal changes of the interval of the high lightning activity becomes similar to changes in the day duration in the Northern hemisphere, and the daily increase of the thunderstorm intensity in Asia starts in-phase with the sunrise in Europe. The decrease of the global activity occurs around sunset time at European observatories.

An impact of the day-night interface on the SR oscillations was discussed for the first time in the papers by Madden and Thompson (1965) and Bliokh et al. (1968). The results of analysis were summarized in the monographs by Bliokh et al. (1977, 1980) and Nickolaenko and Hayakawa (2002), and in dissertations by Rabinowicz (1988) and Pechony (2007). The terminator effect in the SR band was not proven experimentally, and the model computations indicated that the day–night ionosphere transition has a minor impact at the lower resonance modes because the radio wave is able to multiply circle the globe. The pulse launched by a lightning stroke a few times passes the day and the night hemispheres prior to its complete attenuation. Thus, 'averaging' takes place as if the signal propagates in some effective uniform cavity. Depending on the particular model, the day–night non-uniformity is able to cause the amplitude modulations of ±5 % in the cavity with the smooth terminator and up to 10 % when the day–night transition is sharp or step-like (Rabinowicz 1988; Nickolaenko and Hayakawa 2002; Pechony 2007). Some special experimental technique should be elaborated for detection of such tiny variations in the natural radio noise on the background of the 2.5-fold daily variations caused by the source intensity.

To demonstrate the generality of the lens-like pattern associated with the intensity of global thunderstorms, we compare in Fig. 5.20 the observational data from the Mitspe Ramon observatory (Pechony 2007; Pechony et al. 2007) with the model computations. The annual-diurnal variations are shown in Fig. 5.20 in a form of the customary contour maps relevant to the vertical electric field component. Computations for the uniform and non-uniform models of the Earth-ionosphere cavity were performed with the help of 2D telegraph equations (Pechony et al. 2007).

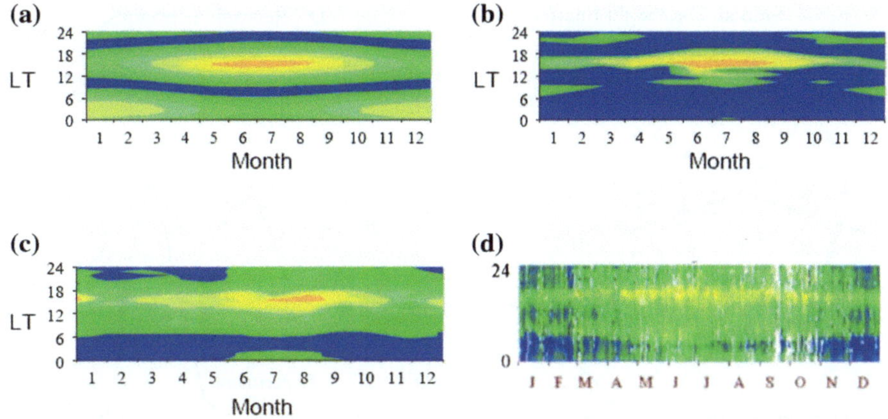

Fig. 5.20 Comparison of the model and experimental data for the Mitspe Ramon observatory. **a** Uniform model and point source that follows the Sun with 3 h delay. **b** Uniform model and point source active over land following the Sun with 3 h retard. **c** Uniform model and OTD monthly average distribution over 1999–2000 **d** Observations

Three different source models were used in particular computations in the uniform cavity. The left upper frame shows the results computed for the equatorial point source of constant amplitude. The source is positioned at the latitude retarded from the noon point by 3 h (15 h local time). One may recognize the characteristic increase in the signal amplitude when the source approaches the observer antipode. The upper right frame corresponds to a more realistic source model when the thunderstorms of the same point source become active only over the land. The antipodal peak vanishes as the source is positioned over the Pacific Ocean in this case. The lower left map was obtained in the framework of the source model based on the OTD data averaged over 1999–2002 period. The right lower map reproduces the experimental data from Mitspe Ramon observatory.

We notice from Fig. 5.20 that familiar lens-shape patterns appear again in the computations even for the uniform cavity model caused by the source properties. In this case, however, the different formal approach was used based on the 2D telegraph equations, but still the model results are similar to each other and to observations. If so, is there any terminator effect that might be expected in the SR records? The answer to this question is positive: the effect really exists, but it is small. We show relevant data in Fig. 5.21.

Two plots are shown in Fig. 5.21. The left one presents the result by Rabinowicz (1988). The smooth terminator model is used in computations, which means that the wave propagation conditions depend on the solar zenith angle α_S as cos (α_S). The plot shows the relative vertical electric field disturbance (in %) versus angular distance α of the "parallel" propagation path at the 8 Hz frequency. The source–observer distance is 5 Mm, and they occupy the latitudes of 22.5° N and 22.5° S at the same meridian, so that the distances from the source and the observer to the terminator line are always equal. The left plot demonstrates that the

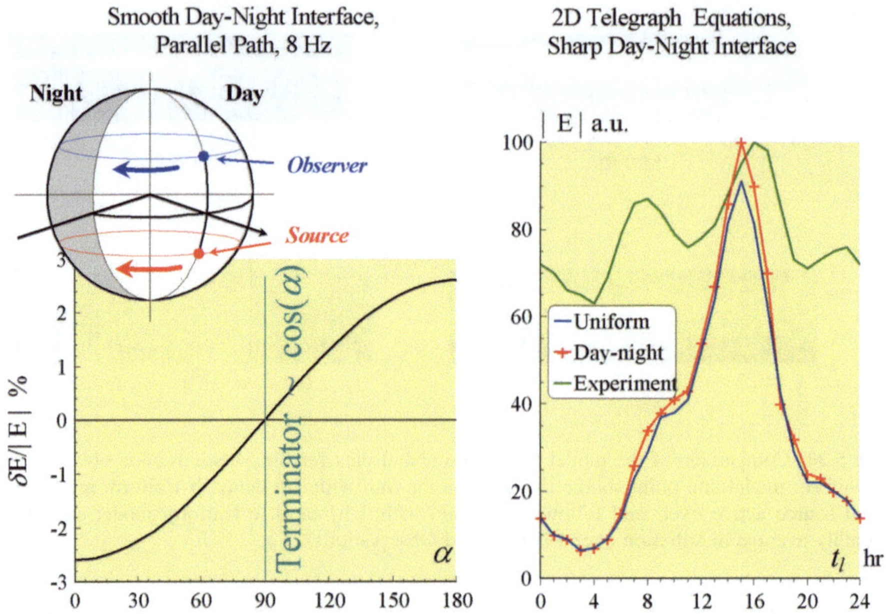

Fig. 5.21 Impact of terminator in the SR amplitude

smooth terminator causes very small amplitude modulation at the first SR mode (see also Nickolaenko and Hayakawa 2002). The right plot in Fig. 5.21 presents the data of modeling with the 2D telegraph equations (Pechony 2007) based on the OTD spatial distribution of lightning strokes. The local time of the Mitspe Ramon observatory is plotted on the abscissa, and the vertical electric field amplitude is shown on the ordinate in arbitrary units. Three graphs are shown here. The blue line is the computational result for the uniform cavity with the OTD sources. While, the red line with crosses shows the data of similar computations when the sharp day-night transition is taken into account. We note that the utmost modification due to the sharp day-night asymmetry is about 10 %, while the daily variations reach an order of magnitude in both the uniform and non-uniform cavities. The green line in the right frame depicts the results of observations.

Thus, the expected impact of terminator in the SR range is rather small, that is, it is a few times smaller than variations caused by variations in the thunderstorm activity. There is additional feature present in Fig. 5.21: the range of observed amplitude variations is much smaller than that predicted by model computations. Observational data leave an impression that the diurnal curve is "elevated" over the abscissa due to some "podium" signal. We will return to this issue in Chap. 10.

References

Ahlquist JE (1982) Normal mode global Rossby waves. J Atmos Sci 39:193–202

Balser M, Wagner CA (1960a) Observation of Earth-ionosphere cavity resonances. Nature 188:638–641

Balser M, Wagner C (1960b) Measurement of the spectrum of radio noise from 50 to 100 c/s. J Res NBS 64D:415–418

Bannister PR (1999) Further examples of seasonal variations of ELF radio propagation parameters. Radio Sci 34:199–208

Belyaev GG, Schekotov AYu, Shvets AV, Nickolaenko AP (1999) Schumann resonances observed using poynting vector spectra. J Atmos Solar Terr Phys 61:751–763

Bliokh PV, Nickolaenko AP, Filippov YuF (1968) Diurnal variations of the natural frequencies of the Earth-ionosphere resonator in relation to the eccentricity of the geomagnetic field. Geomagn Aeron 8(2):198–206

Bliokh PV, Bormotov VN, Kontorovich VM, Nickolaenko AP, Sapogova NA, Shulga VF, Filippov YF (1971) On the degeneracy lift in the spherical Earth-ionosphere cavity, Preprint of Institute of Radio-Physics and Electronics of Ukrainian Acadamy of Science, No. 10, Kharkov (in Russian)

Bliokh PV, Nickolaenko AP, Filippov YuF (1977) Global electromagnetic resonances in the Earth-ionosphere cavity. Naukova Dumka, Kiev (in Russian)

Bliokh PV, Nickolaenko AP, Filippov YF (1980) Schumann resonances in the Earth-ionosphere cavity. In: Jones DL (ed) Peter Peregrinus, Oxford, New York, Paris

Boccippio DJ, Williams ER, Heckman SJ, Lyons WA, Baker IT, Boldi R (1995) Sprites, ELF transients and positive ground strokes. Science 269:1088–1091

Burke CP, Jones DL (1992) An experimental investigation of ELF attenuation rates in the Earth-ionosphere duct. J Atmos Terr Phys 54:243–254

Burke CP, Jones DL (1995) Global radiolocation in the lower ELF frequency band. J Geophys Res 100:26263–26271

Burke CP, Jones DL (1996) On the polarity and continuing currents in unusually large lightning flashes deduced from ELF events. J Atmos Terr Phys 58:531–540

Chrissan DA, Fraser-Smith AC (1996) Seasonal variations of globally measured ELF/VLF radio noise. Radio Sci 31:1141–1152

Christian HJ, Blakeslee RJ, Boccippio DJ, Boeck WL, Buechler DE, Driscoll KT, Goodman SJ, Hall JM, Koshak WJ, Mach DM, Stewart MF (2003) Global frequency and distribution of lightning as observed from space by the optical transient detector. J Geophys Res 108(D1):4005 doi: 10.1029/2002JD002347

Danilov DL, Zhiglyavsky AA (1997) Principal Component of the Time Series: the Caterpillar Method. St.-Petersburg state University, St.-Petersburg, Russia, p 307 (in Russian)

Danilov DL (1996) Proceedings of statistical computing section of the American statistical association, p 156

Fraser-Smith AC, McGill PR, Bernardi A, Helliwell RA, Ladd ME (1991) Global measurements of the low frequency radio noise. In: Kikuchi H (ed) Environmental and space electromagnetics. Springer, Tokyo, pp 191–200

Fraser-Smith AC, Bannister PR (1998) Reception of ELF signals at antipodal distances. Radio Sci 33:83–88

Füllekrug M (1995) Schumann resonances in magnetic field components. J Atmos Terr Phys 57:7479–7484

Füllekrug M, Fraser-Smith AC (1996) Further evidence for a global correlations of the Earth-ionosphere cavity resonances. Geophys Res Lett 23:2773–2776

Füllekrug M, Fraser-Smith AC (1997) Global lightning and climate variability inferred from ELF magnetic field observations. Geophys Res Lett 24:2411–2414

Füllekrug M, Fraser-Smith A, Reising SC (1998) Ultra-slow tails of sprite associated lightning flashes. Geophys Res Lett 25:3495–3498

Füllekrug M, Reising SC (1998) Excitations of Earth-ionosphere cavity resonances by sprite-associated lightning flashes. Geophys Res Lett 25:4145–4148

Füllekrug M, Fraser-Smith AC, Bering EA, Few AA (1999) On the hourly contribution of global cloud-to-ground lightning activity to the atmospheric electric field in the Antarctic during December 1992. J Atmos Solar Terr Phys 61:745–750

Füllekrug M, Constable S (2000) Global triangulation of lightning discharges. Geophys Res Lett 27:333–336

Füllekrug M (2000) Dispersion relation for spherical electromagnetic resonances in the atmosphere. Phys Lett A 275:80–89

Füllekrug M, Mareev EA, Rycroft MJ (eds) (2006) Sprites, elves, and intense lightning discharges. Springer, Dordrecht

Galejs J (1970) Frequency variations of Schumann resonances. J Geophys Res 75:3237–3251

Galejs J (1972) Terrestrial propagation of long electromagnetic waves. Pergamon Press, New York

Gendrin R, Stefant R (1962) Effect of high altitude thermonuclear explosion on July 1962 on the resonance in the Earth-ionosphere cavity, experimental results. CR Acad Sci 255:2273–2275 (in French)

Greifinger C, Greifinger P (1978) Approximate method for determining ELF eigen-values in the Earth-ionosphere waveguide. Radio Sci 13:831–837

Golyandina N, Nekrutkin V, Zhiglyavsky A (2001) Analysis of time series structure. Chapman and Hall, CRC, Boca Raton

Harth W (1982) Theory of low frequency wave propagation. In: Volland H (ed) Handbook of atmospherics, vol 2. CRC Press, Boca Raton, pp 133–302

Hayakawa M (2001) Final report, NASDA's earthquake remote sensing frontier research, seismo-electromagnetic phenomena in the lithosphere, atmosphere and ionosphere. The University of Electro-Communications, Chofu-city, Tokyo, March 2011

Hayakawa M, Sekiguchi M, Nickolaenko AP (2005) Diurnal variation of electric activity of global thunderstorms deduced from OTD data. J Atmos Electr 25(2):55–67

Hayakawa M, Nickolaenko AP, Shvets AV, Hobara Y (2011) Recent studies of Schumann resonance and ELF transients. In: Wood MD (ed) Lightning: properties, formation and types, Chap. 3. Nova Science Publishers, pp 39–71

Heckman SJ, Williams ER, Boldi R (1998) Total global lightning inferred from Schumann resonance measurements. J Geophys Res 103:31775–31779

Hobara Y, Iwasaki N, Hayashida T, Tsuchiya N, Williams ER, Sera M, Ikegami Y, Hayakawa M (2000a) New ELF observation site in Moshiri, Hokkaido Japan and the results of preliminary data analysis. J Atmos Electr 20(2):99–109

Hobara Y, Yamaguchi H, Akinaga Y, Watanabe T, Koons HC, Rieder JL, Hayakawa M (2000b) New ULF/ELF measurement in Seikoshi, Izu peninsula, Japan, Abstracts and Program of International Workshop on Seismo Electromagnetics, The University of Electro-Communications, Chofu-city, Tokyo, 19–22 September, p 9

Hobara Y, Iwasaki N, Hayashida T, Hayakawa M, Ohta K, Fukunishi H (2001) Interrelation between ELF transients and ionospheric disturbances in association with sprites and elves. Geophys Res Lett 28:935–938

Hobara Y, Harada T, Ohta K, Sekiguchi M, Hayakawa M (2011) A study of global temperature and thunderstorm activity by using the data of Schumann resonance observed at Nakatsugawa, Japan. J Atmos Electr 32(2):11–19

Huang E, Williams E, Boldi R, Heckman S, Lyons W, Taylor M, Nelson T, Wong C (1999) Criteria for sprites and elves based on Schumann resonance observations. J Geophys Res 104:16943–16964

IEEE Trans, Com-22, No. 4, (1974) Special issue on the 'Sanguine' project

Jones DL (1969) The apparent resonance frequencies of the Earth-ionosphere cavity when excited by a single dipole source. J Geomagn Geoelectr 21:679–684

Jones DL (1970a) Numerical computations of terrestrial ELF electromagnetic wave fields in the frequency domain. Radio Sci 5:803–809

Jones DL (1970b) Propagation of ELF pulses in the Earth-ionosphere cavity and application to slow tail sferics. Radio Sci 5:1153–1163

Jones DL, Kemp DT (1970) Experimental and theoretical observations of Schumann resonances. J Atmos Terr Phys 32:1095–1108

Jones DL, Kemp DT (1971) The nature and average magnitude of the sources of transient excitation of Schumann resonances. J Atmos Terr Phys 33:557–566

Jones DL (1999) ELF sferics and lightning effects on the middle and upper atmosphere. In: Stuchly MA (ed) Modern radio science 1999. Published for URSI by Oxford University Press, pp 171–191

Leblanc F, Aplin KL, Yair Y, Harrison RG, Lebreton JP, Blanc M (eds) (2008) Planetary Atmospheric Electricity, vol 30. Springer, New York. doi:10.1007/978-0-387-87664-1

Lyons WA, Armstrong RA, Bering EA III, Williams ER (2000) The hundred years hunt for the sprite. EOS 81:373–377 (Aug 15)

Madden T, Thompson W (1965) Low frequency electromagnetic oscillations of the Earth-ionosphere cavity. Rev Geophys 3:211–254

Marple SL Jr (1987) Digital spectral analyses with applications. Prentice-Hall, Englewood Cliffs, p 581

McGorman DR, Rust WD (1998) The electrical nature of storms. Oxford University Press, Oxford

Melnikov A, Price C, Sátori G, Füllekrug M (2004) Influence of solar terminator passages on Schumann resonance parameters. J Atmos Solar Terr Phys 66 (13-14 SPEC. ISS.):1187–1194. doi:10.1016/j.jastp.2004.05.014

Molchanov OA, Hayakawa M, Miyaki K (2001) VLF/LF sounding of the lower ionosphere to study the role of atmospheric oscillations in the lithosphere-ionosphere coupling. Adv Polar Upper Atmos Res 15:146–158

Nelson PH (1967) Ionospheric perturbations and Schumann resonance data, Ph.D. Thesis, MIT, Cambridge Mass

Nickolaenko AP (1986) Scattering of the ELF radio waves by the global non-uniformities of the Earth-ionosphere cavity. Izv VUZov Radiofizika 29(1):33–40 (in Russian)

Nickolaenko AP, Rabinowicz LM (1982) Possible global electromagnetic resonances on the planets of the solar system. Translated from Kosmicheskie Issledovaniya, vol 20(1). Plenum Publishing Corporation, pp 82–88

Nickolaenko AP, Rabinowicz LM (1987) Applicability of ultralow—frequency global resonances for investigating lightning activity on Venus, Translated from Kosmicheskie Issledovaniya, vol 25(2). Plenum Publishing Corporation, pp 301–306

Nickolaenko AP, Rabinowicz LM (1995) Study of the annual changes of global lightning distribution and frequency variations of the first Schumann resonance mode. J Atmos Terr Phys 57(11):1345–1348

Nickolaenko AP, Korol MA, Shvets AV, Kudintseva IG (1996) Effective parameters of low frequency antennas. J Atmos Electr 16:81–88

Nickolaenko AP (1997) Modern aspects of the Schumann resonance studies. J Atmos Solar Terr Phys 59(7):805–816

Nickolaenko AP, Rabinowicz LM (1998) Seasonal drift of the global thunderstorms and Schumann resonance frequencies. In: Proceedings of 7th international conference on mathematical modeling in electromagnetic theory, MMET'98, Kharkov, p 294

Nickolaenko AP, Sátori G, Zieger B, Rabinowicz LM, Kudintseva IG (1998) Parameters of global thunderstorm activity deduced from long term Schumann resonance records. J Atmos Sol Terr Phys 60:387–399

Nickolaenko AP, Hayakawa M (2002) Resonances in the Earth-ionosphere cavity. Kluwer Acadamic Publisher, Dordrecht, p 380

Nickolaenko AP, Hayakawa M, Sekiguchi M (2006a) Variations in global thunderstorm activity inferred from the OTD records. Geophys Res Lett 33:L06823. doi:10.1029/2005GL024884

Nickolaenko AP, Pechony O, Price C (2006b) Model variations of Schumann resonance based on optical transient detector maps of global lightning activity. J Geophys Res 111:D23102. doi:10.1029/2005JD006844

Nickolaenko AP, Yatsevich EI, Shvets AV, Hayakawa M, Hobara Y (2011) Universal and local time variations deduced from simultaneous Schumann resonance records at three widely separated observatories, Radio Sci 46:RS5003. doi:10.1029/2011RS004663

Ogawa T, Tanaka Y, Miura T, Yasuhara M (1966) Observations of natural ELF and VLF electromagnetic noises by using ball antennas. J Geomagn Geoelectr 18:443–454

Ogawa T, Tanaka Y, Yasuhara M (1968) Diurnal variations of resonance frequencies in the Earth-ionosphere cavity. Contr Geophys Inst Kyoto University 8:15–20

Ogawa T, Murakami Y (1973) Schumann resonance frequencies and the conductivity profiles in the atmosphere. Contr Geophys Inst Kyoto Univ 13:13–20

Ogawa T, Otsuka S (1973) Comparison of observed Schumann resonance frequencies with the single dipole source approximation theories. Contr Geophys Inst Kyoto University 13:7–11

Ogawa T, Kozai K, Kawamoto H (1979) Schumann resonances observed with a balloon in the stratosphere. J Atmos Terr Phys 41:135–142

Orville RE (1981) Lightning detection from space. In: Volland H (ed) Handbook of atmospherics, vol 2. CRC Press, Boca Raton, pp 80–96

Orville RE, Henderson RW (1986) Global distribution of midnight lightning: September 1977 to August 1978. Mon Weather Rev 114:2640–2653

Pechony O, Price C (2004) Schumann resonance parameters calculated with a partially uniform knee model on Earth, Venus, Mars, and Titan. Radio Sci 39:RS5007. doi:10.1029/2004RS003056

Pechony O (2007) Modeling and simulations of schumann resonance parameters observed at the mitspe Ramon field station (study of the day-night asymmetry influence on Schumann resonance amplitude records), Ph.D. Thesis, Tel-Aviv University, Israel, March 2007, p 92

Pechony O, Price C, Nickolaenko AP (2007) Relative importance of the day-night asymmetry in Schumann resonance amplitude records. Radio Sci 42:RS2S06. doi:10.1029/2006RS003456

Polk C (1969) Relation of ELF noise and Schumann resonances to thunderstorm activity. In: Coroniti SC, Hughes J (eds) Planetary electrodynamics, vol 2. Gordon and Breach, New York, pp 55–83

Price C, Melnikov A (2004) Diurnal, seasonal and inter-annual variations in the Schumann resonance parameters. J Atmos Solar Terr Phys 66:1179–1185. doi:10.1016/j.jastp.2004.05.004

Price C, Finkelstein M, Starobinets B, Williams E (1999) A new Schumann resonance station at the Negev desert for monitoring global lightning activity. In: Proceedings of 11th international conference on atmospheric electricity, Guntersville, Alabama, 7–11 June 1999, pp 695–697

Rabinowicz LM (1986) On the impact of the day-night non-uniformity of the ELF fields. Radiofizika Izv VUZov 29:635–644

Rabinowicz LM (1988) Global electromagnetic resonances in non-uniform and anisotropic Earth-ionosphere cavity, Ph.D. Thesis, Kharkov (in Russian)

Rakov VA, Uman MA (2003) Lightning: physics and effects. Cambridge University Press, Cambridge

Sao K, Yamashita M, Tanahashi S (1971) Day to day variations of Schumann resonance frequency and occurrence of Pc1 in view of solar activity. J Geomagn Geoelectr 23:411–415

Sao K, Yamashita M, Tanahashi S, Jindoh H, Ohta K (1973) Experimental investigations of Schumann resonance frequencies. J Atmos Terr Phys 35:247–253

Sátori G, Szendröi J, Verö J (1996) Monitoring Schumann resonances—I. Methodology. J Atmos Terr Phys 58:1475–1482

Sátori G (1996) Monitoring Schumann resonances—II. Daily and seasonal frequency variations. J Atmos Terr Phys 58:1483–1488

Sátori G, Zieger B (1996) Spectral characteristics of Schumann resonances observed in Central Europe. J Geophys Res 101:29663–29669

Sátori G, Zieger B (1999) El Niño related meridional oscillation of global lightning activity. Geophys Res Lett 26:1365–1368

Sátori G, Williams ER, Zieger B, Boldi R, Heckman S, Rothkin K (1999) Comparison of long-term Schumann resonance records in Europe and North America. In: Proceedings of 11th international conference on atmospheric electricity, Guntersville, Alabama, 7–11 June 1999, pp 705–708

Sátori G (2003) On the dynamics of the North-South seasonal migration of global lightning. In: Proceedings 12th international conference on atmospheric electricity, Versailles, France (Global Lightning and Climate). p 1

Sátori G, Williams ER, Boccippio DJ (2003) On the dynamics of the north-south seasonal migration of global lightning, AGU Fall Meeting, San Francisco, 8–12 December 2003. p AE32A-0166

Sátori G, Neska M, Williams E, Szendröi J (2006) Signatures of the day-night asymmetry of the Earth-ionosphere cavity in high time resolution Schumann resonance records. Radio Sci 42:RS2S10. doi:10.1029/2006RS003483

Sátori G, Mushtak V, Williams E (2009) Schumann resonance signatures of global lightning activity. In: Betz HD, Schumann U, Laroche P (eds) Lightning: principles, instruments and applications. Springer, Dordrecht, pp 347–386

Schlegel K, Füllekrug M (1999) Schumann resonance parameter changes during high-energy particle precipitation. J Geophys Res 104(A5):10111–10118

Schlegel K, Füllekrug M (2000) Diurnal harmonics in Schumann resonance parameters observed on both hemispheres. Geophys Res Lett 27:2805–2808

Shimakura S, Yamamoto T, Hayakawa M (1991) On the short and long periodicities in whistler occurrence rate and their implication. Res Lett Atmos Electr 11:23–36

Sekiguchi M, Hayakawa M, Hobara Y, Nickolaenko A, Williams E (2004) Links of Schumann resonance intensity with average global land temperature. Radiophys Electron 9(2):383–391 (in Russian)

Sekiguchi M, Hayakawa M, Nickolaenko AP, Hobara Y (2006) Evidence of a link between the intensity of Schumann resonance and global surface temperature. Ann Geophysicae 24:809–817

Sentman DD (1990) Approximate Schumann resonance parameters for a two-scale height ionosphere. J Atmos Terr Phys 52:35–46

Sentman DD, Fraser BJ (1991) Simultaneous observation of Schumann resonances in California and Australia: evidence for intensity modulation by local height of D region. J Geophys Res 96(9):15973–15984

Sentman DD (1995) Schumann resonances. In: Volland H (ed) Handbook of atmospheric electrodynamics, vol 1. CRC Press, Boca Raton, London, Tokyo, pp 267–298

Sentman DD (1996) Schumann resonance spectra in a two-scale height Earth-ionosphere cavity. J Geophys Res 101:9474–9487

Special Issue of Journal of Atmospheric and Solar-Terrestrial Physics (2003), J Atmos Solar-Terr Phys 65(2) (Special issue on sprites)

Uman MA (1987) The lightning discharge. Academic Press, Oxford

Whitley T, Füllekrug M, Rycroft M, Bennett A, Wyatt F, Elliott D, Heinson G, Hitchman A, Lewis A, Sefako R, Fourie P, Dyers J, Thomson A, Flower S (2011) Worldwide extremely low frequency magnetic field sensor network for sprite studies. Radio Sci 46:RS4007. doi:10.1029/2010RS004523

Williams ER (1992) The Schumann resonance: a global tropical thermometer. Science 256:1184–1188

Williams ER (1994) Global circuit response to seasonal variations in global surface air temperature. Mon Weather Rev 122:1917–1929

Williams ER, Mushtak VC, Boldi R, Dowden RL, Kawasaki ZI (2007) Sprite lightning heard round the world by Schumann resonance methods. Radio Sci 42:RS2S20. doi:10.1029/2006RS003498

WMO (1956) World distribution of thunderstorm days, part 2: tangles of marine data and world maps, OMM-No. 21, TP. 21

Yatsevich EI, Nickolaenko AP, Shvets AV, Rabinowicz LM (2006) Two component source model of Schumann resonance signal. J Atmos Electr 26(1):1–10

Yatsevich EI, Nickolaenko AP, Pechony OB (2008) Diurnal and seasonal variations in the intensities and peak frequencies of the first three schumann-resonance modes. Radiophys Quantum Electron 51(7):528

Yatsevich EI (2009) Schumann resonance monitoring and source models, Ph.D., Kharkov, IRE NASU (in Russian)

Yatsevich EI, Nickolaenko AP, Shvets AV, Hayakawa M, Hobara Y (2011) Schumann-resonance records at three observatories and ULF universal- and local-time variations. Radiophys Quantum Electron 53(12):706–716. doi:10.1007/s11141-011-9263-y

Zieger B, Sátori G (1999) Periodic variations of solar and tropospheric origins in Schumann resonance. In: Proceedings of 11th international conference on atmospheric electricity, Guntersville, Alabama, 7–11 June 1999, Alabama, pp 701–703

Chapter 6
Disturbances in SR

This chapter deals with an impact of the ionospheric non-uniformities to the SR oscillation. Resonances demonstrate a higher sensitivity to the 'symmetric' ionosphere modifications that occur simultaneously at the both polar regions. This is why the solar proton events (SPEs) are readily observed in the SR records. We address the published experimental data on SPEs and model these by using modification of the typical exponential conductivity profile of atmosphere. We show that relatively weak events 'improve' the quality of cavity resonator: the signal intensity, peak frequencies, and resonance Q-factors simultaneously increase. In contrast, the powerful modifications reducing the ionosphere height, cause the polar cap absorption (PCA) which results in a decrease in all the resonance parameters. We present the multi-point SR records during the Bastille Day SPE, and compare these with the model computations and demonstrate that the event has doubled the effective surface impedance of the polar ionosphere. While, SR is less sensitive to the 'asymmetric' modifications such as the day-night non-uniformity. By using resonance monitoring, we nevertheless demonstrate that a huge, abrupt, the 20 km reduction of the dayside ionosphere by the galactic gamma-ray flare has caused the simultaneous short reduction of the observed peak frequencies on December 27, 2004. Finally, unusual SR signals regularly appear in Japan in association with the Taiwan earthquakes. We demonstrate that these modifications might arise from the scattering of natural ELF radiation from the ionospheric reduction over the epicenter zone.

6.1 Solar Proton Events

An impact of natural ionosphere disturbances on the SR spectra was studied and modeled since the middle of 60s (Madden and Thompson 1965; Nelson 1967; Sao 1967). There are two types of disturbances in the Earth–ionosphere cavity, and both of them have the cylindrical symmetry. One of them is called the Sudden Ionosphere Disturbance (SID) and the other is the Polar Cap absorption (PCA), and the role of these modifications at ELF was addressed by Madden and

A. Nickolaenko and M. Hayakawa, *Schumann Resonance for Tyros*,
Springer Geophysics, DOI: 10.1007/978-4-431-54358-9_6,

Thompson (1965) and Nelson (1967). The SID events are the abrupt changes in the lower ionosphere caused as a rule by the solar X and gamma flares. The resulting modifications have the "odd" symmetry of the day–night type, which usually cannot produce substantial changes in the resonant frequencies of the cavity (Bliokh et al. 1977, 1980; Nickolaenko and Hayakawa 2002). This formal conclusion is supported by observations (Sao 1967; Sao et al. 1971, 1973; Cannon and Rycroft 1982). As we show below, the galactic gamma ray and X-ray flares are able to noticeably modify the cavity eigen values (resonance frequencies), provided that their flux exceeds that of the solar flares by many orders in magnitude (Nickolaenko et al. 2010, 2012).

Disturbances of the second type (PCA) arise when the charged particles arrive at the Earth from the Sun. The charged particles are deflected by the geomagnetic field and the precipitation is observed in the sub-polar regions that may penetrate into the lower ionosphere. We will not consider the interaction of the charged particles with magnetosphere, but we only note that the ionosphere D region goes down at the both polar regions, and the event is observed as temporal radio blackouts at the sub-polar propagation paths. The PCA also has the cylindrical geometry (but of "even" type) as it is symmetric with respect to the geomagnetic equator. The even symmetry enhances the impact of such a disturbance on the SR frequencies (Bliokh et al. 1977, 1980; Nickolaenko and Hayakawa 2002).

Usually, the solar flares start with the X-ray and gamma radiation. The 150 million kilometer distance is covered by them in about 8 min. These hard photons are followed with a short delay by the relativistic solar proton event (SPE) or electron event (SEE). The proton energy may exceed the level of 10–100 MeV, and that of electrons exceeds a few MeV. These abrupt changes are accompanied in a couple of days by modifications in the solar wind, which lead to the geomagnetic storms. We are going to demonstrate that PCA and SID events appear in different ways in the SR records, although both of them are associated with an increase of the ionization in the lower ionosphere caused by the solar ionization.

6.2 Ionosphere Disturbances in the Framework of Greifinger Model

The work by Schlegel and Füllekrug (1999) examined in detail the modifications in the SR parameters caused by SPE and SEE events. They processed several years of records collected at the Arrival Heights observatory (78° S and 167° E) of Stanford University (Fraser-Smith et al. 1991). The nine strongest precipitations formed the data ensemble covering the interval from March 1989 to February 1994, and all events were associated with the sharp simultaneous increase of the frequency, amplitude and quality factor of the first SR mode. The data were accumulated as the waveforms of 52 s duration recorded every 30 min. Resonance parameters were found by using the Prony method (see e.g. Marple 1987) and the so-called Z–transform.

This kind of transform was applied in the SR studies for the first time by Tanahashi (1976). The physical background of Z–transform is simple: it is the Fourier transform of the complex frequencies, i.e. with exponential time functions of increasing amplitude. Such an orthogonal basis allows for compensating the attenuation factor present in the initial data and thus tapering the resonance line. Tightening of resonance peaks would increase the accuracy of measurements of peak frequencies. The final goal of Z–transform in the work by Tanahashi (1976) was disclosing of the SR line splitting, which should become obvious when the resonance curves are narrow.

Shortcomings of Z–transform are clear from simple physical considerations. Procedure compensating the losses of resonance energy will work correctly until the moment when the introduced increment is equal to the existing natural decrement. The oscillations start to grow after compensating the decrement, and the processing results become unstable. We have to apply the word "unstable" in its general meaning, because nobody knows in advance what a particular form the instability would acquire in the spectral estimates of natural radio noise. The delicate and key moment of Z–transform is finding the situation when the losses are completely compensated, i.e., when the previously attenuating sinusoid becomes the sinusoid of a constant amplitude. We tentatively accept in the spectral processing that the initial signal is a "stable" one, so that the signal processed must be "stationary". However, we know from Sect. 1.4 that ordinary SR spectra become stable after their considerable averaging over the ensembles corresponding to the time intervals of a few minutes at least. One must remember this when applying any processing worsening the stability of spectral estimates; otherwise the final results might become simply unreliable, irreproducible. Therefore, one cannot expect that the results of Z–transform will be stable and consistent. For instance, how one can prove that the fine structure appearing in the amplitude spectrum is conditioned by the hidden resonance line splitting rather than a simple instability? Besides, what is the criterion for choosing the unknown increment rate?

An experimental study by Schlegel and Füllekrug (1999) showed that all SPE and SEE events were followed by an increase in the peak amplitudes, frequencies, and the quality factors (Q–factors). They suggested an interpretation of observations based on the Greifinger's model (Greifinger and Greifinger 1978) and made a conclusion that experimental data are in contradiction to the theoretical results by Bliokh et al. (1977, 1980), who associated the polar cap absorption (PCA) with a decrease in resonance frequencies. We must note immediately that observations by Schlegel and Füllekrug (1999) are also in conflict with the experimental data used in the work by Madden and Thompson (1965) and observed by Nelson (1967). In what follows, we demonstrate that there is no contradiction between the observations of 60s and 80s, and that the Greifinger's model explains both the increase and the decrease in observed resonance parameters, all depending on intimate details of the ionosphere modifications.

Clearly, the observational data by Schlegel and Füllekrug (1999) correspond to improvements in the ELF propagation conditions, because the peak frequency grew and the wave attenuation decreased. Hence, the ionosphere profile was disturbed in such a way that it reduced the wave attenuation at the first resonance at least. We conclude that all SPE and SEE events observed in the Antarctica were associated with anti-polar cap absorptions (anti-PCA) when the wave attenuation decreased instead of increasing. We suspect that this result was conditioned by the desire of experimentalists to extend the data ensemble by considering relatively weak events. It is also possible that events of 80s would pass unnoticed in 60s owing to an increase in sensitivity of space-borne measurements. We show below that the weak precipitations cause the anti-PCA events, while the stronger fluxes induce the wave absorption as before. In other words, observations were in accord with the models and theories, since particle precipitations can cause both increase and decrease in the ELF radio wave absorption. To demonstrate this we explain observations by Schlegel and Füllekrug (1999) with the help of modified Greifinger model.

From the physical point of view, particle precipitations increase the ionosphere plasma density. Two limiting types of modifications might be mentioned. The first one is a classical PCA event when the whole ionosphere profile moves downward. Modifications of the second kind occur in the higher, the E region of ionosphere and do not reach the D layer, and the conductivity remains unchanged at the lower edge of plasma. It is obvious that a disturbance of the second kind simply alters the tilt of the conductivity profile, so that reflectivity increases of the plasma boundary and the wave attenuation decreases.

Let us turn to the modified Greifinger's model with the upper reference height that was used in Chap. 5. The reference height is $G_h = 89$ km, the reference frequency $F_g = 10$ kHz, and the scale height $\zeta = 3.8$ km. This means that the height conductivity profile is described by the function $\sigma(z) = \sigma_0 \exp[(z - G_h)/\zeta]$. The conductivity σ_0 is equal to the effective conductivity of displacement currents at frequency F_g: $\sigma_0 = 2\pi F_g\ \varepsilon_0$, where $\varepsilon_0 = 8.859{\cdot}10^{-12}$ F/m is the dielectric constant of vacuum. By using these parameters of conductivity profile, we calculate the electric height h_E, which is the altitude where conductivity currents are equal to the displacement currents at an arbitrary frequency f:

$$h_E(f) = G_h + \zeta \ln(f/F_g) \tag{6.1}$$

One uses the $f \approx 8$ Hz in computations relevant to the first SR mode and obtains the electric height around 50–60 km. The upper, magnetic height h_M, is the height of magnetic field diffusion into the ionosphere over the electric height:

$$h_M(f) = h_E(f) - 2\zeta \ln(2k\zeta) \tag{6.2}$$

The magnetic height is the 100–110 km at the first SR mode. The peak frequency in the resonance spectrum is found from Nickolaenko and Rabinowicz (1982):

$$f_n = \frac{c}{2\pi a}\sqrt{n(n+1)}\sqrt{\frac{h_E}{h_M}} \tag{6.3}$$

where c is the light velocity, a is the Earth's radius, and n is the mode number.

The particular parameters of profile provide the first peak frequency of 7.88 Hz (see Fig. 5.9 and Table 5.4). The classical PCA event corresponds to the downward shift of the conductivity profile, which means that only the reference height G_h is reduced, all other parameters remaining unchanged. The disturbance is formally introduced by the substitution $G_h \to G_h - S$ where S denotes the downward shift. As a result, every characteristic altitude is reduced by the S value, and changes in the frequency are computed from Eqs. (6.1)–(6.3). The following relation connects the frequency modification with the height reduction and undisturbed ionosphere parameters:

$$\delta f_n = -S \frac{h_M - h_E}{2\, h_E h_M} \tag{6.4}$$

Obviously the reduction of reference height lowers the resonance frequency and the Q-factor (as it increases the wave attenuation factor). Such a modification corresponds to the PCA event in the common sense. However, it occurs only at the sub-polar regions, so that an increase of attenuations will be noted at the polar propagation paths, and the reduction in the peak frequency will be smaller than that formula (6.2) predicts for a uniform reduction over the whole globe. Since we consider the first SR mode, the effect will be smaller by the ratio of the area occupied by the disturbance to the area of the whole globe. Of course, the solution we discuss is an approximate one, and the exact solutions of such and similar problems were published in the literature including the resonance line splitting (Madden and Thompson 1965; Nelson 1967; Nickolaenko and Hayakawa 2002; Nickolaenko et al. 2008).

In contrast to the above, the phenomenon of anti-PCA is observed when the energetic particles reduce only the upper, the magnetic ionosphere height h_M and leave the electric lower height h_E undisturbed. Exactly this kind of modification was suggested by Schlegel and Füllekrug (1999) when interpreting the Antarctic data. Such a disturbance obviously makes the conductivity profile steeper and improves its reflection ability.

We speak about some "median" conductivity profile. The real precipitations generate complicated spatial structures in the surrounding undisturbed plasma. One may recall the photographs of aurora containing many swaying "curtains" and "streams". An impact of such small details in the ionosphere structure might be noticeable at higher frequencies. The ELF wavelength is so great that the plasma might be treated as a horizontally uniform medium having some effective properties.

Let the precipitations of the second kind reduce the reference height by the same quantity so that $G_h \to G_h - S$. However, the height scale ζ is also reduced in such a way that the lower (electric) characteristic height does not vary:

$$h_E = const = G_h + \zeta \cdot \ln(f/F_g) = G_h - S + \zeta_d \cdot \ln(f/F_g) \tag{6.5}$$

One can readily find that the disturbed height scale is found from the following equation:

$$\zeta_d = \zeta - \frac{S}{\ln(F_g/f)} \tag{6.6}$$

In these conditions, the upper (magnetic) height (6.1) will be modified by the value:

$$\Delta\, h_M \cong \frac{2S}{\ln(F_g/f)}[1 + \ln(2k\zeta)] \tag{6.7}$$

The anti-PCA event "rotates" the conductivity profile around the fixed point h_E, as the plasma density grows faster with altitude (see Fig. 6.1). Such a disturbance reduces losses in the Earth-ionosphere cavity. Indeed the h_E height remains constant, while the altitude h_M is reduced, so that it follows from (6.3) that we expect an increase in the peak frequency. In terms of effective ionosphere height and the effective surface impedance, the classical PCA event lowers the effective height, but does not vary the surface impedance of the ionosphere. The dimensionless parameter *Z*/*kh* that determines the cavity eigen-values increases, so that the resonance frequency reduces and the wave attenuation grows. The anti-PCA event leaves the ionosphere height as it was and simultaneously reduces the effective surface impedance. As a result, the parameter *Z*/*kh* becomes smaller, which results in the increase in resonance frequency and Q-factor.

Computations confirm the above conclusions (see Fig. 6.1). We used the uniform distribution of thunderstorms over the globe to avoid the dependence on the source–observer distance. Three conductivity profiles were used as shown in the upper panel of Fig. 6.1. The abscissa depicts the conductivity (S/m) on logarithmic scale in this panel. The regular profile corresponds to $G_h = 89$ km and $\zeta = 3.8$ km. All plots corresponding to the regular profile are shown in Fig. 6.1 by the smooth green lines. The second profile is shown by the blue line with open wedges, which corresponds to the uniform 10 km reduction of the ionosphere. This is the case of PCA event when the reference height becomes $G_h = 79$ km while the scale height remains the same. The anti-PCA disturbance is shown by the red curves with filled triangles, which suggests both the reduced reference height $G_h = 79$ km and the scale height $\zeta = 2.4$ km. During anti-PCA, the lower characteristic height h_E remains invariable at frequencies around the first SR mode. This area is outlined by the orange ellipse in the upper plot of Fig. 6.1. The magnetic heights h_M of all three profiles are in the violet ellipse of the upper panel.

One may observe that the green (the regular ionosphere) and the red curves (the anti-PCA disturbance) practically coincide in frequencies of the first SR mode. The electric height goes down by 10 km during the PCA event (blue line). Concerning the magnetic height h_M, all three modifications are well separated here. The upper green line occupies the highest position. During the PCA, the height h_M is shifted downward by 10 km (blue line with wedges). During the anti-PCA event, the magnetic height acquires additional 6–11 km reduction shown by red

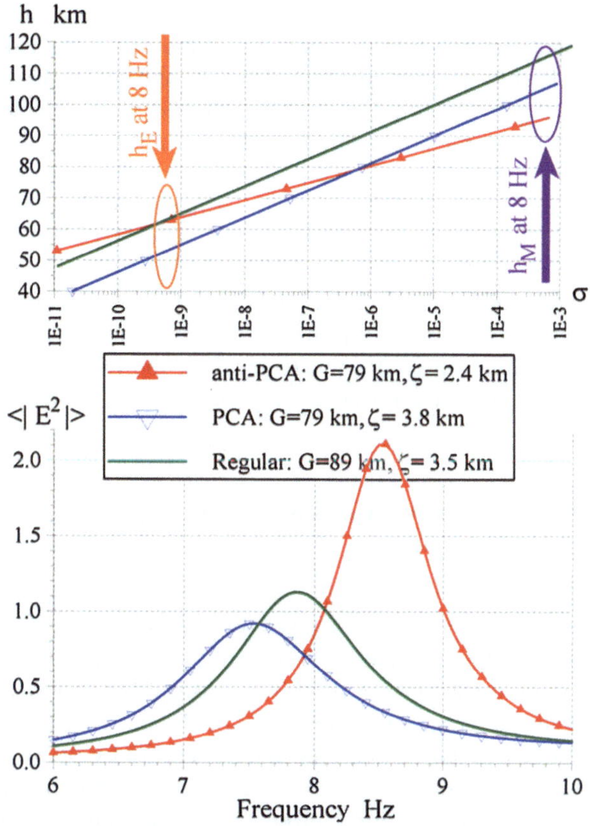

Fig. 6.1 The PCA and anti-PCA models comprising all the ionosphere. The characteristic heights (the *upper frames*) and the power spectra for the global uniform source distribution

line with triangles. This particular effect results from the "rotation" of the conductivity profile, which is clearly seen in the upper panel. Thus, the electric heights practically coincide for the regular and the anti-PCA cases (green and red lines in the left ellipse), while the anti-PCA magnetic height is severely reduced due to reduction in the height scale (lines in the violet ellipse).

One can find the propagation constant by substituting the profile characteristics into Eqs (5.9) and (5.10) and afterwards, to compute the power spectrum of the vertical electric field component. The lower panel in Fig. 6.1 demonstrates three spectra in the vicinity of the first SR mode. The regular spectrum is shown by green line with the peak frequency of ~7.9 Hz. The "classic" PCA event (blue line) reduces the peak frequency and the resonance intensity, while the width of spectral peak increases, which is obvious from Fig. 6.1. The anti-PCA event is related to fluxes that modify only the E region of the ionosphere and leave the D region unchanged. The resonance properties are improved, and the spectral intensity increases together with the peak frequency and the resonance Q-factor (red curve).

We must remark that besides explaining the published data, the model computations interpret the increase of the signal intensity also observed experimentally, but not explained by Schlegel and Füllekrug (1999). Thus, we can deduce that precipitation events they selected were relatively weak and therefore had improved the propagation conditions in the sub-polar region. This assertion could be checked by the signal processing at higher resonance modes: their frequencies, intensities, and Q-factors must increase together with those of the first SR mode.

The long-term observations of natural ELF radio signals in the Antarctica provided the data on the SR modifications by the SPE. Observations are consistently interpreted by the classical approach describing the modifications in the vertical conductivity profile. It is important that precipitations shifting the whole polar ionosphere downward result in the PCA events, whereas the modifications not reaching the lower ionosphere edge cause the opposite, the anti-PCA reaction.

6.3 Experimental Record of the PCA Caused by Bastille Day SPE

A question arises as a result of the above discussion: Are the usual PCA still observed in the modern SR records, or they are absent nowadays? Figure 6.2 gives the positive answer to this question. The classical disturbances persist being detected as an abrupt decrease of the peak frequencies during powerful modifications of the lower ionosphere. The links of SR with the space weather and magnetosphere activity was discussed in the literature for a long time. Many publications contained proof of magnetosphere influence on the global electromagnetic resonance (e.g. Kleimenova 1965; Zybin and Kleimenova 1965; Sao 1967; Sao et al. 1971, 1973; Zieger and Sátori 1999).

We demonstrate in Fig. 6.2 the impact of a very strong polar perturbation, the PCA event caused by the powerful SPE that is regarded in the literature as Bastille Day SPE, because it occurred around the UT noon on July14, 2000. The time is shown on the abscissa of Fig. 6.2 in days of July. The peak frequencies are shown on the ordinates. The upper panels show an abrupt fall of the first and the second SR mode frequencies observed at the Lehta observatory, Karelia, Russia (64° N and 34° E). The vertical red arrow denotes the onset of the SPE. One may note the rapid decrease of the peak frequencies of 0.4 Hz at the first, and 0.8 at the second modes. The tendency of subsequent recovery of the resonance frequencies is also seen in the figure.

Data are shown in two lower frames of Fig. 6.2 recorded at the Kamchatka Karimshino observatory, Russia (53° N and 158° E). Changes are depicted in the peak frequencies of two orthogonal components of the horizontal magnetic field: components H and D (Nickolaenko et al. 2004, 2008). The H–component lies in the plane of magnetic meridian, and it is close to the geographic H_{SN}–component.

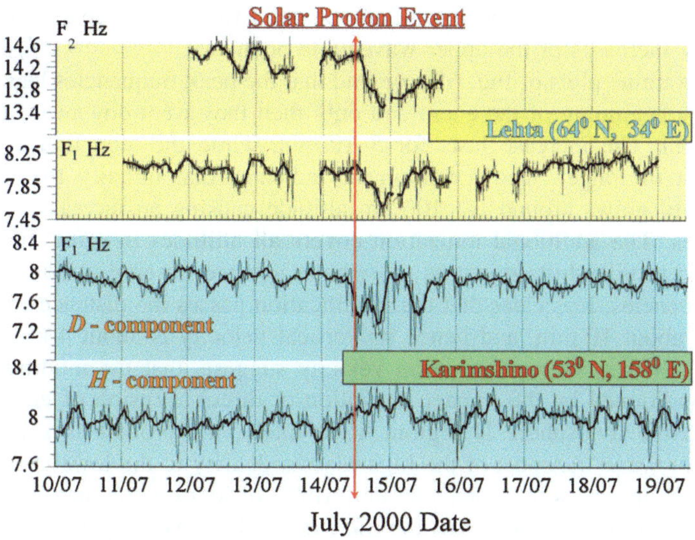

Fig. 6.2 Abrupt changes in the peak frequencies caused by the Bastille Day SPE

The *D*–component is perpendicular to this plane and is close to the H_{EW}–component.

Plots in Fig. 6.2 are shown by thin and thick lines. The data of monitoring are shown by thin black lines depicting the peak frequencies of the power spectra accumulated every 10 min. The thick lines show the moving average of quickly varying data. The width of moving window was about an hour.

Development of a classical PCA is obvious. The fall in the SR frequencies and the global charter of modification become apparent when comparing records at Karelia and Kamchatka. The drops of peak frequency practically coincide with the arrival of solar protons to the Earth. Widening of resonance peaks was detected (not shown in the plots) during the PCA event, which directly indicates the increase in the wave absorption. The sharp decrease of peak frequencies in different field components was combined with the global registration of simultaneous alterations points to the global ionosphere modification. The first SR frequency decreases by $\sim$0.4 Hz at both the observatories and afterwards it returned to its regular value (Roldugin et al. 1999, 2001, 2003; Nickolaenko et al. 2008).

We already know that particle precipitations during the SPE can cause an abrupt change in the peak frequency reaching a fraction of a hertz. Particular variation, either increase or decrease, depends on the character of interaction between the polar ionosphere and the precipitating particles. The peak frequencies go down when the precipitation is strong enough to reduce the ionosphere as the whole. They increase when the magnetospheric particles cannot reach the lower ionosphere edge and the plasma conductivity increases at the E region and higher, while that remains unchanged at the D region. The conductivity profile becomes steeper at the altitude range important for the ELF radio propagation, and the reflectivity increases of the upper waveguide boundary.

If we examine plots of Fig. 6.2, we find that the peak frequencies had increased at the very beginning of the event and only then they went downward (Roldugin et al. 2001). The increase lasts for 5–10 min, while the reduction is observed during 1 h or longer. The development indicates that initially the ionosphere conductivity grows around the 100 km altitude causing an increase in the peak frequencies. The additional ionization covers all altitudes in a few minutes, the ionosphere profile shifts down as the whole, and we observe a reduction in the peak frequencies. It is clear that the modification passes the distance from 100 to 60 km in about 10 min, and hence its vertical velocity is about 4 km/minute or ~70 m/s. Such a slow propagation velocity suggests that the modification was hardly conditioned by the impact ionization or direct intrusion of energetic particles into the ionosphere D region. Most likely, the process is some kind of diffusion or wave transport of the high plasma density to the lower heights.

The SPE generates the polar non-uniformity of ionosphere, and such a disturbance must cause the splitting of the SR eigen-values. We exploit formulas from the paper by Nickolaenko et al. (2008) [see also Chap. 2 in Nickolaenko and Hayakawa (2002)]. The first SR mode frequency is described by the following equation:

$$f_1^m = f(n = 1, m) = f_1 + \Delta f \cdot S_1(m) \tag{6.8}$$

where $f_1 = 10.6$ Hz is the SR frequency for the ideal cavity, $\Delta\Delta f$ is the complex correction of the frequency due to finite conductivity of the ionosphere boundary and the function $S_1(m)$ accounts for spatial structure of non-uniformity, which depends on the azimuthal index $m \in [-1, +1]$. In the uniform Earth—ionosphere we have $\mathrm{Re}\{\Delta f\} = 2.6$ Hz.

One obtains the following equations for the splitting function (Nickolaenko and Hayakawa 2002; Nickolaenko et al. 2008):

$$S_1(m) = \begin{cases} \int\limits_{-1}^{1} (1 + {\rho_n}^2 x^2)(1 - x^2)\, dx = \frac{2}{3}\left(1 + \frac{{\rho_n}^2}{5}\right) & m = 0 \\ \frac{1}{2}\int\limits_{-1}^{1} (1 + {\rho_n}^2 x^2)(1 + x^2)\, dx = \frac{2}{3}\left(1 + \frac{2{\rho_n}^2}{5}\right) & m = \pm 1 \end{cases} \tag{6.9}$$

We use the symmetric (even) polar non-uniformity described by the $Z/kh = const\,[1 + \rho_n^2 \cos^2\theta]$ angular function where Z and h are correspondingly the effective surface impedance and the effective height of the lower ionosphere boundary, and θ is the co-latitude, and ρ_n^2 is the non-uniformity index.

The first SR mode splitting due to the polar non-uniformity is shown in Fig. 6.3. A parameter ρ_n^2 is negative when the losses in the polar region are reduced (this is the anti-PCA event) and the left portion of the plot corresponds to such disturbances. The classical PCA increases the sub-polar attenuation, the parameter ρ_n^2 becomes positive in this case, and the relevant dependence is shown in the right part of frame. When $\rho_n^2 = 0$, the cavity is uniform, and its peak frequency is 8 Hz regardless the wave azimuthal index $m = 0$ or $|m| = 1$. In this case one speaks about the frequency degeneracy in the spherically symmetric cavity (Bliokh et al. 1977, 1980; Nickolaenko and Hayakawa 2002).

The index of polar non-uniformity might be readily estimated if we use the experimentally observed frequency variation of +0.4 for anti-PCA and −0.4 during the Bastille Day PCA. The parameter is $\rho_n^2 = 0.4$ when the index $|m| = 1$, and $\rho_n^2 = 0.8$ when the index $m = 0$. Hence, the SPEs modified the ionosphere effective surface impedance (or the effective height) by 40–80 %. These estimates were obtained for a particular $Z/kh = const\ [1 + \rho_n^2 \cos^2\theta]$ dependence, however, they will not change crucially if one applies any other reasonable model. It is important to note that Fig. 6.3 demonstrates that the line splitting is a few tenths of a hertz, so that it is not easy to discover such a splitting around 8 Hz frequency when the spectral peak has the width of ~2 Hz.

Summarizing the above, we admit that thc Bastille Day SPE of 2000 was associated with approximately 50 % modification of the ionosphere at the poles. Smaller frequency changes would be interpreted by progressively minor parameters of the polar non-uniformity corresponding to less significant modifications. The frequency falls observed experimentally might be attributed to the sole effective surface impedance of the ionosphere, to the effective height, or a combined effects of the two, as the resonance frequencies depend on the dimensionless parameter Z/kh (e.g. Nickolaenko and Hayakawa 2002, Chap. 2). This small parameter of electromagnetic problem defines both changes of peak frequency and the line splitting. Both these effects might be evaluated by using Fig. 6.3.

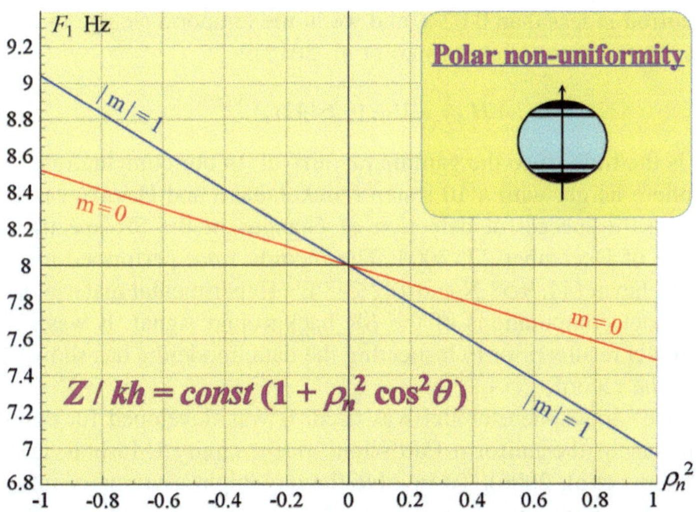

Fig. 6.3 Line splitting caused by the polar non-uniformity of the type: $Z/kh = const[1 + \rho_n^2 \cos^2\theta]$

6.4 Impact of Powerful Galactic Gamma-Flare on SR Background Spectrum

The orbital X- and gamma ray observatories recorded two especially powerful gamma busts that arrived from the extra-solar sources (Inan et al. 1999, 2007; Dowden et al. 1999; Terasawa et al. 2005; Hurley et al. 2005; Tanaka et al. 2008). Records of remote VLF transmitters indicated that the ionosphere depression was caused by the gamma rays covering almost the entire hemisphere. An attempt was unsuccessful in finding changes in the SR records caused by the *SGR* 1900 + 14 gamma flare on August 27, 1998 (Price and Mushtak 2001). We consider an impact of the powerful gamma ray flare from *SGR* 1806-20 that occurred on December 27, 2004 and the detailed description of the event might be found in the papers by Hurley et al. (2005), Terasawa et al. (2005), Inan et al. (2007), and Tanaka et al. (2011). We mention only the minimal information here. The flare occurred around 21 h 30 min and 26.5 s UT when the hard X/gamma rays arrived at the dayside of the Earth. Radiation came from a neutron star 30–40 thousand light years away. The peak flux exceeded the most intense solar flares by five orders of magnitude, which was 100 times greater than the *SGR* 1900 + 14 gamma flare of 1998. The disturbance was centered above the Pacific Ocean, at the geographic coordinates, 146.2° W and 20.4° S, i.e. at a distance of ≈450 km from the center of the dayside hemisphere. Monitoring of different trans-Pacific VLF transmissions at the "Palmer" Antarctic station allowed for the following interpretation of observations (Inan et al. 2007). The ionosphere depression covered all aquatory of the Pacific Ocean up to angular distances $\psi = 60°$ from the center of disturbance. The gamma flare lowered the dayside ionosphere by ~20 km, and modification lasted for more than 1 h. An abrupt reduction in the ionosphere height occurred in less than 0.02 s, and we fit the temporal change of the height by the following function (Nickolaenko et al. 2012):

$$dH = -19 \cdot (9.6441t)^{-0.1501}. \tag{6.10}$$

Here t is the time since the gamma ray arrival. In our computations, we change the ionosphere height with a 10 s step (Nickolaenko and Hayakawa 2010).

We discuss a successful detection of changes in the SR spectra during the gamma flare of December 27, 2004. The records were performed at the Moshiri observatory, Japan (44.365° N and 142.24° E). Experimental material is compared with the model computations of the SR background signal. It was important to have the model results prior to inspecting the data, modeling has shown us what to look for in the record.

The "knee" ionosphere model was used. It was developed for computing the SR parameters in the uniform Earth–ionosphere cavity (Mushtak and Williams 2002; Williams et al. 2006). To involve the day–night asymmetry into modeling, one exploits the partially uniform knee (PUK) ionosphere model (Pechony and Price 2004) (see Chap. 14). Similarly to the previous sub-section, both the models imply two characteristic altitudes: the electric and magnetic heights, but these parameters are defined in a slightly different way. The "knee model' implies two exponential conductivity profiles, and their intersection forms a "knee" at some altitude. The height scale ζ is smaller above the knee altitude and greater below this altitude, and this forms the necessary fracture of the curve. We postulate, for simplicity, that only the 'knee altitude' is reduced by gamma rays. Therefore, the 'electric' height of the Earth–ionosphere cavity goes down, while the 'magnetic' height remains undisturbed.

The physical effect of a gamma flare on SR is conditioned by a huge reduction of the ionosphere height, which alters the eigen-values of the Earth–ionosphere cavity. As we know from the previous sub-section, such a modification simultaneously reduces the peak amplitudes, peak frequencies, and the quality factors of all SR modes. In the most adequate modeling, the day-night non-uniformity should be accounted for, especially as the 20 km reduction of the ionosphere height occurred at the dayside of the globe, and this was done by using the PUK model (Pechony and Price 2004). Bearing in mind the global nature of SR, we applied the 10 km depression over the whole globe in the model of a uniform Earth–ionosphere cavity (Nickolaenko and Hayakawa 2010). We compare the results of both approaches (Nickolaenko et al. 2012) and see that the day-night asymmetry has a minor impact on the quantitative effect of the gamma rays. These models of the Earth–ionosphere cavity were used combined with the OTD data on the global lightning source distribution.

The 'physical' parameters of the standard and disturbed ionospheres are listed below. The air conductivity varies with the height h above the ground in the following way:

$$\sigma(h) = \sigma_0 \exp[(h - H_{KNEE})/\zeta] \tag{6.11}$$

We adopt the knee altitude $H_{KNEE} = 55$ km for the regular model, and conductivity σ_0 corresponds to the knee frequency $F_{KNEE} = \sigma_0/(2\pi\cdot\varepsilon_0) = 10$ Hz. This means that conductivity and displacement currents at the frequency of 10 Hz become equal at the 55 km altitude. Thus, $\sigma_0 = 20\pi\cdot\varepsilon_0$. The vertical conductivity profile has the scale heights $\zeta_a = 2.9$ km above and $\zeta_b = 8.3$ km below the knee altitude. The upper ionosphere is characterized by the magnetic height $H_{mp} = 96.5$ km and $F_{mp} = 8$ Hz, which logarithmically varies with frequency. Conductivity profile at the 'magnetic' altitude has a definite scale height also depending on frequency (Mushtak and Williams 2002; Williams et al. 2006). The globally uniform ionosphere disturbance relevant to the gamma rays flare reduces the altitude H_{KNEE} from 55 to 45 km, and all other model parameters remain the same (Nickolaenko and Hayakawa 2010). Data of such a disturbed profile are used for obtaining the propagation constant of ELF radio waves and computing the resonance spectra.

We use the PUK model (Pechony and Price 2004; Pechony et al. 2007) when accounting for the cavity day-night asymmetry. The ionosphere is subdivided into the day and night hemispheres, and the knee model is used for each of them having

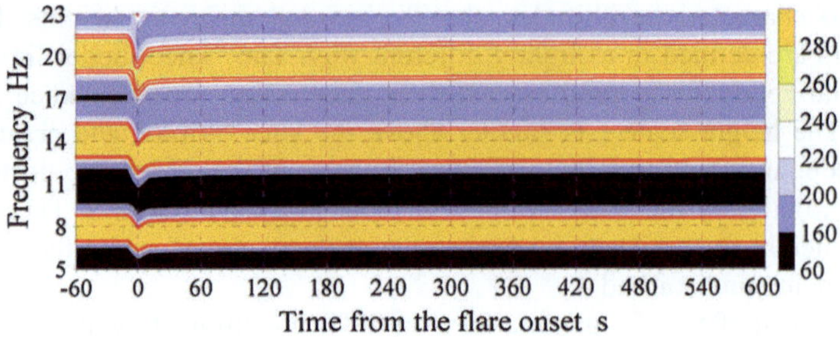

Fig. 6.4 Abrupt modifications of the SR spectra produced by the gamma ray flare of December 27, 2004

different parameters. In particular, $F_{KNEE} = 13$ Hz and $F_{mp} = 6$ Hz are fixed everywhere. The dayside is characterized by $H_{KNEE} = 54$ km, $\zeta_a = 2.7$ km, $\zeta_b = 7.5$ km, and $H_{mp} = 97.5$ km. The night-side has the values of $H_{KNEE} = 60$ km, $\zeta_a = 3.8$ km, $\zeta_b = 9.1$ km, and $H_{mp} = 99.0$ km. The gamma flare reduced the dayside knee height by 20 km (to 34 km). These parameters were used in the procedure of resolving the 2D telegraph equations for SR (e.g. Pechony and Price 2004; Pechony et al. 2007; Kirillov 1993, 1996; Kirillov and Kopeykin 2002).

For computing the SR spectra, one has to know the source distribution. Two spatial distributions of thunderstorms were used. The simple one has the lightning strokes uniformly distributed worldwide. A more realistic second model implies the global distribution acquired by the OTD (Christian et al. 2003). This space-borne sensor recorded optical flashes from the lightning strokes worldwide during 5 years of observations. The final results were presented as global maps of lighting activity corresponding to every hour UT and each month of a year. Since the gamma rays arrived at 21:30 UT, we had to average the OTD maps for 21 and 22 h in our computations. The more detailed description might be found in Nickolaenko and Hayakawa (2010).

Figure 6.4 shows the expected sonogram of SR around the moment of the gamma ray flare $t = 0$. The knee height modification was described by Eq. (6.10) with the initial global reduction of 10 km instead of 19 km. Thunderstorms were positioned in accordance with the OTD data. Resonance spectra of vertical electric field component were computed with the 10 s time step.

An abrupt downward shift of the SR pattern is clearly seen in Fig. 6.4 at the onset of gamma flare, and spectra slowly return to the regular outline afterwards. Obviously, one has to look for a sudden drop in experimental sonograms at the time of the gamma burst occurrence. To stress the outline of modification, we draw the levels 240 and 260 a.u. by thick lines in Fig. 6.4. The contours will be helpful when analyzing observational data.

Figure 6.5 compares the impact of gamma rays in both the uniform Earth–ionosphere cavity (upper plot) and in the non-uniform cavity (lower plot). The spectra were computed for the vertical electric field component by using the 2D

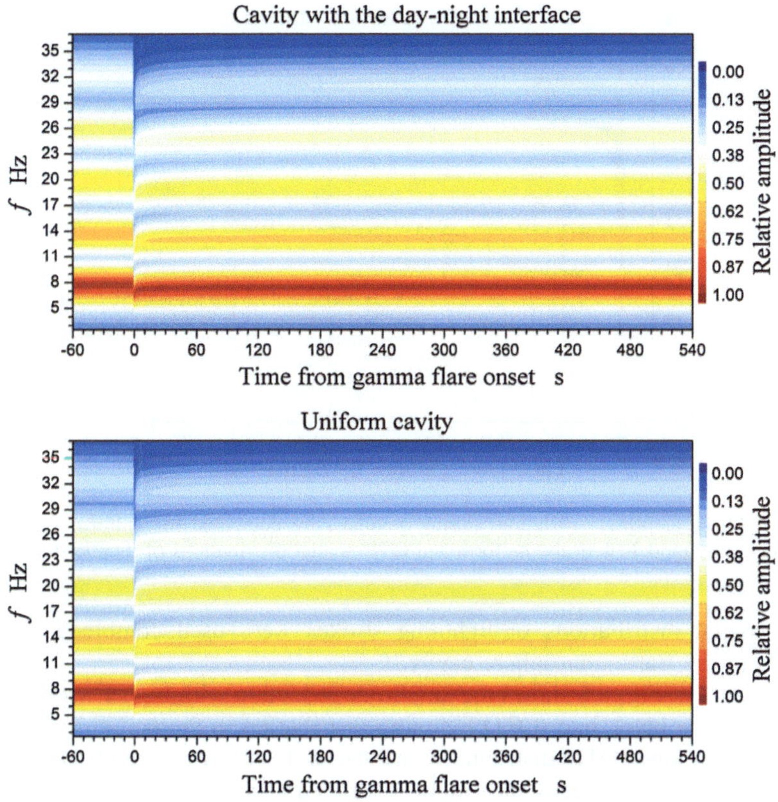

Fig. 6.5 Changes in the SR sonograms at the moment of gamma flare computed for the Earth–ionosphere cavity with day–night interface (*upper frame*) and the uniform resonator (*lower frame*)

telegraph equations with the PUK day-night model (Pechony et al. 2007; Nickolaenko et al. 2010, 2012). The OTD source distribution was used in both cases. The upper plot of Fig. 6.5 practically repeats the sonogram of Fig. 6.4 computed for the global modification of the uniform Earth–ionosphere cavity. We must emphasize that spectra of Fig. 6.4 were computed by using the field expansions into zonal harmonic series representation, and spectra of Fig. 6.5 were computed with the help of 2D telegraph equations. Thus, different models and separate formal solutions provide similar results: dynamic spectra of SR contain an abrupt drop in the resonance pattern. It is clear now in what a way ionosphere modification could reveal itself caused by the gamma flare.

Amplitude variations are given in Fig. 6.6 obtained with 2D telegraph equations. We plot here modulations in the uniform and non-uniform Earth–ionosphere cavities. The observer is placed at 'Palmer' Antarctic station where the ionospheric impact of gamma flare was detected in VLF transmissions. Both uniform and the

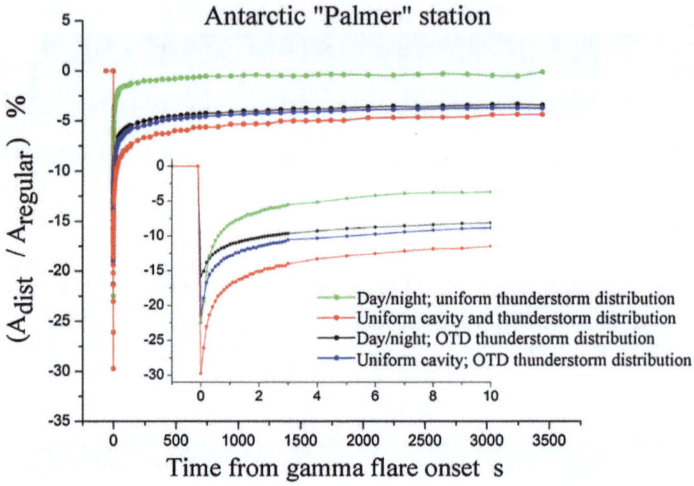

Fig. 6.6 Sharp amplitude variations during gamma flare computed for uniform and the non-uniform cavity models with the uniform and OTD spatial distributions of thunderstorms

OTD spatial distributions of lightning strokes were used combined with two models of the Earth–ionosphere cavity.

The inset in Fig. 6.6 shows the amplitude reduction with higher resolution. By comparing the black line (OTD lightning distribution in the non-uniform cavity) with the blue one (OTD distribution and the uniform drop), we observe that the impact of the ionosphere day–night asymmetry is insignificant. The highest deviation of curves is pertinent to the uniform distribution of lightning strokes. Except this systematic deviation, the general behavior remains the same: an abrupt reduction reaching 30 % and a gradual recovery afterwards. We conclude that computations predict essentially the same modification regardless the particular model: gamma rays abruptly reduce all resonance parameters. It is important that sudden alterations occur simultaneously at all resonance modes, and hence we must look for the similar discontinuity pattern in experimental ELF sonograms around the occurrence time of gamma flare.

It is obvious now that uniform and non-uniform cavity models predict similar reaction of global electromagnetic resonance on the gamma flare regardless the particular distribution of thunderstorms. All resonance parameters instantly reduce at all modes, and the sonogram shows a characteristic discontinuity at the moment of arrival of gamma rays. This distinctive pattern must be sought in observations data.

Experimental ELF records are collected permanently at the Moshiri observatory (Japan) (geographic coordinates, 44.365° N and 142.24° E) since 1996. Waveforms of vertical electric, east–west (EW), and north–south (NS) horizontal magnetic field components are recorded here with the sampling frequencyof 4 kHz (Ando et al. 2005). The ELF data from Moshiri were used in the studies of sprites

(Hobara et al. 2001; Hayakawa et al. 2004) and the global lighting activity (Sekiguchi et al. 2006; Hayakawa et al. 2011b). We have chosen a fragment of ELF record in the vicinity of the gamma ray event for the spectral processing. Dynamic spectra of H_{NS} and H_{EW} components are shown in Fig. 6.7 (the electric signal was spoiled by local interference). Time is shown on the abscissa in seconds, and the zero time is linked to the moment of gamma ray arrival. The ordinate depicts the frequency with the step of 6 Hz. We present two survey sonograms in this figure.

Apparently, data processing must satisfy two contending conditions. Since computations indicate that the abrupt shift of resonance pattern in a sonogram is about 1 Hz, the frequency resolution must be 0.1 Hz at least. Hence, the segments of 10 s duration must be used in the FFT procedure. On the other hand, the ionosphere disturbance driven by the gamma flare decreases significantly in 10 s, so that inpact of modification will be reduced in the amplitude spectra composing the sonograms. Therefore, the separate spectra of the sonogram were obtained with the 1 s step. Thus, for $t = 0$, we process the segment from −5 to +5 s. For the time $t = 1$, the segment is [−4; 6 s], etc. The time shown on the abscissa corresponds to the middle of a record 10 s long.

Figure 6.7 shows that we immediately stuck with the above-mentioned instability of experimental SR spectra. Owing to the stochastic nature of lightning strokes, the observational sonograms are highly structured. The smooth resonance lines appear after averaging of many elementary spectra. The procedure usually takes 5–10 min (30 or 60 elementary spectra), and so one cannot expect the 'regular' experimental resonance pattern like those obtained in the modeling (Figs. 6.4 and 6.5). Spectra in Fig. 6.7 are spontaneous in both the field components and this is the major obstacle in finding an impact of gamma rays: the expected outline of modification will be obscured by random fragments relevant to the global thunderstorms. In this aspect, experimental spectra are similar to data obtained by Price and Mushtak (2001).

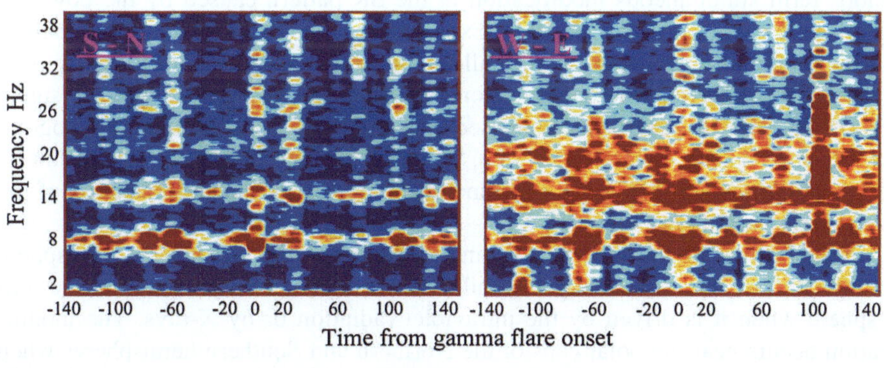

Fig. 6.7 Experimental sonograms of SR recorded at Moshiri observatory around the gamma flare

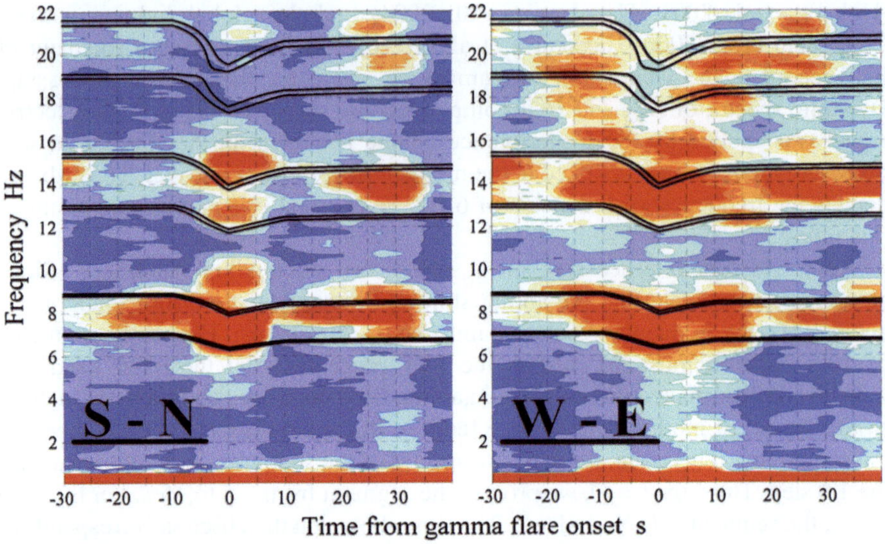

Fig. 6.8 Comparison of observed data with the model dynamic spectra in the vicinity of gamma flare. Modification is evident of resonance pattern

To reveal modification, we depict a smaller portion of experimental sonograms and combine them in Fig. 6.8 with the spectra of Fig. 6.4. The thick black lines depict the levels emphasized in the model sonogram of Fig. 6.4. These lines are superimposed on the color experimental sonogram corresponding to the interval ±30 s. The expected simultaneous reduction in the resonance pattern becomes apparent in both frames of Fig. 6.8 in spite of highly unstable experimental spectra. The time of sudden modification matches well with the arrival moment of gamma rays to the Earth. Even the tendency toward recovery might be perceived in the plots of Fig. 6.8.

Direct comparison of computational and experimental sonograms reveals a short- term simultaneous modification in the SR pattern caused by the powerful gamma flare of December 27, 2004. Experimental data itself, without relationship to the model spectra, would hardly allow for convincing detection of the modulation. To single out the modification, we also had to apply a special signal processing: the 10 s segments were used in the FFT procedure, while the sonogram was formed by the spectra of 1 s step. These results agree with the conclusion by Price and Mushtak (2001) that random nature of spectra impedes detection of the short-term gamma ray effect.

Irregularities in the cosmic ionizing radiation result in different ionosphere modifications. The disturbance is similar to the day–night asymmetry of the ionosphere when it is driven by the ultraviolet radiation or by X-rays. The modification occurs near the polar caps of the Northern and Southern hemispheres when the ionization is caused by the flux of charged particles since the geomagnetic field deflects these particles towards the poles. Of course, this is an oversimplified picture, because we do not mention the Earth's magnetosphere at al. From the viewpoint of SR, the "symmetric" disturbances of the second kind provide more noticeable modifications of resonance parameters (Bliokh et al. 1977, 1980; Nickolaenko and Hayakawa 2002). This is why the SPE is clearly seen in the SR records, while an impact of the solar chromosphere flares is dubious if present at all. Even the giant gamma flare of December 27, 2004 did not cause the contrast and durable modifications in the SR spectra comparable with the SPE impact described in the previous sub-section.

6.5 Disturbances Associated with Seismic Activity

Electromagnetic waves emerging from the seismic activity have been reported during the last few decades, and the idea is usually accepted with skepticism. Simultaneously, it is commonly accepted that detection of seismogenic electromagnetic emissions (if any) is facilitated at low frequencies, as these attenuate to the lower extent in the conducting soil (e.g. Hayakawa and Molchanov 2007; Hayakawa et al. 2008, 2010). In the present subsection, we demonstrate the experimental data first and then suggest a possible interpretation.

The anomalies were noted in the Japanese ELF records during a few years that were clearly associated with the earthquakes (EQs) in Taiwan (Hayakawa et al. 2005), i.e., at the distance of about 3,000 from the epicenter. Unusual signals were noted initially in the records of Nakatsugawa observatory (35.45° N and 137.3° E), which were associated with the powerful Taiwan EQ on September 21, 1999, the Chi–Chi EQ. The anomaly appeared as an amplitude increase in the vicinity of the fourth (sometimes, the third) SR mode. This preliminary result was confirmed by the statistical analysis of Nakatsugawa records for the Taiwan EQs with magnitude exceeding 5, which occurred during 6 years (Ohta et al. 2001, 2002, 2006, 2009; Hayakawa et al. 2008, 2011a). Similar modifications were found also in the records of Moshiri observatory during the Ping-tong EQ in Taiwan on December 26, 2006 (Hayakawa et al. 2008).

We show typical experimental data and the details might be found in papers by Hayakawa et al. (2010, 2011a) and references therein. The observatory Nakatsugawa monitors the magnetic field components south–north and west–east together with the vertical component. The sampling frequency is 100 Hz, and data are recorded on the computer hard disc drive. The receiving equipment is described in the papers by Ohta et al. (2001, 2006) and Hayakawa et al. (2005). The spectral processing used the FFT procedure with 1,024 samples, so that the segment duration was 10.24 s and the frequency resolution was $\sim$0.097 Hz. The following anomalies are mentioned in the paper by Hayakawa et al. (2005) associated with the Chi–Chi EQ.

1. An increase was noted in the amplitude of the fourth (sometimes third) SR mode with simultaneous shift of the peak frequency by approximately 1 Hz.
2. Unusual signals appeared about a week before the EQ, they lasted a few days, and they were present during the aftershock activity.

A dynamic spectrum containing seismic anomalies is shown in Fig. 6.9. The time interval is shown on the abscissa in days (UT), covering the September of 1999. The ordinate shows the frequency in Hz, and the signal amplitude is shown by color. The color scale is shown at the bottom of plot. The first and the second (marked as n = 1 and n = 2 in the plot) resonance peaks are easily recognized as horizontal trails around 8 and 14 Hz frequency. The third mode is also well resolved. One may observe the regular diurnal modulation of the signal amplitude with the pulsed interference superimposed in regular trend. Interferences are seen as the vertical light strips.

One may observe that the anomaly is an abrupt increase of the signal around the fourth SR mode frequency. The upward shift of ~1 Hz is also noticeable. The "pulsed" radio signals are also observed on the background of a general amplitude increase, which have the 25 Hz "carrier" frequency, i.e., around the peak of the fourth resonance mode. Sometimes, the pulsed component is absent, and the regular "stationary" increase is observed (Ohta et al. 2001, 2002, 2006, 2008, 2009).

Changes in spectral amplitude were observed from a week to a few days prior to the main shock, and this is illustrated by Fig. 6.9 where the ELF activity starts on September 16 while the EQ took place on September 21st. The wave arrival direction might be estimated by using two horizontal orthogonal magnetic components, which appeared to be close to the direction toward Taiwan.

The spectral structure was unusual, so a statistical analysis was performed of the Nakatsugawa records six years long. The goal of this study was verification of a link between the SR anomalies observed in Japan with the EQs occurring in Taiwan. The anomalies were present in Japanese records appearing in a week to the main shock of the land Taiwan EQs of magnitude higher than 5 (Ohta et al. 2006, 2009). The number of events was 27, and statistical processing indicated a statistically significant correlation between the observations at Nakatsugawa and

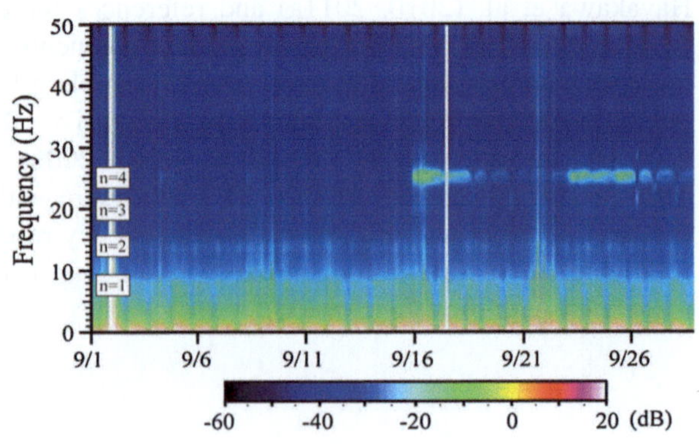

Fig. 6.9 Anomalous dynamic spectrum at the Nakatsugawa observatory recorded in September of 1999

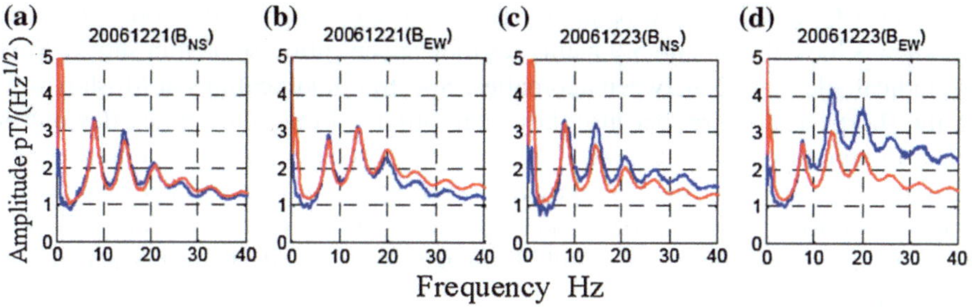

Fig. 6.10 Resonance spectra at the Moshiri station on December 21 and December 23, 2006 recorded in the interval 20–22 h UT

EQs in Taiwan. Concerning the EQ under the sea, only two events were accompanied by SR anomaly. One of them was the most powerful, and the other occurred nearby the sea shore. Observations at Nakatsugawa played an important role in explaining how the ionosphere disturbance originates due to seismic activity (Molchanov and Hayakawa 2008), but we will not go in this direction. It is worth mentioning that the closer data examination allowed us to discover also the different kinds of abnormal traces of some Japanese EQs (Hayakawa et al. 2011a).

Anomalies were also found in the records of Moshiri observatory, Hokkaido (44.29° N and 142.21° E). We demonstrate these data first and then we discuss a possible mechanism of generating the anomaly in the SR data due to the radio wave scattering by the ionosphere disturbance over Taiwan. The EQ called Ping-Tong took place in Taiwan on 12:26 UT on December 26, 2006. Its magnitude was 7.3 and the depth was 10 km. The anomaly was observed in the Moshiri record. Unfortunately, the Nakatsugawa equipment was modernized at the moment, so that simultaneous records are not available. Records in the three UT intervals were analyzed: 8 ± 1 h (Asian thunderstorms), 15 ± 1 h (African thunderstorms), and 21 ± 1 h (American thunderstorms). An anomaly was noted only in the third time interval of the daily records relevant to thunderstorms in America. We show the SR spectra relevant to the time interval 21 ± 1 h UT in an ordinary day of December 21. 2006 and in the day of disturbance (December 23). Figure 6.10 contains four frames: frame (a) depicts the spectrum of H_{SN} component on December 21, frame (b) shows the H_{WE} spectrum of component on December 21, frame (c) corresponds to H_{SN} component on December 23, and frame (d) depicts the spectrum of H_{WE} on December 23. The frequencyis shown on the abscissa in Hz, and the ordinate shows the spectral amplitude density in pT/(Hz)$^{1/2}$. The blue curves demonstrate the undisturbed amplitude spectra that were averaged in the interval of ±15 days around the December 21 date. The red curves shows the disturbed spectra averaged in the same ±15 day interval around the December 23 date.

Spectral anomalies of both Japanese observatories have similar features, but deviations are also present. The similarity is found in the following: Anomalies are found in the records of different years, which were associated in time with the powerful EQ in Taiwan. Anomalies were observed at higher SR modes. Unusual SR signals have the precursory nature, as they appear prior to the min shock. Next, we indicate deviations between anomalies: Spectral modifications at Moshiri were of the short-term nature, because they were observed only during one day and in the UT interval of 21 $\pm$ 1 h. At this time, the maximum of global thunderstorm activity is found at the South America. Moshiri anomaly was a precursor in such a way that it was present prior to the EQ shock and was absent afterwards. The Moshiri anomaly occupied the wide frequency band, while that at Nakatsugawa was narrow-banded.

The interference of the direct wave and that reflected from the ionospheric disturbance placed over the EQ is a reasonable explanation of spectral anomaly (Hayakawa et al. 2005). The constructing or destructing interference of direct and reflected waves depends on the mutual phase shift, and this latter depends on positions of the propagation and scattering paths. The frequency at which the interference becomes constructing directly depends on the geometrical difference of propagation path lengths (Nickolaenko et al. 2006). Let us assume that thunderstorms are concentrated at Africa, America, or Asia. This will allow for evaluating the anomaly on different times of the day. We list in Table 6.1 parameters or propagation paths calculated for three compact thunderstorm centers in Asia, Africa and America.

There is an obvious reason why the thunderstorms of South America must play an outstanding role in observations in Japan: these sources are positioned not very far from the observer antipode (Hayakawa et al. 2005). For example, the antipode of the Nakatsugawa observatory is positioned at a point with the coordinates 35.45° S and 42.7° W, not far from the South America. The situation is similar with Moshiri. Keeping in mind a finite width of antipode minimum, one may speak about higher sensitivity of resonance records in Japan to the South American thunderstorms. Since the observer—disturbance distance is 3–4 Mm, the high amplitude of the wave coming to Japan suggests that the amplitude is also high of the wave scattered by the Taiwan irregularity.

An important feature is the back-scattering of the ELF radio waves coming from American sources pertinent to the Taiwan non-uniformity. The geometry of radio propagation paths is illustrated in a diagram in Fig. 6.11. An increase of the wave scattering in the backward direction was noted in theoretical works by Nickolaenko (1986, 1994), Nickolaenko and Hayakawa (2002) and Hayakawa et al. (2005). Influence of a localized ionosphere disturbance associated with the Taiwan EQ increases when the signal arrived from the South America as seen in Fig. 6.11.

It is rather simple to evaluate the frequency at which the sum reaches maximum of the direct and the reflected from the ionosphere disturbance wave. The difference in path length for American source is about the 5 Mm value, therefore maximum interference is expected at the third–fourth SR mode. Similar speculations indicate that interference of the direct and reflected waves arriving from the other thunderstorm centers will increase the amplitude at higher frequencies, beyond the confident detection of separate SR peaks.

Table 6.1 Geometric parameters of three possible configurations source–observer–non-uniformity

Moshiri observatory (44.365° N and 142.24° E)			
Point source position	Source–observer distance (Mm)	Observer–disturbance distance (Mm)	Source–disturbance distance (Mm)
Asia (10° N and 105° E)	5.22	3.68	2.34
Africa (0° N and 25° E)	12.12	3.68	8.90
America (0° N and 75° W)	13.85	3.68	16.55
Nakatsugawa (35.45° N and 137.3° E)			
Point source position	Source–observer distance (Mm)	Observer–disturbance distance (Mm)	Source–disturbance distance (Mm)
Asia (10° N and 105° E)	4.32	3.17	2.34
Africa (0° N and 25° E)	12.00	3.17	8.90
America (0° N and 75° W)	14.84	3.17	16.55

We introduce the seismogenic ionosphere disturbance in the simplest way: we move downward by 20 km the part of the regular "knee" conductivity profile at heights below the 80 km. The regular and disturbed knee profiles are shown in Fig. 6.12 by blue and red lines with corresponding marks. Such a modification does not modify the upper, "magnetic" ionosphere height, with all other parameters of profile also remaining unchanged.

Detailed description of ELF radio wave scattering by a localized ionosphere disturbance might be found in Nickolaenko (1994) and Nickolaenko and Hayakawa (2002). Here we just outline the way of obtaining the solution. The vertical electric field component is considered because of the simplicity of relevant solutions: the field orientation is independent of positions of the source, observer and

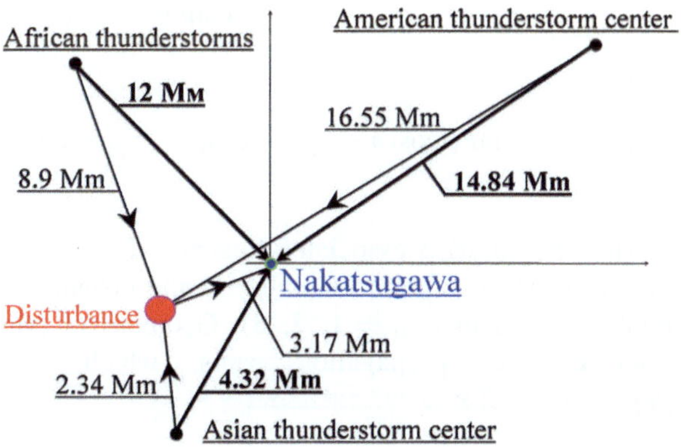

Fig. 6.11 Propagation and scatter paths geometry for three global thunderstorm centers, observer at Nakatsugawa, and disturbance over Taiwan

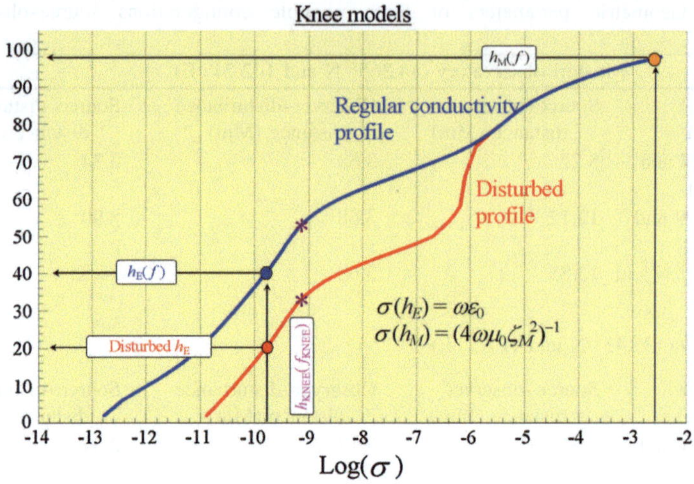

Fig. 6.12 Knee vertical profiles of atmospheric conductivity and relevant characteristic height

scatterer. The complete field at the observatory is the sum of direct E_1 and scattered E_2 radio waves:

$$E = E_1 + E_2 \tag{6.12}$$

The direct (primary) wave propagates from the source to an observer in the uniform Earth—ionosphere cavity and is found from the following relation:

$$E_1(\omega) = \frac{M_C(\omega)}{4ha^2\varepsilon_0}\frac{i\nu(\nu+1)}{\omega}\frac{P_\nu[\cos(\pi - \theta_H)]}{\sin \pi\nu} \tag{6.13}$$

Here $M_C(\omega)$ is the source current moment, $\nu(\omega)$ is the propagation constant, $P_\nu(x)$ is Legendre function, and θ_H is the source–observer angular distance. The relative field disturbance is found as the following ratio:

$$B_D = \frac{E_2}{E_1} = \frac{\int \sin\theta \, d\theta d\varphi \, Q_i \delta \, C_\nu^2}{4 \sin \pi \, \nu \, P_\nu[\cos(\pi - \theta_H)]} \tag{6.14}$$

where

$$Q_i = \nu(\nu+1)P_\nu[\cos(\pi-\gamma_i)]P_\nu[\cos(\pi-\gamma)] - dMP_\nu^1[\cos(\pi-\gamma_i)]P_\nu^1[\cos(\pi-\gamma)] \tag{6.15}$$

Here $P_\nu^1(\cos\theta)$ is associated Legendre function, γ is the angular distance from the observer to disturbance, and γ_i is the distance from the disturbance to a particular field source (i = 1, 2, 3). Geometrical parameter dM accounts for the orientation of propagation paths, which is found from the relation $dM = \frac{\partial\gamma_i}{\partial\gamma} = \frac{\sin\theta_i \cos\gamma_i \cdot \cos\alpha_i - \sin\gamma_i \cos\theta_i}{\sin\gamma}$.

ELF radio propagation in the Earth–ionosphere cavity is characterized by the Brillouin waves, and the regular complex cosine function is introduced as (Wait 1962):

$$C_\nu^2 = 1 - \frac{\nu(\nu+1)}{(ka)^2} \tag{6.16}$$

where a is the Earth's radius, and k is the free space wave number.

The disturbance of this cosine function involved in Eq. (6.14) is caused by the ionosphere modification, which is the difference of the regular and disturbed values:

$$\delta\, C_\nu^2 = C_\nu^2|_{\text{Disturbed}} - C_\nu^2|_{\text{Regular}} = \frac{\nu_B(\nu_B+1) - \nu(\nu\ +1)}{(ka)^2} \tag{6.17}$$

Values of the regular propagation parameter ν and of thc disturbed parameter ν_B are computed through regular and disturbed conductivity profiles. We also accept that the δC_ν^2 parameter varies with the radius of disturbance by the "Gauss" law:

$$\delta C_\nu^2 = \Delta C_\nu^2 \exp\left(\frac{\cos\beta - 1}{d^2}\right) \tag{6.18}$$

where ΔC_ν^2 is the maximum value at the center of disturbance, β is the angular distance from the center of disturbance to the current integration point in (6.14), and d is the characteristic size of disturbance that depends on the earthquake magnitude.

By integrating Eq. (6.14) with the Gauss non-uniformity, we obtain the following formula:

$$B_D = \frac{E_2}{E_1} = \frac{\pi d^2}{2} \frac{\Delta C_\nu^2 \cdot Q_i}{\sin(\pi\nu) P_\nu[\cos(\pi - \theta_H)]} \tag{6.19}$$

Computational data confirm (see Figs. 6.13 and 6.14) that a disturbance in the lower ionosphere over Taiwan can cause a noticeable change of the ELF field in Japan. Plots in Fig. 6.13 correspond to situations when a point source is positioned at Asia, Africa, and America. The left panels show results for Moshiri and the right—for Nakatsugawa. Blue curves correspond to the uniform cavity and the red lines with dots show data for the cavity with localized ionosphere modification positioned over Taiwan and having the characteristic diameter $d = 2$ Mm. One may observe that the modification substantially depends on frequency: its amplitude increases with frequency when the global lightning occurs at the point source. Data presented in Fig. 6.13 demonstrate that the concept of a seismogenic localized ionosphere non-uniformity successfully explains the observations. The realistic source distribution is much more complicated, so that we perform computations for the models of spatially distributed sources based on orbital observations by OTD (Christian et al. 2003).

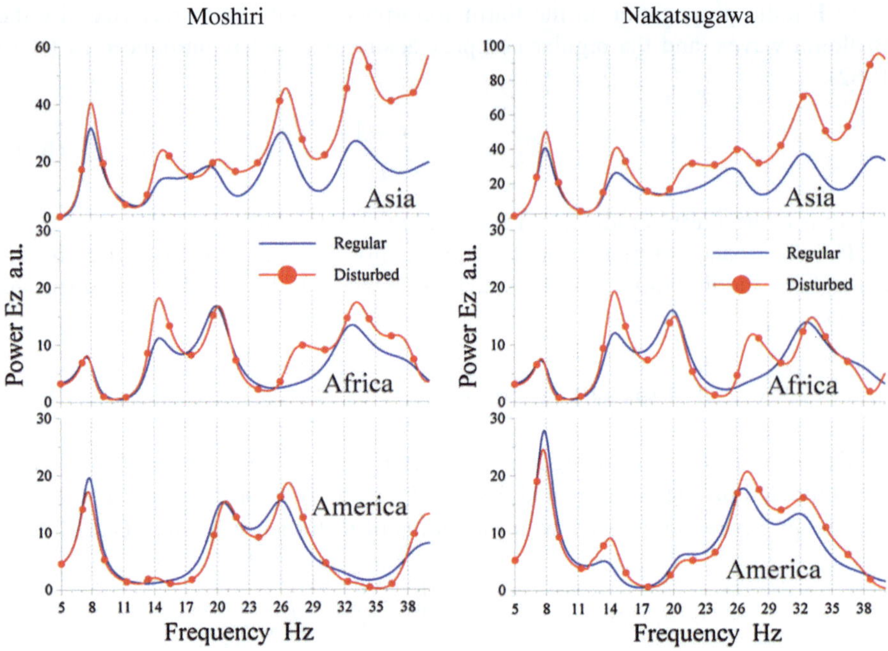

Fig. 6.13 Power resonance spectra corresponding to separate compact sources. *Blue curves* show the data for regular cavity and the *red curves with dots* describe the presence of ionosphere modification of characteristic diameter 2 Mm

Initial data contain 24 maps of spatial distribution of optical flashes from lightning strokes all over the world. Each map corresponds to a particular time of day (UT). The lightning activity is given on the flash rate per square kilometer in the cells of 2.5° × 2.5° size. This quantity was transformed into integer number of flashes in the cell, and the size of these cells was increased to 10° × 10°. Such an extension passes unnoticed in SR computations, because the size of a cell remains much smaller than the wavelength, but the computational speed grows by an order of magnitude since the number of cells is reduced by the factor of 16.

Centers of cells have the longitudes: −175°, −165°, ..., +175° (from west to east) and the latitudes: −85°, −75°, ..., +85° (from south to north). Computations were organized in the following way. The UT hour is fixed and the relevant OTD map is chosen of the global lightning activity. In computations, the program moves from one cell to the other and finds the source–observer distance for the cell center θ_H and the distance γ_i from the cell center to the current point of ionosphere disturbance by using Eq. (6.15). All other propagation parameters are also computed together with the current parameter of derivatives *dM*. By using formula (6.13), we compute the direct wave E_1 at the given frequency. It is supposed that every cell has the unit current moment $M_C(\omega) = 1$. Simultaneously, the complex relative field disturbance is computed B_D, so that the absolute disturbed field $E_D = E_1 + E_2$ is equal to:

$$E_D = E_1(1 + B_D) = \frac{i\nu(\nu+1)}{4ha^2\varepsilon_0\omega}\frac{P_\nu[\cos(\pi - \theta_H)]}{\sin\pi\nu}(1 + B_D). \tag{6.20}$$

Contribution from a given cell into the power spectrum in the uniform cavity is directly proportional to the number of flashes N_i in this cell:

$$P_i^{(U)} = N_i\left|\frac{i\nu(\nu+1)}{4ha^2\varepsilon_0\omega}\frac{P_\nu[\cos(\pi - \theta_H)]}{\sin\pi\nu}\right|^2 \tag{6.21}$$

The same holds for the cavity with the disturbance:

$$P_i^{(D)} = P_i^{(U)} \cdot |1 + B_D|^2 \tag{6.22}$$

Here $P_i^{(U)}$ is the contribution of i-th cell into power spectrum, and $P_i^{(D)}$ is its contribution in the non-uniform cavity. Cumulative power spectram in the regular and disturbed cavities are the sum contributions of all the cells:

$$P_U(f, UT) = \sum_i P_i^U \tag{6.23}$$

$$P_D(f, UT) = \sum_i P_i^D \tag{6.24}$$

As we already remarked, such a procedure suggests that random lightning discharges follow the Poisson law. In this case, the stochastic mutual time delay between separate pulses is distributed exponentially (Erlang law), so that pulses do not interfere in the frequency domain, and the cumulative power spectrum is just the sum of the power spectra of individual pulses. Thus, we obtain the cumulative power spectra (6.22) and (6.23) at a given frequency, while the universal time (UT) is a parameter. By varying the frequency and performing computations all over the global cells, we obtain the resonance power as the function of frequency for the fixed time of day.

Model spectra are illustrated in Fig. 6.14 for UT = 10, 16, and 20 h. The blue curves depict the regular spectra computed for both the observatories, while anomalous spectra are shown by red curves with dots. The times chosen correspond to the maximum activity concentrated in Asian, African, and American centers. Signal frequency is shown on the abscissa ranging from 5 to 29 Hz. The power density is plotted on the ordinate in arbitrary units. We confirm from computations taking into account the realistic distribution of thunderstorms over the planet the idea that the seismogenic ionosphere reduction over Taiwan is able to noticeably modify the SR amplitude (power) spectra in Japan at the modes with $n \geq 4$. Model seismogenic effects are higher at Moshiri than at Nakatsugawa.

Computations show that the resonance intensity and the spectral pattern vary during the day due to the motion and alteration in intensity of the global

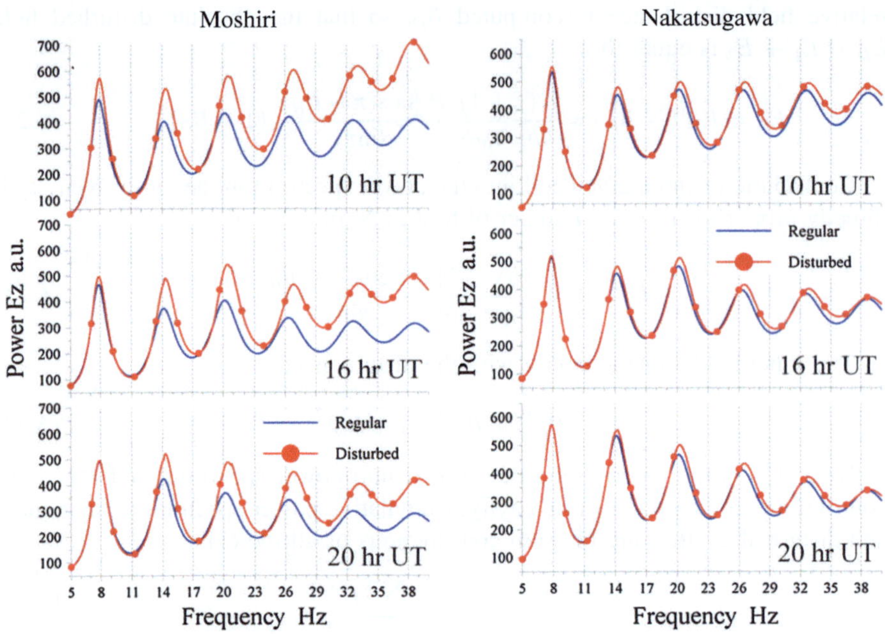

Fig. 6.14 Seismogenic alterations in the SR spectra observed at times of maximum thunderstorm activity at Asian, African, and American centers

thunderstorms. Since we do not know exactly what source distribution took place on a particular day, we must consider Fig. 6.14 as a mere estimate of possible effects caused by the seismogenic ionosphere modification. However, if one compares the model spectra of Fig. 6.14 with the corresponding observations at the Moshiri site on December 21 and 23, 2006, he/she will find a striking similarity in their outlines. This indicates that data interpretation is rather realistic, and the suggested mechanism of the ELF wave scattering by ionospheric disturbance was rather feasible. It is important to note that the modification involved in our interpretations is beyond the reach of higher frequencies, and no man-made devices fly at its altitudes, so the SR records remain so far the only "witness" of modifications over Taiwan.

There is an alternative explanation of origin of the observed anomalous ELF radio signals (Hayakawa et al. 2010, 2011a). It is based on the detection of ionospheric anisotropic waves, which are originated in the anisotropic plasma under the influence of the narrow-band ELF radio emission (15–20 Hz) coming from beneath the ground. This seismogenic emission causes the currents in the plasma that polarize the non-uniformity in the ionosphere conductivity. Computations show that such gyrotropic waves can propagate along the lower ionosphere boundary with the velocity from a few tens to a few hundred kilometers per second (Sorokin and Pokhotelov 2005; Sorokin et al. 2009; Sorokin and Hayakawa 2008). The details of how such waves generate the radio signals with the spectrum resembling observational data might be found in the paper by Hayakawa et al. (2010). We must stress that the causative mechanisms are different: radiation from global thunderstorms versus radiation from underground,—the SR modification appears only when the ionospheric disturbance is present over the future EQ.

SR monitoring revealed anomalous signals in Japan associated with the distant EQs in Taiwan. The electromagnetic disturbances in the frequency range of global electromagnetic resonance are of the precursory nature, because they appear a few days prior to the main shock of EQ. Experimental data is well explained by the sub-ionospheric radio wave scattering from the localized plasma modification over the future epicenter or by the scattering from the same non-uniformity of the seismogenic radio emission coming from the crust. Obviously, observations should go on aiming not only at extending the existing statistics, but looking for some other pairs of the EQs and remote observatories with the SR anomalies.

References

Ando Y, Maltsev P, Sukhynyuk A, Goto T, Yamauchi T, Hobara Y, Sekiguchi M, Ikegami Y, Sera M, Korepanov V, Hayakawa M (2005) New ELF observation system at Moshiri, Japan and assessment of acquired data. J Atmos Electr 25(1):29–39

Bliokh PV, Nickolaenko AP, Filippov YuF (1977) Global electromagnetic resonances in the Earth-ionosphere cavity. Naukova Dumka, Kiev (in Russian)

Bliokh PV, Nickolaenko AP, Filippov YuF (1980) Schumann resonances in the Earth-ionosphere cavity. London, Paris, Peter Perigrinus, New York 168 p

Cannon PS, Rycroft MJ (1982) Schumann resonance frequency variations during sudden ionospheric disturbances. J Atmos Terr Phys 44:201–206

Christian HJ, Blakeslee RJ, Boccippio DJ, Boeck WL, Buechler DE, Driscoll KT, Goodman SJ, Hall JM, Koshak WJ, Mach DM, Stewart MF (2003) Global frequency and distribution of lightning as observed from space by the optical transient detector. J Geophys Res 108(D1):4005. doi:10.1029/2002JD002347

Dowden RL, Thomson NR, Rodger CJ, Hayakawa M (1999) VLF perturbation caused by gamma flare. Paper Presented at URSI General Assembly (Tronto), G2, p 2

Fraser-Smith AC, McGill PR, Bernardi A, Helliwell RA, Ladd ME (1991) Global measurements of the low frequency radio noise. Electromagnetics, Springer, Tokyo, 191–200

Greifinger C, Greifinger P (1978) Approximate method for determining ELF eigenvalues in the Earth-ionosphere waveguide. Radio Sci 13:831–837

Hayakawa M, Molchanov OA (2007) Seismo-electromagnetics as a new field of radiophysics: electromagnetic phenomena associated with earthquakes. Radio Sci Bull 320:8–17

Hayakawa M, Nakamura T, Hobara Y, Williams E (2004) Observation of sprites over the Sea of Japan and conditions for lightning-induced sprites in winter. J Geophys Res 109:A01312. doi:10.1029/2003JA009905

Hayakawa M, Ohta K, Nickolaenko AP, Ando Y (2005) Anomalous effect in Schumann resonance phenomena observed in Japan, possibly associated with the Chi-chi earthquake in Taiwan. Ann Geophysicae 23:1335–1346

Hayakawa M, Nickolaenko AP, Sekiguchi M, Yamashita K, Ida Y, Yano M (2008) Anomalous ELF phenomena in the Schumann resonance band as observed at Moshiri (Japan) in possible association with an earthquake in Taiwan. Nat Hazards Earth Syst Sci 8:1309–1316. www.nat-hazards-earth-syst-sci.net/8/1309/2008/

Hayakawa M, Ohta K, Sorokin VM, Yaschenko AK, Izutsu J, Hobara Y, Nickolaenko AP (2010) Interpretation in terms of gyrotropic waves of Schumann-resonance-like line emissions observed at Nakatsugawa in possible association with nearby Japanese earthquakes. J Atmos Solar-Terr Phys 72:1292–1298

Hayakawa M, Hobara Y, Ohta K, Izutsu J, Nickolaenko AP, Sorokin V (2011a) Seismogenic effects in the ELF Schumann resonance band. IEEE J Trans Fundam Mater 131(9):684–690. doi:10.1541/ieejfms.131.684

Hayakawa M, Nickolaenko AP, Shvets AV, Hobara Y (2011b) Recent studies of Schumann resonance and ELF transients. In: Wood MD (ed) Lightning: properties, formation and types. Nova Sci Pub, pp 39–71

Hobara Y, Iwasaki N, Hayashida T, Hayakawa M, Ohta K, Fukunishi H (2001) Interrelation between ELF transients and ionospheric disturbances in association with sprites and elves. Geophys Res Lett 28(5):935–938

Hurley K, Boggs SE, Smith DM, Duncan RC, Lin R, Zoglauer A, Krucker S, Hurford G, Hudson H, Wigger C, Hajdas W, Thompson C, Mitrofanov I, Sanin A, Boynton W, Fellows C, von Kienlin A, Lichti G, Rau A, Cline T (2005) An exceptionally bright flare from SGR 1806–20 and the origins of short-duration big gamma-ray bursts. Nature 434:1098–1103

Inan US, Lehtinen NG, Lev-Tov SJ, Johnson MP, Bell TF, Hurley K (1999) Ionization of the lower ionosphere by γ–rays from a magnetar: detection of a low energy (3–10 keV) component. Geophys Res Lett 26:3357–3360

Inan US, Lehtinen NG, Moore RC, Hurley K, Boggs S, Smith DM, Fishman GJ (2007) Massive disturbance of the daytime lower ionosphere by the giant γ –ray flare from magnetar SGR 1806–20. Geophys Res Lett 34:L08103. doi:10.1029/2006GL029145

Kirillov VV (1993) Parameters of the Earth-ionosphere waveguide at ELF. Prob Dffraction Wave Propag 25:35–52 (in Russian)

Kirillov VV (1996) Two-dimensional theory of electromagnetic wave propagation in the ELF range in the Earth-ionosphere waveguide, Izv VUZov Radiophyz 39:1103–1112 (in Russian)

Kirillov VV, Kopeykin VN (2002) Solving a two dimensional telegraph equation with anisotropic parameters. Izv VUZov Radiophyz 45:1011–1024 (in Russian)

Kleimenova NG (1965) Some observations of the natural electromagnetic field in the 1-20 Hz band at the Tiksi and Lovosero polar stations. Fizika Zemli 3:97–102 (in Russian)

Madden T, Thompson W (1965) Low frequency electromagnetic oscillations of the Earth-ionosphere cavity. Rev Geophys 3:211–254

Marple SL Jr (1987) Digital spectral analyses with applications. Prentice-Hall, Englewood Cliffs

Molchanov OA, Hayakawa M (2008) Seismo electromagnetics and related phenomena: history and latest results. TERRAPUB, Tokyo 189 p

Mushtak VC, Williams ER (2002) ELF propagation parameters for uniform models of the Earth-ionosphere waveguide. J Atmos Solar-Terr Phys 64:1989–2001

Nelson PH (1967) Ionospheric perturbations and Schumann resonance data. Ph.D. thesis, MIT, Cambridge Mass

Nickolaenko AP (1986) Scattering of the ELF radio waves by the global non-uniformities of the Earth-ionosphere cavity. Izv VUZov Radiofizika 29(1):33–40 (in Russian)

Nickolaenko, AP (1994) ELF radio wave propagation in a locally non-uniform Earth-ionosphere cavity. Radio Sci 29(5):1187–1199

Nickolaenko AP, Hayakawa M (2002) Resonances in the Earth-ionosphere cavity. Kluwer Academic Publishers, Dordrecht-Boston-London

Nickolaenko AP, Hayakawa M (2010) Model disturbance of Schumann resonance by the *SGR* 1806–20 γ –ray flare on December 27, 2004. J Atmos Electr 30(1):1–11

Nickolaenko AP, Rabinowicz LM (1982) Possible global electromagnetic resonances on the planets of the solar system. Translated from Kosmicheskie Issledovaniya, vol 20(1). Plenum Publishing Corporation, pp 82–88

Nickolaenko AP, Rabinowicz LM, Shvets AV, Schekotov AYu (2004) Polarization characteristics of low frequency resonances. Izv VUZov Radiofizika 47(4):267–291 (in Russian)

Nickolaenko AP, Hayakawa M, Sekiguchi M, Ando Y, Ohta K (2006) Model modifications in Schumann resonance intensity caused by a localized ionosphere disturbance over the earthquake epicenter. Ann Geophys 24:567–575

Nickolaenko AP, Rabinowicz LM, Shvets AV (2008) Polar non-uniformity of ionosphere related to solar proton events. Telecommun Radio Eng 67(5):413–435

Nickolaenko AP, Kudintseva IG, Pechonaya O, Hayakawa M, Nakamura T, Hobara Y, Tanaka Y (2010) Impact of a gamma-ray burst on the Schumann resonance. Radiophys Quantum Electron 53(9–10):542–556. doi:10.1007/s11141-011-9249-9

Nickolaenko AP, Kudintseva IG, Pechony O, Hayakawa M, Hobara Y, Tanaka YT (2012) The effect of a gamma ray flare on Schumann resonances. Ann Geophys 30:1321–1329. doi:10.5194/angeo-30-1321-2012

Ohta, K, Umeda K, Watanabe N, Hayakawa M (2002) Relationship between ELF magnetic fields and Taiwan earthquakes. In: Hayakawa M, Molchanov OA (eds) Seismo electromagnetics: lithosphere–atmosphere–ionosphere coupling. TERRAPUB, Tokyo, pp 233–237

Ohta K, Umeda K, Watanabe N, Hayakawa M (2001) ULF/ELF emissions observed in Japan, possibly associated with the Chi–Chi earthquake in Taiwan. Nat Hazard Earth Syst Sci 1:37–42 http://www.nat-hazards-earth-syst-sci.net/1/37/2001/

Ohta K, Watanabe N, Hayakawa M (2006) Survey of anomalous Schumann resonance phenomena observed in Japan, in possible association with earthquakes in Taiwan. Phys Chem Earth 31:397–402

Ohta K, Izutsu J, Hayakawa M (2009) Anomalous excitation of Schumann resonances and additional anomalous resonances before the 2004 Mid-Niigata prefecture earthquake and the 2007 Noto Hantou Earthquake. Phys Chem Earth 34(6–7):441–448 (Special issue, In: Hayakawa M, Liu JY, Hattori K, Telesca L (eds) Electromagnetic phenomena associated with earthquakes and volcanoes)

Pechony O, Price C (2004) Schumann resonance parameters calculated with a partially uniform knee model on Earth, Venus, Mars, and Titan. Radio Sci 39:RS5007. doi:10.1029/2004RS003056

Pechony O, Price C, Nickolaenko AP (2007) Relative importance of the day-night asymmetry in Schumann resonance amplitude records, Radio Sci 42:RS2S06. doi:10.1029/2006RS003456

Price, C, Mushtak V (2001) The impact of the August 27, 1998, γ-ray burst on the Schumann resonances. J Atmos Solar-Terr Phys 63:1043–1047

Roldugin VC, Maltsev YuP, Vasiljev AN, Vashenyuk EV (1999) Changes of the first Schumann resonance frequency during relativistic solar proton precipitation in the 6 November 1997 event. Ann Geophysicae 17:1293–1297

Roldugin VC, Maltsev YP, Petrova GA, Vasiljev AN (2001) Decrease of the first Schumann resonance frequency during solar proton events. J Geophys Res 106, 18555–18562

Roldugin VC, Maltsev YP, Vasiljev AN, Shvets AV, Nickolaenko AP (2003) Changes of Schumann resonance parameters during the solar proton event of 14 July 2000. J Geophys Res 108(A3):1103. doi:10.1029/2002JA009495

Sao K (1967) Correlation between solar activity and atmospheric the atmospheric potential gradient at the Earth's surface in polar regions. J Atmos Terr Phys 29:2047–2053

Sao K, Yamashita M, Tanahashi S (1971) Day to day variations of Schumann resonance frequency and occurrence of Pc1 in view of solar activity. J Geomagn Geoelectr 23:411–415

Sao K, Yamashita M, Tanahashi S, Jindoh H, Ohta K (1973) Experimental investigations of Schumann resonance frequencies. J Atmos Terr Phys 35:247–253

Schlegel K, Füllekrug M (1999) Schumann resonance parameter changes during high-energy particle precipitation. J Geophys Res 104(A5):10111–10118

Sekiguchi M, Hayakawa M, Nickolaenko AP, Hobara Y (2006) Evidence on a link between the intensity of Schumann resonance and global surface temperature. Ann Geophysicae 24:1809–1817

Sorokin VM, Hayakawa M (2008) On the generation of narrow-banded ULF/ELF pulsations in the lower ionospheric conducting layer. J Geophys Res 113:A06306. doi:10.1029/2008JA013094

Sorokin VM, Pokhotelov OA (2005) Gyrotropic waves in the mid-latitude ionosphere. J Atmos Solar–Terr Phys 67:921–930

Sorokin VM, Sergeev IYu, Pokhotelov OA (2009) Low latitude gyrotropic waves in a finite thickness ionospheric conducting layer. J Atmos Solar–Terr Phys 71:175–179

Tanahashi S (1976) Detection of the line splitting of Schumann resonance from ordinary data. J Atmos Terr Phys 38:132–142

Tanaka YT, Terasawa T, Yoshida M, Horie T, Hayakawa M (2008) Ionospheric disturbances caused by SGR 1900 + 14 giant gamma ray flare in 1998: constraints on the energy spectrum of the flare. J Geophys Res 113:A07307. doi:10.1029/2008JA013119

Tanaka YT, Hayakawa M, Hobara Y, Nickolaenko AP, Yamashita K, Sato M, Takahashi Y, Terasawa T, Takahashi T (2011) Detection of transient ELF emission caused by the extremely intense cosmic gamma-ray flare of 27 December 2004. Geophys Res Lett 38:L08805. doi:10.1029/2011GL047008

Terasawa T, Tanaka YT, Takei Y, Kawai N, Yoshida A, Nomoto K, Yoshikawa I, Saito Y, Kasaba Y, Takashima T, Mukai T, Noda H, Murakami T, Watanabe K, Muraki Y, Yokoyama T, Hoshino M (2005) Repeated injections of energy in the first 600 ms of the giant flare of SGR 1806–20. Nature 434:1110–1111. doi:10.1038/nature03573

Wait JR (1962) Electromagnetic waves in Stratified media. Pergamon Press, Oxford

Williams ER, Mushtak V, Nickolaenko AP (2006) Distinguishing ionospheric models using Schumann resonance spectra. J Geophys Res 111:D16107. doi:10.1029/2005JD006944

Zieger B, Sátori G (1999) Periodic variations of Solar and tropospheric origins in Schumann resonance. In: Proceedings of 11th international conference on atmospheric electricity, Guntersville, Alabama, 7–11 Jun 1999, pp 701–703

Zybin KYu, Kleimenova NG (1965) Amplitude spectrum of micropulsations in the 1-20 Hz band. Geomagn Aeron 5:1125–1126 (in Russian)

Chapter 7
Coherence of SR

Global nature of SR provides deep impression, and it permanently stimulates attempts to test whether the natural ELF radio signal is really of the global nature. We mention a few early experiments. Similarity of natural ELF background was checked in the trans–Atlantic measurements of resonance amplitude. A conclusion was made that a continuous SR record reflects the motion of global thunderstorms. Another investigation used the synchronous records of ELF transients (Q–bursts) at a global network of remote observatories. The waveforms were readily recognized in every record, and the onset occurred practically simultaneously at all sites. The signal processing included a comparison of the wave arrival angles and measurements of mutual time delays of the pulses. It allowed for obtaining the coordinates of the parent strokes concurrently with evaluating the propagation constant of ELF radio waves. The globality becomes most obvious in the coherence measure of the records acquired simultaneously at the field-sites separated by a great distance. We describe the results of such cross-correlation measurements in the vertical electric field component. It allowed for separating the coherent signal from the incoherent one and we demonstrate that the incoherent signal has also the resonance peaks. Coherence might also be found of the two orthogonal horizontal magnetic field components. Their cross-correlation is linked to the polarization matrix, i.e., the field polarization. We describe the procedure of obtaining the field polarization and demonstrate obvious advantages of this characteristic when finding the resonance phenomena.

7.1 Coherence Measure of SR Background Data

We describe classical results on the coherence measure of the two SR signals recorded at distant sites (Bormotov et al. 1971; Bliokh et al. 1971, 1977a, 1980; Lazebny et al. 1987). Two observatories simultaneously and coherently recorded the vertical electric field component in the frequency band of SR. A few pairs of sites were used. The major data were accumulated in the measurement campaigns of 1969, 1972, and 1974. The pairs of sites were: Kharkov, Ukraine (50° N and

A. Nickolaenko and M. Hayakawa, *Schumann Resonance for Tyros*,
Springer Geophysics, DOI: 10.1007/978-4-431-54358-9_7,

37° E)—Irkutsk, Baikal Lake (50° N and 105° E); and Severomorsk, Kola Peninsula (66° N and 34° E)—Kara-Dag, Crimea (44° N and 34° E). The analog spectral processing of the records on magnetic tapes allowed obtaining the power spectra of both the observatories and the relevant complex cross-spectra of SR oscillations.

The measurements were aiming at three goals. The first one was the suppression of local radio noise always present at the site. As a result, the measurement accuracy was increased of SR spectra, its frequencies, amplitudes and Q-factors. This objective was achieved due to the global nature of resonance, which conditioned its coherence at distant sites. The local interference (noise) is incoherent, so that it is reduced in the cross-spectra. Integration of the cross-spectra at a given frequency f during the accumulation time T_A causes the proportional increase in the coherent power, while the incoherent signals interfere destructively and their intensity does not increase at all. As a result, the ratio of coherent amplitude to incoherent amplitude varies as $(T_A)^{1/2}$.

The second aim was to obtain quantitative estimates for the coherence measure M_{COH} of radio signal that characterizes the portion of coherent signal in a record. The coherence measure is defined as:

$$M_{\mathrm{COH}} = \frac{\langle E_1(f)E_2^*(f)\rangle}{\sqrt{\left\langle |E_1(f)|^2 + |n_1(f)|^2\right\rangle\left\langle |E_2(f)|^2 + |n_2(f)|^2\right\rangle}} = \frac{\langle A_1A_2\rangle\langle\cos\alpha\rangle}{\sqrt{1+1/R_1^2}\sqrt{1+1/R_2^2}} \tag{7.1}$$

Here $E_1(f)$ and $E_2(f)$ are the complex spectra of coherent SR signals, $n_1(f)$ and $n_2(f)$ are the spectra of local interferences at site 1 and site 2, A_1 and A_2 are the amplitudes of coherent signals, $\alpha(f)$ is the phase difference of these signals, and R_1 and R_2 are the signal to noise ratios of each observatory.

One can see from Eq. (7.1) that the local interference is found only in the denominator of the coherence measure. The amplitude decreases $|M_{\mathrm{COH}}| \to 0$ when the local noises play the dominant role, and it approaches the unity $|M_{\mathrm{COH}}| \to 1$ when the noise disappears. The records become completely coherent in the latter case.

The third goal of the experiment was an attempt to detect the line splitting of the SR oscillations, which is caused by the geomagnetic field. We will return to this problem in the next Chap. 8.

The average spectrum of the coherence measure is given in Fig. 7.1 by the red curve with dots. The blue line with squares demonstrates the spectrum of incoherent noise. Experimental data were recorded at the base Kharkov—Irkutsk in August 1969. The abscissa shows the frequency in Hz and the coherence measure is shown on the left ordinate. Plots of coherence measure vary from day to day and from hour to hour. These deviations are conditioned by the random nature of global thunderstorm activity, and on the other hand, alterations are also linked to

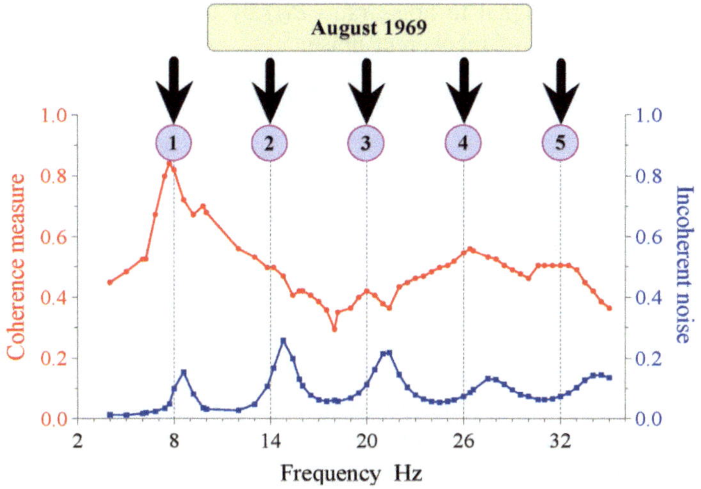

Fig. 7.1 Averaged coherence measure obtained experimentally at the Kharkov—Irkutsk base in August 1969

changes in the level of the local noise. We demonstrate the result averaged over the whole measurement campaign of August 1969.

Numbers in circles and the vertical arrows mark the positions of SR modes. As we observe, the coherence measure of vertical electric field component reaches ~0.8 at the first SR mode. The highest level observed in this series of observations was 0.9. We may state that electromagnetic radiation from the global thunderstorm activity in the vicinity of the first SR mode is high, and it reaches on average 80 % of the signal intensity. The 20 % left at this frequency are relevant to the incoherent noise. We must remark here that the experiment was performed with an analog equipment, and the signal from Siberia was translated to Ukraine via the ordinary telephone line by phase modulation of the 1.8 kHz carrier frequency. This is the reason why those experiments provided a lower estimate for the real coherence measure. Modern measurements exploiting updated equipments and recent technologies will probably show a somewhat higher coherence. On the other hand, the local interferences have grown everywhere since that time, so that obtaining pure natural signals is a hard task nowadays.

One can estimate the spectrum of incoherent noise by using the coherence measure. The following assumptions are necessary for this purpose. Let the phase difference α of coherent signals be negligible at the remote observation points and let the resonance amplitude substantially exceed that of the incoherent noise at both the sites. In a case when the signal to noise ratio is the same at two observatories $R_1 = R_2 = R$, we obtain:

$$R^2 = R_1^2 = R_2^2 \cong \frac{1}{1 - |M_{\mathrm{COH}}|} \qquad (7.2)$$

We can derive the signal to noise ratio $R(f)$ by using Eq. (7.2) and compute afterwards the spectrum of incoherent noise:

$$n(f) = \frac{S(f)}{R(f)} \tag{7.3}$$

The function $n(f)$ is shown by the blue line relevant to the right ordinate in the lower part of Fig. 7.1. It appears that the incoherent noise behaves in an unexpected way. Its spectrum contains the well-resolved peaks centered at the SR frequencies. This feature seems odd, especially in the context of SR as a global phenomenon.

The paradox is resolved if we recall that the field sources occupy the wide areas of the globe. Let us consider two remote observatories and a single area where all lightning strokes take place. One can compute the position of the nodal lines of a particular mode relevant to each observatory. These lines will intersect with the source area, and it is possible to compute the portion of strokes in the vicinity of nodal line. Obviously, there will be thunderstorms found in the nodal vicinity of the first observatory, but simultaneously positioned far away from the nodes of the second observatory. Such sources will compose the two subsets: each of them participates in the signal at one observatory, but plays a minor role in the signal at the other site. These subsets are responsible for the incoherent resonance signals. Thus, the strokes widely distributed in space always form the resonance signals being incoherent when recorded at two distant observatories. The absence of coherence does not mean the cancellation of resonance properties in this case. We conclude that the incoherent noise at two separated observatories may have minor relevance to the local noise.

We considered the coherence measure of the vertical electric field component. Similar characteristics might be introduced for the horizontal magnetic field components. In this case, the orientation should be taken into account, but the general approach remains the same. There exist two possibilities here. One may use the cross-spectra of the orthogonal field components H_X and H_Y recorded at the same observatory. These spectra are used in establishing the field polarization. We will show that according to measurements, about one-half of the recorded horizontal magnetic field belongs to the coherent component, while the second half is incoherent (it is depolarized). The second possibility is the cross-spectra of the records at different observatories.

The cross-coherence was studied by Holtham and McAskill (1988) in the resonance signals recorded in the polar region. The two sites were positioned in the Canadian Arctic separated by 1,100 km. Both the observatories exploited similar digitized measurement equipments and the synchronization signal was picked from the time codes of GOES satellites with the accuracy of ±0.5 ms. A very high coherence measure was found reaching 0.94 at 8 Hz and 0.82 at 13 Hz, and these values remained stable during the measurement session. The authors did not find any relation of their data with the planetary geomagnetic index K_p or with the index of solar activity. Simultaneously, no effects were found that might be attributed to the motion of the global lightning activity.

Exclusively high coherence of the sub-polar records agrees well with the resonance nature of incoherent ELF radio noise. Indeed, the neighborhoods of the nodal lines closely overlay for two sub-polar observatories, and only the thunderstorms beyond this joint area take part in forming the recorded signals. This specific feature of the experimental setup is associated with the exclusively pure records of the sub-polar sites, so that the resulting coherence becomes very high. Concerning the diurnal motion of the global thunderstorm activity, it causes slow changes in the median wave arrival angle while there is no variation in the source–observer distances. Such a motion combined with the relatively small partition of observatories cannot substantially change the signal coherence. Moreover, the spectral patterns must remain stable as well. The only varying parameter will be the signal intensity that reflects approximately the two-fold daily variation of the global thunderstorm activity.

In the paper by Füllekrug and Fraser-Smith (1996) the records were compared of a single field component performed at three observation sites positioned at Arrival Heights, Antarctica (78° S and 167° E) Sønderstrømfjord, Greenland (67° N and 51° E), and Stanford, California (37° N and 122° W). Amplitudes of the first and the second SR modes showed pronounced diurnal variations and considerable variability from day to day during the observation period from Jan. to Apr., 1990. Changes from day to day were similar at all three points. These variations had a period of 20–30 days definitely linked to the solar rotation period. The cross-correlation coefficients for the pairs Antarctica—Greenland, Antarctica—Stanford, and Stanford—Greenland were correspondingly 0.75, 0.71 and 0.7 at the first mode and 0.23, 0.41, and 0.81 at the second mode. Thus, the link was proven between the ELF natural radio emissions recorded at an arbitrary pair of observatories on the Earth, especially at the first SR mode.

Cross-spectra of two orthogonal horizontal magnetic field components recorded at the same site are used in the study of field polarization. An analysis of the Lehta observatory records showed that about 50 % of the resonant magnetic field component are depolarized, and only one-half of resonance signal is polarized (Nickolaenko et al. 2004). This fact is conditioned by the angular pattern of a horizontal magnetic antenna, which makes them sensitive to radiation from different thunderstorm centers. On the other hand, the absence of coherence might arise from the "out of time" lightning strokes taking place at the well-known centers of global thunderstorm activity or from the strokes randomly "scattered" all over the globe. These two reasons explain the depolarized component of magnetic field. The concepts of "out of time" or the "globally scattered" lightning activity simultaneously explain the "podium" observed in the diurnal variations of resonance intensity and strokes (Nickolaenko et al. 2001, 2005a, b, 2006, 2008).

7.2 Polarization of Magnetic Field at SR Frequencies

Experimental studies of the field polarization were initially performed in the SR band by Sentman (1987), and later investigations were continued in the works by Sentman (1989) and Labendz (1998).

The meaning of the above word "polarization" should be explained, as the term is already occupied. We know that the waves of transverse electromagnetic polarization (TEM waves) propagate in the frequencies of SR (8, 14, 20, 26 etc. Hz) in the air slab bounded by the ground and the lover ionosphere plasma. The radio wave of TEM polarization has the non-zero vertical electric field E_r and the horizontal magnetic field H_ϕ. There is also the horizontal electric component E_θ, but it turns to zero on the perfectly conducting surface of the Earth and is therefore very small everywhere owing to the great wavelength. The TEM polarized waves are efficiently excited by the vertical lightning strokes. The wave of this polarization allows for the limiting transition to zero frequency $\omega \to 0$: the horizontal magnetic field vanish in this case $H_\phi \to 0$, and the vertical electric field turns into the electrostatic field of the Earth—ionosphere capacitor. This field is regarded in the literature as the fair weather field, which has the amplitude about 120 V/m. The fair weather field is an element of global electric circuit, which is maintained by the global thunderstorm activity (Bliokh et al. 1977a, 1977b, 1980; Holzworth et al. 1984; Holzworth and Norwille 1992; Ogawa 1985, 1995; Roble and Tzur 1986; McGorman and Rust 1998; Rycroft et al. 2000, 2008; Rycroft 2006; Rycroft and Harrison 2011), although some additional sources are feasible (Bespalov and Chugunov 1994).

Thus the term "polarization" is employed, because we deal with the Earth–ionosphere cavity at ELF with the waves of TE or TEM polarization. Besides, the polarization is usually associated with the time–space behavior of the electric field vector of a wave, but in the global cavity this vector has the single projection E_r. Nevertheless, the "polarization" term is used in the SR band in association with the horizontal magnetic field vector (Fowler et al. 1967).

It is usually accepted that the Earth is a perfectly conducting sphere, which looks reasonable at ELF. One speaks about a uniform isotropic Earth–ionosphere cavity when the parameters of the ionosphere plasma are independent of the direction (isotropic) and of angular coordinates (the origin of the spherical coordinate system $\{r, \theta, \varphi,\}$ is positioned at the center of the planet). The polar axis $\theta = 0$ is directed in such a cavity to the field source (vertical discharge) positioned at the point with the geography coordinates $M_S\{a, 0, 0\}$. The observer is found on the ground surface $r = a$ at the point $M_0\{a, \theta_H, 0\}$. The choice is natural of the coordinate system with the source at the pole and the observer at the zero meridian when the cavity is uniform and isotropic. Then, the field components are the functions of circular frequency ω and the angular source–observer distance θ_H. The E_r field is always directed vertically (along the radius), and the complete magnetic field H_ϕ of the uniform isotropic cavity is directed along the normal to the great circle connecting the source and receiver.

When treating the real problems, one has to use the geography coordinates with the $\theta = 0$ axis directed to the North Pole and the source and observer occupying the points $M_S\{a, \theta_S, \varphi_S\}$ and $M_O\{a, \theta_O, \varphi_O\}$ correspondingly (see Fig. 7.2). Global thunderstorms are concentrated over the continents in the tropics in the zone where the local time ranges between 15 and 18 h. Field sources "ride" around the globe during the day following the sun, so that the direction varies in time from a fixed point toward the source. This is why one has to use three antennas at the field site: the vertical electric and two orthogonal horizontal magnetic antennas. Most often, the magnetic antennas are oriented along the geographic cardinal directions, and the field projections are denoted as H_X and H_Y. The horizontal magnetic field component H_X is directed along the parallel from west to east: $H_X \equiv H_{WE} \equiv H_\varphi$, while the H_Y component is oriented along the meridian from south to north: $H_Y \equiv H_{SN} \equiv H$. Simple expressions link the complete vector and its components:

$$H_X = H_\phi \cos(Az - 180°) = -H_\phi \cos(Az) \tag{7.4a}$$

$$H_Y = H_\phi \sin(Az - 180°) = -H_\phi \sin(Az) \tag{7.4b}$$

Here Az is the source geographic azimuth counted clockwise from the local direction to the north as seen in Fig. 7.2.

When the ionosphere plasma is uniform and isotropic, the Earth–ionosphere cavity is regarded as a uniform isotropic cavity, and the field components vary in time synchronously, so that the tip of magnetic field vector moves along the straight line in time. We say that the horizontal magnetic field is linearly polarized. The Earth–ionosphere cavity is regarded as anisotropic one when we take into account the anisotropy of ionosphere plasma placed in the geomagnetic field. Now, the polar axis $\theta = 0$ is directed to the geomagnetic Pole. It turns up that the geography coordinate system is used again, provided that we neglect the eccentricity and the tilt of the geomagnetic dipole.

There exist five non-zero field components in the anisotropic cavity, and only the radial magnetic field is absent $H_r = 0$. However, the same three components

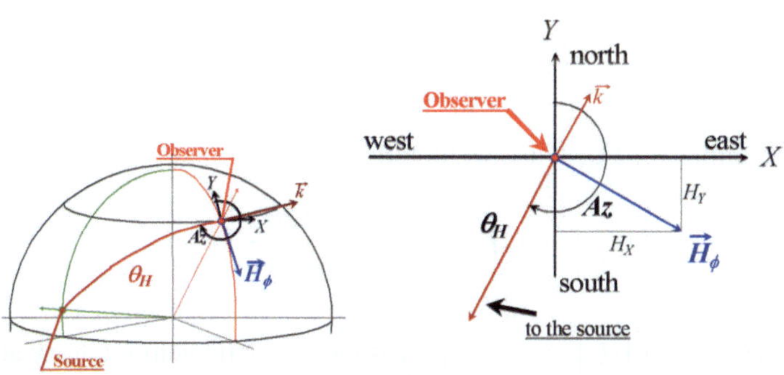

Fig. 7.2 Geography coordinate system at a field site, source bearing, and magnetic field components

E_r, H_θ, and H_φ are present at the well-conducting surface of the ground. The horizontal magnetic field components arise in a natural way as an element of solution of the electrodynamic problem, the meridional component H_θ is directed from north to the south and the component H_φ is directed along the parallel from west to the east. This is the cardinal distinction of magnetic field components H_θ, and H_φ from the projections H_X and H_Y in the isotropic cavity. The sole azimuthal H_ϕ field is present in the isotropic cavity, and this is decomposed into two projections for an arbitrary source–observer geometry.

The anisotropy of the ionosphere forming the upper boundary of the Earth–ionosphere cavity must lead to the splitting of its eigen-values: bring to the lift of resonance degeneracy (Bliokh et al. 1977a, b, 1980; Nickolaenko and Hayakawa 2002). The frequency dependent phase shift appears as a result between the field components $H_X \equiv H_\varphi$ and $H_Y \equiv -H_\theta$. Thus the magnetic field polarization arises deviating from the linear one. The elliptic polarization of horizontal magnetic fields of an ELF radio wave indicates the anisotropy of ionosphere plasma and the relevant splitting of SR frequencies as seen in Chap. 8.

The elliptic polarization was discovered in the first experiments by Sentman (1987, 1989). We will show below the typical measurements results on the magnetic field polarization at SR frequencies. One has to compute the coherence matrix (Fowler et al. 1967) when studying the field polarization:

$$J = \begin{Vmatrix} j_{XX} & j_{XY} \\ j_{YX} & j_{YY} \end{Vmatrix} = \begin{Vmatrix} \langle H_X H_X^* \rangle & \langle H_X H_Y^* \rangle \\ \langle H_Y H_X^* \rangle & \langle H_Y H_Y^* \rangle \end{Vmatrix} \tag{7.5}$$

Here the angular brackets denote averaging over the ensemble. It is clear from Fig. 7.2 that $H_X \equiv H_\varphi$ and $H_Y \equiv -H_\theta$, so that the coherence matrix is given by,

$$J = \begin{Vmatrix} \left\langle H_\varphi H_\varphi^* \right\rangle & \left\langle -H_\varphi H_\theta^* \right\rangle \\ \left\langle -H_\theta H_\varphi^* \right\rangle & \left\langle H_\theta H_\theta^* \right\rangle \end{Vmatrix} \tag{7.6}$$

The outline and orientation of polarization ellipse depends on the angle β, which is found from the following relation:

$$\sin 2\beta = \frac{2\,\mathrm{Im}(j_{XY})}{\sqrt{(j_{XX} - j_{YY})^2 + 4 j_{XY} j_{YX}}} \tag{7.7}$$

By accounting for Eq. (7.6), we obtain:

$$\sin 2\beta = -\frac{2\mathrm{Im}\left(\langle H_X H_Y^* \rangle\right)}{\left\langle |H|^2 \right\rangle} \tag{7.8}$$

Here $\left\langle |H|^2 \right\rangle = \left\langle |H_\varphi|^2 \right\rangle + \left\langle |H_\theta|^2 \right\rangle$ is the magnetic field intensity at the given frequency. For a monochromatic wave we have: $H_X = A_1 \exp[i(\omega t + \eta_1)]$ and $H_Y = A_2 \exp[i(\omega t + \eta_2)]$, so that $\left\langle |H|^2 \right\rangle = A_1^2 + A_2^2$, therefore:

$$\sin 2\beta = \frac{2A_1A_2}{A_1^2 + A_2^2}\sin(\eta_1 - \eta_2) \tag{7.9}$$

The term "ellipticity" (polarization coefficient) is introduced as $e_l = \tan\beta$, which is equal to the ratio of the small and great axes of the ellipse outlined by the tip of magnetic field vector in time. Ellipticity is characterized by the deviation of polarization from the linear one. For example, let the component H_Y retard in phase in comparison with the field H_X by a quarter of the oscillation period, and then the complete field vector rotates counter-clockwise in time. The signal is said to have the left-hand polarization (LHP). In the opposite situation when the rotation occurs clockwise, we have the right-hand circular polarization (RHP). This issue is also discussed in Chap. 8.

The left frame in Fig. 7.3 illustrates the model results for the spectral amplitude of the field component H_φ (see Nickolaenko et al. 2004; Nickolaenko and Sentman 2007). The hedgehog model was used for the geomagnetic field, which is dem onstrated by nomogramme in Fig. 7.3. The hedgehog geomagnetic field is radially directed, has the constant amplitude, and abruptly changes its direction when crossing the equator. The degree of ionosphere plasma anisotropy is characterized by the dimensionless parameter ρ_m, which is equal to the ratio of electron gyro-frequency ω_r to the collision frequency of electrons with the other particles ν_e: $\rho_m = \omega_r/\nu_e$. The value $\rho_m = 1$ was used in computations. Data in the left plot correspond to the minimal source–observer distance when the equatorial source is set at the zero meridian and the observer occupies the same meridian, but at the 50° latitude in the Northern hemisphere. The abscissa shows the signal frequency in Hz. The ordinate indicates the amplitude in arbitrary unites of the horizontal magnetic field H_φ. The narrow lines depict the frequency response (the functions $|(R_{11})|^{-1}$, $|(R_{10})|^{-1}$, and $|(R_{1-1})|^{-1}$) of the individual sub-levels into which the first SR mode has been split. These latter correspond to different values of azimuthal index $m = +1$, $m = 0$, or $m = -1$ (see Chap. 8 for details). Vertical; bars at the frequency axis denote the real parts of the eigen-frequencies in the anisotropic cavity. The thick upper line in the left frame shows the complete amplitude of the field component H_φ obtained for the particular position of the source and observer. As one may see, the splitting of eigen-values reaches a few tenths of a hertz. This quantity is too small in comparison with the width of resonance lines, so that outline of the resulting resonance curve does not show any signs of the line splitting (the lift of degeneracy).

We must remark here that an equatorial source excites only two sub-levels in the vicinity of the first mode, those corresponding to $m = +1$ and $m = -1$ (see Table 8.1). These two functions correspond to sub-levels having the greatest possible frequency separation; the line splitting remains still invisible in the amplitude spectrum.

The right frame in Fig. 7.3 depicts the frequency dependence of the magnetic field ellipticity e_L. Computations show that the polarization coefficient is positive around the first SR frequency, so that the vector of horizontal magnetic field rotates counter-clockwise, and we observe the LHP wave. This result immediately

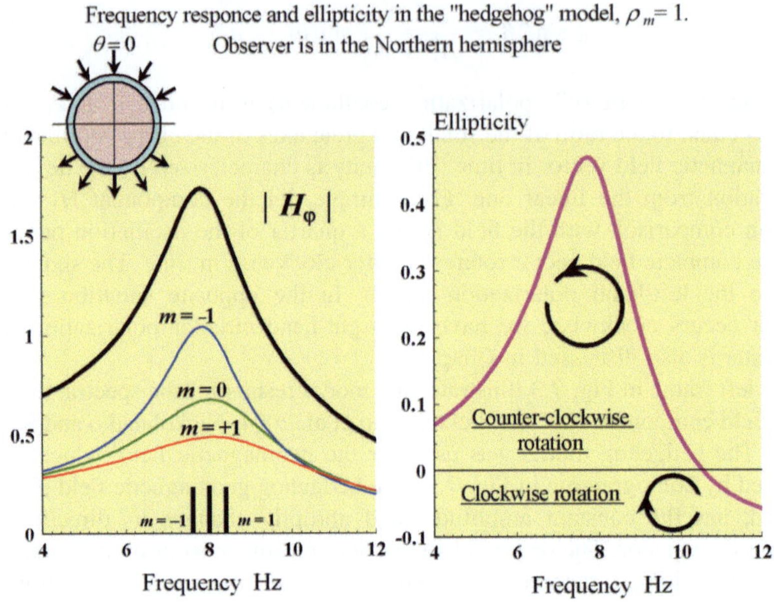

Fig. 7.3 Amplitude spectra of the sub-levels and complete H_φ component (*left frame*) and of the signal ellipticity e_l (*right frame*)

follows from the plots in the left frame of this figure. We see here that the frequency response of the eigen-function with $m = -1$ is much higher than that of the function with $m = +1$. Similar property of the sub-functions was noted in computations of the vertical electric field component by Nickolaenko and Rabinowic (1974) and Bliokh et al. (1977a, b, 1980): the partial wave with $m = -1$ dominates in the cavity at the first resonance frequency. The wave traveling from the west to east corresponds to this eigen-function formally described as $\exp[i(\omega t - \varphi)]$. The standing wave (an actually resonance regime) is observed in a close vicinity of the source in this case.

Computations predict that the wave traveling from west to the east ($m = -1$) dominates around the first peak frequency of SR, and we expect the counter-clockwise rotation of the horizontal magnetic field vector. We will demonstrate below that the similar rotation is a characteristic property of all peak frequencies, while the opposite rotation might occur at intermediate frequencies, i.e., between the resonance peaks.

The ellipticity behaves in a more complicated way when the frequency grows. Its absolute value decreases with frequency, and it might change the sign for particular positions of the source. This change of sign is associated with the spatial field structure that is linked to the point field source, so the alterations of the source–observer distance may alter the direction of the field rotation. Model computations showed that a decrease in the wave attenuation might lead to multiple changes of sign of ellipticity e_l at higher modes, but the sign permanently remains positive in the vicinity of the first resonance mode. Such behavior is relevant to the middle latitude (North) observer and an equatorial source.

Model data agree in part with early observations by Sentman (1987, 1989) and Labentz (1998) who noted that the ellipticity in general repeats the resonance pattern of the power spectrum. However, computational data disagree with the same experimental results claiming that there are periodic (day after day) changes of polarization including that at 8 Hz frequency. Such behavior was considered by the authors as a proof of the resonance splitting of the Earth–ionosphere cavity. As we know, the lift of degeneracy (a line splitting) under the influence of ionosphere anisotropy indeed drives to the elliptical polarization, but the source motion does not influence the sign of polarization in the vicinity of the first mode. We demonstrate below that measurements based on the updated equipment confirm such a conclusion.

Monitoring of ELF radio signals was conducted at Russian observatories Lehta, Karelia, (64° N and 34° E) and Karimshino, Kamchatka (53° N and 157° E). The equipment contained the induction ferromagnetic field sensors with the built-in antenna pre-amplifiers (Belyaev et al. 1999). Signals from antennas were fed by the shielded twisted pair cable 100–200 m long to the input of the receiver that formed the working frequency band and amplified the signal to the level appropriate for the analog–digital converters (ADC). The three-channel ELF receiver included synchronous notch filters locked-in with the industrial 50 Hz interference (Schekotov and Golyavin 1978), scaling amplifiers, low-pass filters with the cut-off frequency $\sim$40 Hz, and the high pass filters with the cut-off frequency $\sim$3 Hz at Lehta and 0.01 Hz at Karimshino. Magnetic antennas at the Lehta observatory were oriented along the south–north and west–east directions in the geographic coordinate system. The antennas at Karimshino were aligned with the geomagnetic coordinates (the geomagnetic field declination is about –6°). The magnetic field sensors were buried to reduce the vibration interference. Besides, the vertical magnetic antenna was used at Karimshino, and the vertical electric antenna was used at Lehta. The electric antenna was an isolated copper sphere of $\sim$40 cm diameter (self-capacitance of $\sim$22 pF) mounted at the 3 m PVC pipe. Positions of antennas at Lehta were shown by a photo in Fig. 2.10.

Signals of the three field components after conditioning (filtering and amplification) are fed to the input of multi-channel ADC in a PC. The PC performs the data accumulation and the preliminary processing in the real time. The current state of measurements is displayed by the PC in a form of the recent waveforms, current spectrum, average and dynamic spectra. The sampling frequencies are 150 Hz at Karimshino (the signal waveform is recorded here) and 172 Hz at Lehta. At Lehta, the following data were stored on the computer hard disk: the power spectra averaged over 5 min (the elementary fragment of record used in FFT is about 12 s) of the field components $\left\langle |E_r|^2 \right\rangle$, $\left\langle |H_X|^2 \right\rangle$, and $\left\langle |H_Y|^2 \right\rangle$, average complex spectra of the Poynting vector $\langle P_X(f) \rangle$ and $\langle P_Y(f) \rangle$, the average complex cross-spectra $\langle H_X H_Y^* \rangle$, and the waveforms of separate ELF transients (Q–bursts) caused by the powerful lightning strokes, provided that their amplitude exceeded a definite threshold (usually, 6 RMS of the current SR background oscillations).

The complex gain of each channel was measured prior to the start of observations with the help of calibrating coils at the magnetic antennas and with the capacitance equivalent to the electric antenna. The regular tests of the channels was performed by a sub-routine generating the calibrating signal in the PC and feeding them to antennas via data acquisition system. The 24-bit ADC are used at Kamchatka with the built-in low pass filter, which increased the dynamic range of the system and allowed us to conduct records without notch filters and scaling amplifiers. The output data of Kamchatka observatory are the continuous records of waveforms detected by two horizontal and one vertical magnetic antennas with the 150 Hz sampling frequency.

The ideal SR background signal at an observatory is a superposition of overlapping pulses that arrived from different directions being radiated by different lightning discharges. The polarized (coherent) field component is observed experimentally together with the non-polarized (incoherent) field. The parameters of ellipticity are relevant to the first component. The polarization degree P_D is the ratio of intensity of the polarized portion of field I_P to the complete intensity of the magnetic field I (Fowler et al. 1967):

$$P_D = \frac{I_P}{I}. \tag{7.10}$$

The polarized intensity is equal to:

$$I_P = \sqrt{(\langle j_{XX}\rangle - \langle j_{YY}\rangle)^2 + 4\langle j_{XY}\, j_{YX}\rangle}, \tag{7.11}$$

and the complete intensity of magnetic field is:

$$I = j_{XX} + j_{YY}. \tag{7.12}$$

The polarization degree (coefficient) belongs to the interval of $0 < P_D < 1$. The depolarization degree of the radio emission is defined as $D_D = 1 - P_D$, and it is relevant to the non-polarized, isotropic part of magnetic field. The horizontal magnetic field at the SR frequencies is partially polarized in both isotropic and anisotropic Earth–ionosphere cavities. In the isotropic cavity, the polarization degree depends on the spatial distribution of the thunderstorm activity. The limiting values of the P_D parameter might arise from the following characteristic source distributions in the isotropic cavity:

1. $P_D \approx 1$ corresponds to the concentration of all thunderstorms in a narrow sector of angles at the observatory.
2. $P_D \approx 0$ corresponds to completely incoherent field components. Such a situation takes place when the pulsed signals arrive from all the directions in the uniform and isotropic cavity, which is the case of globally uniform source distribution. The depolarized signal might appear when we have a pair of

compact incoherent sources positioned at the right angle with respect to the observer. The signals at the antennas will become incoherent in this case.

In the cavity with anisotropic upper boundary, the polarization degree depends on the frequency and on the source–observer geometry. We present in Fig. 7.4 the experimental dynamic spectra of the polarization coefficient of the horizontal magnetic field (upper plot) and the ellipticity (lower plot). Data were used recorded at Lehta from Sep. 18 to Sep. 24, 2000. The date of observations (UT) is shown on the abscissa, the signal frequency is shown on the ordinate in the band from 5 to 25 Hz, and the spectra are shown by the color.

As one may observe from the upper frame in Fig. 7.4, the polarization degree tends to vary simultaneously at all frequencies and reaches the maximum around 15 h UT. During this time, the global thunderstorm activity is concentrated at Africa, i.e. the thunderstorm center closest to the observer. These data agree with the concept of dominance of Africa thunderstorms in the afternoon hours and the activity of other sources is insignificant. The polarization degree is lower at other hours indicating, most probably, the spreading of global thunderstorm activity over the planet.

The lower frame in Fig. 7.4 demonstrates the dynamics of ellipticity e_l. It is clearly seen that the ellipticity remains positive in the vicinity of the first SR frequency where the vector of horizontal magnetic field always rotates counter-clockwise. This observational result is in accord with the model data shown in Fig. 7.3. We compare the model and experimental data in more detail in the next chapter devoted to the degeneracy lift of the SR oscillations.

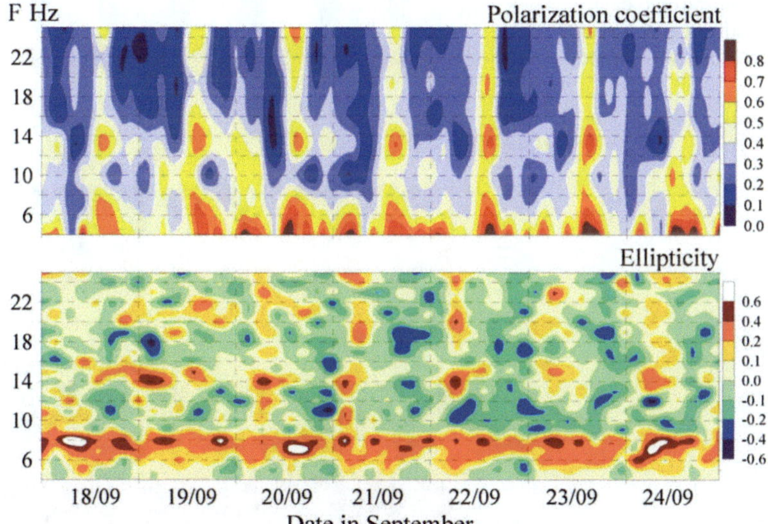

Fig. 7.4 Polarization monitoring of the horizontal magnetic field at the Lehta observatory during September 18–24, 2000

The observed and computed polarizations of magnetic field are compared in Fig. 7.5 in the vicinity of the first SR mode. Seven days (UT) are depicted on the abscissa, and the frequency is shown on the ordinate from 5 to 12 Hz. The ellipticity is shown by color. The upper frame presents the computational data, and two lower frames correspond to observations. The middle frame shows the ellipticity at the Lehta observatory (Karelia) in an interval of 18–24 of September 2000. The lower frame depicts the experimental data of Karimshino (Kamchatka) during July 7–14, 2000, so that periods one week long are shifted in time. Antennas at Karimshino are oriented along the geomagnetic coordinates, but this is insignificant since the magnetic declination at this observatory is about –6.5°. We can accept therefore with minor inaccuracy that the field component *H* at Karimshino is coincident with the south–north projection and the *D*—field is the west–east component. Concerning the sign of polarization, these are opposite at the two observatories. The reason is that the axis *Z* is directed to the zenith in the geographic coordinates, and it is directed to nadir in the geomagnetic coordinates. When plotting experimental data in Fig. 7.5, we transformed the signs of polarization (ellipticity) to the geographic coordinate system.

As one might see, observations at different sites and at different months provide rather similar pictures: dynamics of the field polarization look similar in the vicinity of the first SR mode. It is important to note that the sign of the field polarity is permanently preserved at the fist peak frequency both in computations

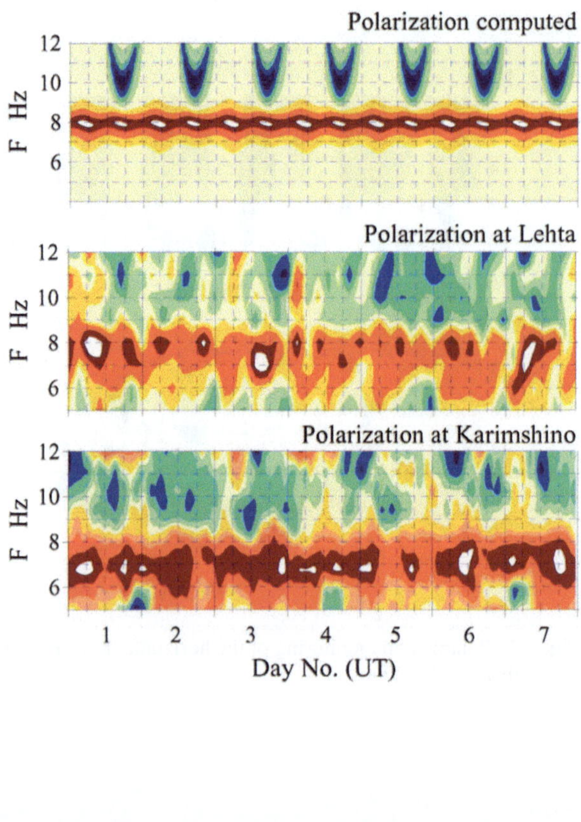

Fig. 7.5 Dynamic spectra of ellipticity computed in the model of anisotropic cavity versus results of experimental observations at two remote sites

and in observations. The observed range of ellipticity variations deviates from that obtained in computations, and these deviations might be used to find a more appropriate value of the parameter of anisotropy ρ_m.

Model computations in the anisotropic cavity used the simplest point source positioned at the coordinates $\{a, \theta_S, \varphi_S\}$. The origin of coordinate system is found at the center of Earth, and the $\theta = 0$ axis points to the North Pole. The exp($i\omega$ t) time dependence is used. The ground surface is perfectly conducting, and the ionosphere anisotropy is accounted for with the help of impedance boundary conditions (e.g. Nickolaenko and Hayakawa 2002):

$$\left[\vec{n} \times \vec{E}\right]\Big|_{S_b} = \hat{z}\left[\vec{n} \times \left[\vec{H} \times \vec{n}\right]\right]\Big|_{S_b} \tag{7.13}$$

where $\vec{E}$ and $\vec{H}$ are the electric and magnetic fields of ELF radio wave, S_b is the lower ionosphere boundary being the sphere of effective radius b, concentric with the Earth's surface $r = a$, $\hat{z}$ is the tensor of the effective surface impedance of ionosphere plasma:

$$\hat{z} = \begin{Vmatrix} z_{\theta\theta} & z_{\theta\varphi} \\ z_{\varphi\theta} & z_{\varphi\varphi} \end{Vmatrix} \tag{7.14}$$

The hedgehog geomagnetic field model is used when the constant magnetic field is directed along the radius and abruptly changes its sign when crossing the equator. For an ionosphere model with sharp boundary, the impedance components acquire the following form (Bliokh et al. 1968, 1977a, b, 1980; Nickolaenko and Hayakawa 2002).

$$z_{\theta\theta} = z_{\phi\phi} = z_1 = \frac{z_0}{\sqrt{2}}\sqrt{1 + \sqrt{1 + \rho_{m^2}}} \tag{7.15}$$

$$z_{\phi\theta} = -z_{\theta\phi} = z_2 \cdot \mathrm{sign}(\cos\theta) = \frac{z_0}{\sqrt{2}} \frac{\rho_m\,\mathrm{sign}(\cos\theta)}{\sqrt{1 + \sqrt{1 + \rho_{m^2}}}} \tag{7.16}$$

Here $\rho_m = |\omega_r/v_e|$ is the parameter of anisotropy, $\omega_r = -\omega_{\mathrm{H}}\cdot\mathrm{sign}(\cos\,\theta)$ is the radial component of electron gyro-frequency, v_e is the effective collision frequency of electrons, $z_0 = \left[i\omega v_e/\omega_0^2\right]^{1/2}$ is the effective surface impedance of the isotropic ionosphere boundary, and ω_0 is the electron plasma frequency. The ionosphere effective height is $h = b{-}a$.

An advanced model was used in computations (Nickolaenko et al. 2004), in which the form (7.14) of the ionosphere surface impedance remained the same, but the terms z_1 and z_2 were found from the realistic frequency dependence of propagation constant $v(f)$ corresponding to experimental data. Fields of the anisotropic cavity are expanded in terms of the eigen-functions of an ideal Earth–ionosphere cavity (Bliokh et al. 1968, 1977a, 1977b, 1980; Nickolaenko and Hayakawa 2002):

$$\vec{H} = \sum_{n,m} \alpha_{n,m}\vec{H}_{n,m} \tag{7.17}$$

$$\vec{E} = \sum_{n,m} \beta_{n,m} \vec{E}_{n,m} \tag{7.18}$$

We will not construct the cumbersome formal solution, which can be found in a paper by Nickolaenko et al. (2004). We note instead that expansions (7.17) and (7.18) allow for obtaining the system of linear algebraic equations for the unknown coefficients α_{nm}, $n \in [1; \infty]$, m $[-n; +n]$, and that contribution from the non-diagonal terms of this system is quadratic with respect to the small parameter of the problem $\mu_{\mathrm{P}} = \frac{\|Z\|}{kh}$, where $\|Z\|$ is the norm of effective surface impedance (this is the absolute value of the impedance in the isotropic cavity), k is the free space wave number, and h is the effective height of ionosphere above the ground. The value of this small parameter depends on the skin-depth δ_S in the ionosphere, which is linked with the quality factor Q of oscillations ($Q \geq 4$). The quality factor is the ratio of the energy lost during one cycle of oscillations to the cumulative energy of the field. Since the losses are concentrated within the skin-depth δ_S, and the effective height of the Earth–ionosphere cavity h is much smaller than the planetary radius a, we obtain a simple relation:

$$Q = \frac{V_{field}}{V_{losses}} = \frac{h}{\delta_S}$$

where V_{field} is the volume occupied by the field and V_{losses} is the volume where losses take place. Since $h \ll a$, the volume occupied by the field is equal to the Earth's surface area multiplied by the height h. The volume of losses is the product of the same area and the skin-depth δ_S. Hence we obtain that the small parameter of the problem $\mu_P \leq$ ¼. Therefore, only diagonal terms should be accounted for in computations, and in the vicinity of the first SR mode ($n = 1$; $m = -1$, $m = 0$, $m = +1$) we can use the relations listed in Table 7.1. The other terms play the role of corrections to our estimates. Certainly the complete system of equations is used in real computations (up to $n = 100$ for $f \approx 8$ Hz), and there the particular symmetry properties of matrix are helpful [see Nickolaenko et al. (2004)].

Each column of Table 7.1 contains separate eigen-functions having different eigen-frequencies in the anisotropic cavity. Deviations of eigen-values arise from the different way of inclusion of the non-zero impedance term z_2 into the dispersion relation. It is absent when $m = 0$, and it is included into the R_{mn} function with the positive sign when $m = +1$ and with negative sign when $m = -1$. So the frequency characteristics of the sub-levels in the horizontal magnetic field component $|R_{nm}(f)|^{-1}$ reach their maximum at different frequencies (see Fig. 7.3).

Table 7.1 shows that field components H_φ and H_θ are combined from different eigen-functions, and the $\pm i$ factor present in the H_θ component indicates the phase shift of $\pm\pi/2$ of this component against the H_φ. We use the $\exp(i\omega\ t)$ time dependence. Let us assume that in some frequency band the functions dominate (with the greater amplitudes) with $n = 1$ and $m = -1$. Here, according to

Table 7.1 The major field components in the vicinity of the first SR mode in the cavity with the anisotropic ionosphereIonosphere

	$n = 1, m = 0$ Standing wave	$n = 1, m = +1$ Wave traveling to west	$n = 1, m = -1$ Wave traveling to east
$-E_r$	$\frac{R^1_{10} \cos\theta_S \cos\theta_0}{R_{10}} \exp(i\omega t)$	$\frac{R^1_{11} \sin\theta_S \sin\theta_0 \exp[i(\omega t + \varphi_0 - \varphi_S)]}{2R_{11}}$	$\frac{R^1_{1-1} \sin\theta_S \sin\theta_0 \exp[i(\omega t - \varphi_0 + \varphi_S)]}{2R_{1-1}}$
H_φ	$\frac{\cos\theta_S \sin\theta_0}{R_{10}} \exp(i\omega t)$	$\frac{-\sin\theta_S \cos\theta_0 \exp[i(\omega t + \varphi_0 - \varphi_S)]}{2R_{11}}$	$\frac{-\sin\theta_S \cos\theta_0 \exp[i(\omega t - \varphi_0 + \varphi_S)]}{2R_{1-1}}$
H_θ	0	$+i\frac{\sin\theta_S \exp[i(\omega t + \varphi_0 - \varphi_S)]}{2R_{11}}$	$-i\frac{\sin\theta_S \exp[i(\omega t - \varphi_0 + \varphi_S)]}{2R_{1-1}}$
R_{nm}	$\omega_1^2 - \omega^2 + \frac{icz_1\omega}{h}$	$\omega_1^2 - \omega^2 + \frac{icz_1\omega}{h} + \frac{3cz_2\omega}{4h}$	$\omega_1^2 - \omega^2 + \frac{icz_1\omega}{h} - \frac{3cz_2\omega}{4h}$
R^1_{nm}	$\omega - \frac{icz_1}{h}$	$\omega - \frac{icz_1}{h} - \frac{3cz_2}{4h}$	$\omega - \frac{icz_1}{h} + \frac{3cz_2}{4h}$

Table 7.1, the component H_θ is multiplied by $-i$, and the field $H_Y = -H_\theta = i\,\frac{\sin\theta_S \exp[i(\omega t - \varphi_0 + \varphi_S)]}{2R_{1-1}}$ retards against the $H_X = H_\varphi = \frac{-\sin\theta_S \cos\theta_0 \exp[i(\omega t - \varphi_0 + \varphi_S)]}{2R_{1-1}}$ component. The complete magnetic field vector rotates counter clockwise in the case and the ellipticity is positive $e_l > 0$. At frequencies where the eigen-function dominates with indices $n = 1$ and $m = +1$, the component H_θ is multiplied by $+i$, and the field H_Y leads against the H_X component. Correspondingly, the rotation direction changes to the opposite one, and the ellipticity becomes negative $e_l < 0$.

We must underline that the right and the left rotations are the result of an agreement. We could direct the vertical axis Z of the coordinate system downward, to nadir instead of zenith. Then, we will look at the observatory not from the space, but from under the ground. The rotation direction will change as a result. The similar effect will take place when the observer moves to the Southern hemisphere where $\cos\theta_0 < 0$ and the radial geomagnetic field has an opposite direction. Finally, we must recall that rotation directions are opposite in radio science and optics. The reason is that an observer in optics looks toward the source, against the propagation direction. In radio science, the observer looks along the propagating wave, so that right and left rotations interchange their places.

When constructing a realistic propagation model, we initially introduce the scalar effective surface impedance of the ionosphere z_0 in such a way that the peak frequencies acquire the observed values in the absence of geomagnetic field (the anisotropy parameter $\rho_m = 0$). The geomagnetic field is "switched on" afterwards, the anisotropy parameter ρ_m becomes a non-zero value, and the surface impedance turns into a tensor (7.14)–(7.16). Let us apply the heuristic frequency dependence $\upsilon(f) = (f - 2)/6 - if/100$ based on the measurements (see Chap. 14). The diagonal elements of the tensor are computed then with the help of the following formula:

$$z_1 = \frac{ih}{\omega}\left\{\frac{c^2}{a^2}\nu(\nu + 1) - \omega^2\right\} \tag{7.19}$$

The elements of the side diagonal are recalculated from the diagonal elements by using the following relation:

$$z_2 = z_1 \frac{\rho_m}{1 + \sqrt{1 + (\rho_m)^2}} \quad (7.20)$$

The resonance fields and the field ellipticity e_l were computed in the anisotropic cavity with $\rho_m = 1$. The frequency ranged from 6 to 24 Hz and covered three resonance modes. The most pronounced effects were observed around the first SR frequency. The observer was positioned at the Greenwich meridian and the latitude of 50° N. The field source was a vertical electric dipole with the current moment independent of the frequency. A dipole is found at the equator near the evening terminator, at the point where the local time is equal to 17 h. Such a source circles the globe during the day periodically coming closer (afternoon UT) and further (after midnight UT) from the observer.

Figure 7.5 demonstrates a high reciprocity between the computational and observational data despite the outstanding simplicity of the field source model. Mutual correspondence was probably conditioned by a low sensitivity of polarization characteristics to the field amplitude. The phase relations depending on the cavity characteristics rather than on the source properties determine the polarization. Character of the frequency dependence of the field ellipticity indicates the successful detection of the resonance frequency splitting, which arises in the Earth–ionosphere cavity under the influence of geomagnetic field.

7.3 Advantages of Polarization Measurements in the Resonance Detection

Dynamic spectra of ellipticity (polarization coefficient) facilitates detection of resonance structures. We demonstrate this property by using the spectra of so-called ionospheric Alfvén resonator (IAR), which is described in more detail in Chap. 12. Data were collected at Karimshino (Kamchatka), and these are relevant to frequencies below the first SR mode. The IAR is observed during the local night since the lower ionosphere has a relatively sharp lower boundary and smaller losses at this time. During the night, there is a decrease in the thickness of ionosphere and it increases before the dawn. The ionosphere changes modulate the observed resonance IAR frequencies, so that the dynamic spectrum acquires a characteristic pattern. The resonances vanish in the morning. The resonance itself is observed as a succession of spectral peaks in the horizontal magnetic field at a few Hz frequencies. Sometimes, the peaks of the IAR are extended up to 10 Hz.

We show in Fig. 7.6 the results of processing of the ELF–ULF records collected at the Karimshino observatory in September of 2000. The figure demonstrates four types of dynamic spectra of the same record. All spectra are shown as 2D maps over the time–frequency plane, and the frequency $f \in [0.1-6]$ Hz is plotted along the ordinates. The dates (local time) are shown along the abscissa.

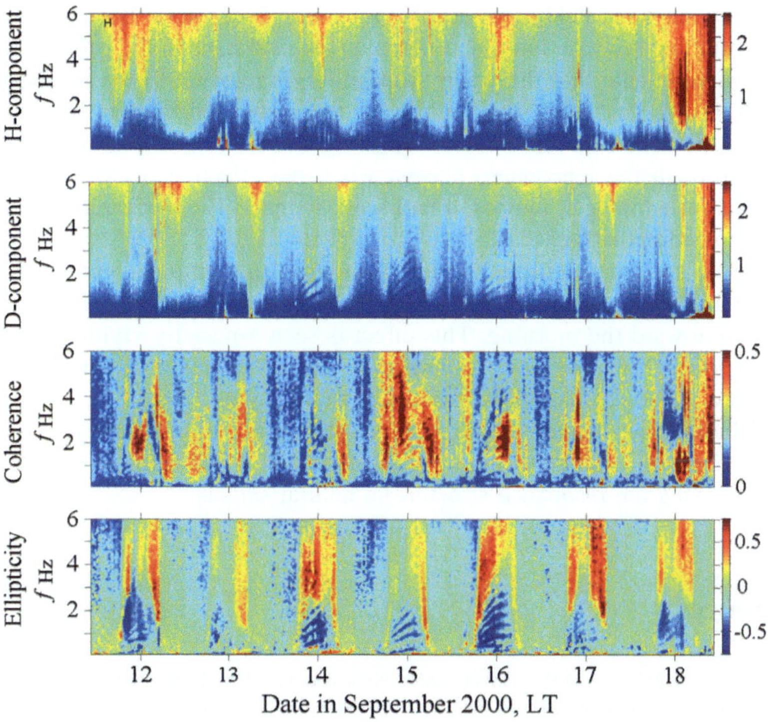

Fig. 7.6 Dynamic spectra of IAR in the frequency band 0.1–6 Hz. Data collected at Karimshino on September 12–18, 2000

Two upper maps demonstrate dynamic spectra of the magnetic field components H_H and H_D. We already mentioned that the geomagnetic coordinate system is used at the site. Since the declination at Karimshino is rather small ($-6.5°$), the following approximate relations are valid: $H_H \approx H_{SN}$ and $H_D \approx H_{WE}$. The third map in Fig. 7.6 demonstrates dynamic spectrum of the coherence function $M_{\mathrm{COH}} = \frac{\langle |H_{\mathrm{H}} \cdot H_{\mathrm{D}}^{*}| \rangle}{\sqrt{\langle |H_{\mathrm{H}}|^2 \rangle + \langle |H_{\mathrm{D}}|^2 \rangle}}$. The lowest map of Fig. 7.6 depicts the field ellipticity e_l. As one may see, the characteristic pattern of IAR is observed in the clearest way at the lowest map. The polarization of horizontal magnetic field shows that the IAR pattern is present almost every night, and this cannot by said about the spectra of separate field components.

Peak frequencies in the polarization plot correspond to the positive ellipticity; the peaks merge into a kind of curved ridges. The valleys correspond to the linear polarization. Therefore, peaks are seen as the dark blue tilted lines in the map. Circular polarization of the ULF signal is quite natural, because the wave comes to the observer through the magneto-active ionosphere plasma. Still, one cannot consider the spectra of ellipticity as trivial ones. Indeed, the circular polarization perceives the traveling wave regime; it is connected with the propagation direction relative the magnetic field of the Earth. However, it does not follow from this fact that the rotation direction must be different at the peaks and in the valleys. Besides, one can hardly state that the wave propagation is relevant to the resonance frequencies at all, since the standing waves are associated with resonance. Most probably, the field maxima observed on the ground are associated with the wave passing through the ionosphere plasma when the effective thickness of plasma layer contains an integer number of half-wave lengths. The mechanism reminds us of the "coated optics" effect.

Dynamic spectra of ellipticity clearly indicate the temporal variations of the effective thickness of the IAR. Initially, it decreases rather fast, then becomes fixed and grows toward the morning. This effect is seen especially vividly in the plots for September 15th and 16th. Polarization properties make it a useful tool in ionosphere studies by using low frequency radio waves.

Concluding the chapter we must admit that application of the cross-spectra, coherence measure and the field polarization associated with them allows for better resolving the resonance structure in natural signals.

References

Belyaev GG, Schekotov AYu, Shvets AV, Nickolaenko AP (1999) Schumann resonances observed using Poynting vector spectra. J Atmos Solar-Terr Phys 61:751–763

Bespalov PA, Chugunov YuV (1994) Rotation of plasmasphere and nature of atmospheric electricity. Doklady RAN 337:467–469 (in Russian)

Bliokh PV, Nickolaenko AP, Filippov YuF (1968) Diurnal variations of the natural frequencies of the Earth-ionosphere resonator in relation to the eccentricity of the geomagnetic field. Geomagn Aeron 8(2):198–206

Bliokh PV, Bormotov VN, Kontorovich VM, Nickolaenko AP, Sapogova NA, Shulga VF and Filippov YuF (1971) On the degeneracy lift in the spherical Earth-ionosphere cavity, Preprint Inst Radio-Phys Electron Ukrainian Acad Sci 10 Kharkov (in Russian)

Bliokh PV, Nickolaenko AP, Filippov YuF (1977a) Global electromagnetic resonances in the Earth-ionosphere cavity. Naukova Dumka, Kiev (in Russian)

Bliokh PV, Galyuk YuP, Hynninen EM, Nickolaenko AP, Rabinowicz LM (1977b) On the resonance phenomena in the Earth-ionosphere cavity. Izv VUZov Radiofizika 20:501–509 (in Russian)

Bliokh PV, Nickolaenko AP, Filippov YuF (1980) Schumann resonances in the Earth-ionosphere cavity. Peter Perigrinus, New York, London, Paris, 168 p

Bormotov VN, Lazebny BV, Nickolaenko AP, Sapogova NA, Shulga VF (1971) Removal of the resonance degeneration in the spherical cavity between the Earth and the ionosphere. In: Proceedings of URSI symposium on the wave propagation, USSR, abstracts of reports, Tbilisi, pp 409–412

Fowler RA, Kotick BJ, Elliott RD (1967) Polarization analysis of natural and artificially induced geomagnetic micropulsations. J Geophys Res 72:2871–2883

Füllekrug M, Fraser-Smith AC (1996) Further evidence for a global correlations of the Earth-ionosphere cavity resonances. Geophys Res Lett 23:2773–2776

Holtham PM, McAskill BJ (1988) The spatial coherence of Schumann activity in the polar cap. J Atmos Terr Phys 50:83–92

Holzworth, RH, Norwille KW (1992) The global circuit as deduced from balloon-borne measurements. In: Proceedings of 9th international conference on atmospheric electricity, St. -Petersburg, Russia, 15–19 Jun 1992, vol 1. pp 14–17

Holzworth RH, Onsager T, Kintner P, Powell S (1984) Planetary scale variability of the fair weather vertical electric field. Phys Rev Lett 53:1398–1401

Labendz D (1998) Investigation of Schumann resonance parameters. J Atmos Solar-Terr Phys 60:1779–1789

Lazebny BV, Nickolaenko AP, Paznukhov VE, Rabinowicz LM, Shulga VF (1987) Evaluation of global lightning activity parameters from the records of the coherence measure. Geomagn Aeron 27:516–518 (in Russian)

McGorman DR, Rust WD (1998) The electrical nature of storms. Oxford University Press, Oxford

Nickolaenko AP, Hayakawa M (2002) Resonance in the Earth-ionosphere cavity. Kluwer Academic Publishers, Dordrecht-Boston-London 380 pp

Nickolaenko AP, Rabinowicz LM (1974) Speeding up the convergence of zonal harmonic series representation in the Schumann resonance problem. J Atmos Terr Phys 36:979–987

Nickolaenko AP, Sentman DD (2007) Line splitting in the Schumann resonance oscillations. Radio Sci 42, RS2S13. doi:10.1029/2006RS003473

Nickolaenko AP, Shvets AV, Yatsevich EI (2001) Deducing variations of the global thunderstorm activity from Schumann resonance data. Radiophys Electron 6(1):71–77 (in Russian)

Nickolaenko AP, Rabinowicz LM, Shvets AV, Schekotov AYu (2004) Polarization characteristics of low frequency resonances. Izv VUZov Radiofizika 47(4):267–291 (in Russian)

Nickolaenko AP, Yatsevich EI, Rabinowicz LM, Shvets AV, Belyaev GG, Schekotov AYu (2005a) Comparison of experimental observations of Schumann resloance with the model of a single global thunderstorm center. Izv VUZov Radiofizika 48(4):283–298 (in Russian)

Nickolaenko AP, Yatsevich EI, Shvets AV, Rabinowicz LM (2005b) Two component model of Schumann resonance signal. Radiophys Electron 10(2):224–232 (in Russian)

Nickolaenko AP, Yatsevich EI, Shvets AV, Rabinowicz LM (2006) Two component source model of Schumann resonance signal. J Atmos Electr 26(1):1–10

Nickolaenko AP, Yatsevich EI, Pechony OB (2008) Diurnal and seasonal variations in the intensities and peak frequencies of the first three Schumann-resonance modes. Radiophys Quantum Electron 51(7):528 (in Russian)

Ogawa T (1985) Fair weather electricity. J Geophys Res 90:5951–5960

Ogawa T (1995) Lightning currents. In: Volland H (ed) Handbook of atmospheric electrodynamics, vol 1. CRC Press, Florida, pp 23–63

Ogawa T, Tanaka Y, Fraser-Smith AC, Gendrin R (1967) Worldwide simultaneity of a Q-burst in the Schumann resonance frequency range. J Geomagn Geoelectr 19:377–384

Polk C (1969) Relation of ELF noise and Schumann resonances to thunderstorm activity. In: Coroniti SC, Hughes J (ed) Planetary electrodynamics, vol 2. Gordon and Breach, New York, pp 55–83

Polk C (1982) Schumann resonances. In: Volland H (ed) Handbook of atmospherics, vol 1. CRC Press, Boca Raton, pp 111–178

Polk C, Fitchen F (1962) Schumann resonances of the earth-ionosphere cavity-extremely low frequency reception at Kingston, R. I. J Res NBS 66D:313–318

Roble RG, Tzur I (1986) The global atmospheric electrical circuit. In: Krider EP, Roble RG (eds) The earth's electrical environment. National Academy Press, Washington, pp 206–231

Rycroft MJ (2006) Introduction to the physics of sprites, elves, and intense lightning discharges. In: Füllekrug M, Mareev EA, Rycroft MJ (eds) Sprites, elves, and intense lightning discharges. Springer, Dordrecht, pp 1–18

Rycroft MJ, Harrison RG (2011) Electromagnetic atmosphere-plasma coupling: the global atmospheric electric circuit. Space Sci Rev 140. doi:10.1007/s11214-011-9830-8

Rycroft MJ, Israelsson S, Price C (2000) The global atmospheric electric circuit, solar activity and climate change. J Atmos Solar-Terr Phys 62:1563–1576

Rycroft MJ, Harrison RG, Nicoll KA, Mareev EA (2008) An overview of Earth's global electric circuit and atmospheric conductivity. Space Sci Rev 137:83–105. doi:10.1007/s11214-008-9368-6

Schekotov AY, Golyavin AM (1978) Locked-in notch filter of industrial frequency and its harmonics. Devices Meas Tech 4:175–178 (in Russian)

Sentman DD (1987) Magnetic elliptical polarization of Schumann resonance. Radio Sci 22:595–606

Sentman DD (1989) Detection of elliptical polarization and mode splitting in discrete Schumann resonance excitations. J Atmos Terr Phys 51:507–519

Williams E, Castro D, Boldi R, Chang T, Huang E, Mushtak V, Lyons W, Nelson T, Heckman S, Boccippio D (1999) The relationship between the background and transient signals in Schumann resonances. In: Proceedings of 11th international conference on atmospheric electricity, 7–11 Jun 1999, Guntersville, Alabamapp, pp 758–761

Chapter 8
SR Line Splitting

We discuss in this chapter the lift of resonance degeneracy (line splitting), which is caused by the influence of ionosphere anisotropy. The anisotropy of ionospheric plasma arises when placed in the outer geomagnetic field. We account for this field in the framework of the hedgehog model: a constant radial magnetic field that abruptly changes its direction when crossing the equator. As a result, the ionosphere surface impedance becomes a tensor, and radio waves acquire different phase velocities and attenuation rate when traveling along the equator from west to east and from east to west. Thus, the first resonance splits into three sub-frequencies, corresponding to the standing wave and to the waves traveling from west to east and from east to west. It is impossible to detect the line splitting by using outline of the resonance curve due to the low quality factor of SR. The lift of resonance degeneracy might be proven only by using the spatial characteristics of different sub-levels: different phase velocities of the west–east and east–west propagating waves, or different sense of polarization of the horizontal magnetic field. We present results of measurements of the line splitting by using the vertical electric field recorded at two longitudinally separated points, and we also describe the measurements of magnetic field polarization. The latter must vary with frequency in a definite way in the presence of the line splitting. Finally, we compare the observational and model data and estimate the effective parameter of ionosphere anisotropy (gyrotropy) by the value of 0.5. This means that the ratio of electron gyrofrequency to the electron collision frequency is about 0.5 in the lower ionosphere plasma.

8.1 General Consideration of Line Splitting

The vertical electric field component was used in the first experiments aiming at the detection of line splitting in the Earth–ionosphere cavity (Bormotov et al. 1971). Later, an attempt was undertaken to detect the line splitting by using the form of resonance line of the power spectrum of the horizontal magnetic field (Tanahashi 1976). Also, the polarization of horizontal magnetic field vector was

A. Nickolaenko and M. Hayakawa, *Schumann Resonance for Tyros*,
Springer Geophysics, DOI: 10.1007/978-4-431-54358-9_8,

used (Sentman 1987, 1989; Labendz 1998). All these works presented purely experimental data with no comparison with the model computations. It was not clear then in what a way the splitting of resonance frequencies must appear in observations. The relevant computations were made later allowing for correct interpretation of experimental data (Nickolaenko et al. 2002, 2004; Nickolaenko and Hayakawa 2002; Nickolaenko and Sentman 2007).

It is impossible to detect the line splitting by using the form of resonance line due to the low quality factor of global electromagnetic resonance oscillations. The peak width of the basic SR mode at 8 Hz is about 2 Hz, while the splitting of eigen-values does not exceed a few tenths of Hz. Besides, as we already know, the observed resonance lines always have a fine structure arising from the random nature of the field source. Averaging the power spectra (see Chap. 1) reduces these fluctuations, so that resonance peaks become smoother. However, the fine structure is never removed completely. The fine structure in the vicinity of the first SR might arise from other reasons, for instance, from the local IAR observed at night (see Chap. 12) or due to the penetration of magnetospheric oscillations into the Earth–ionosphere cavity. This is why detection of the line splitting must imply the special properties of radio signals conditioned exclusively by the lift of degeneracy.

An attempt was undertaken by Tanahashi (1976) to detect directly the resonance line splitting with the help of Z-transform (e.g. Marple 1987) instead of the usual Fourier transform. Z-transform should sharpen the separate sub-peaks in the resonance line by compensating the natural losses at the expense of basic functions growing in time. As we noted in Chap. 6, such an approach reduces the stability of spectral estimates and the problem arises of reliability of data obtained. One can prove the splitting of resonance frequencies by demonstrating the relevant spatial properties of the field components. In particular, transition from one frequency to another causes prevalence of a particular eigen-function. We concentrate on the effects conditioned by difference in the eigen-frequencies of radio waves traveling from west to east and from east to west.

The spherical coordinate system $\{r, \theta, \varphi\}$ is used with the origin at the center of the Earth. When the ionosphere parameters are independent of the angular coordinates (a uniform cavity), the polar axis $\theta = 0$ is directed to the field source, and the observer is positioned at the zero meridian. Such an arrangement simplifies computations, because the solution depends only on the source frequency ω and the source–observer distance θ_H. Only the TEM polarized waves propagate at ELF, so that the vertical (radial) electric E field and the horizontal magnetic H_ϕ component might be detected on the ground surface. The magnetic field is perpendicular to the source–observer great circle arc.

When the diurnal motion of the source is treated, the geography coordinate system becomes convenient having the polar axis $\theta = 0$ directed to the North Pole. In this case, the observer and the receiver acquire the coordinates: $M_S(a, \theta_S, \varphi_S)$ and $M_0(a, \theta_0, \varphi_0)$. Global thunderstorms circle the globe during the day, so that both the source–observer distance and the source azimuth vary in time. An observatory usually records the vertical electric and two orthogonal horizontal magnetic field components. The magnetic antennas are oriented, as a rule, along the west–east and the south–north directions. The field components are denoted as: $H_X = H_{EW}$ directed from the west to east and $H_Y = H_{NS}$ directed from south to north.

In the uniform and isotropic cavity, the temporal variations of the H_X and H_Y fields occur in phase at any given frequency, and the tip of horizontal magnetic field vector moves along the straight line. We speak of linear polarization in this case. Actually, the term "polarization" is already occupied: we treat the TEM polarized waves. Still, the expression "linear polarization" is more compact than the "motion of the vector tip along the straight line"; therefore the term "polarization" is used in the literature and also in the subsequent text.

The ionosphere plasma becomes anisotropic under the influence of geomagnetic field, and the polar axis of coordinate system must be coincident with the direction of geomagnetic dipole. Since we neglect the tilt of geomagnetic dipole and its eccentricity, it appears that the same geography coordinates are used. The non-zero field components E_r, H_θ, and H_φ are present on the perfectly conducting ground surface in the anisotropic Earth–ionosphere cavity. We emphasize that the field components $H_\theta = -H_Y$ and $H_\varphi = H_X$ appear from the solution of the electromagnetic problem itself, not as the mere projections of the same "arbitrary oriented" H_ϕ. vector.

Equations indicate that H_θ, and H_φ components acquire a mutual phase shift depending on frequency due to the lifting of oscillation degeneracy under the impact of geomagnetic field. The elliptic polarization appears of the horizontal magnetic field vector that was used in the pioneering works by Sentman (1987, 1989). However, computations were not available in those times for oscillations in the anisotropic Earth–ionosphere cavity, and only simplistic considerations were used when discussing the observational data. We are able to use the model data now and discuss the splitting of the first SR mode (n = 1) into a triplet. Three different eigen-frequencies of this triplet correspond to the three eigen-functions with the azimuthal numbers $m = -1$, $m = 0$, and $m = +1$.

We will not go into cumbersome solution of the problem with the anisotropic ionosphere, because it might be found in many works (e.g. Wait 1962, 1965; Large and Wait 1966; Large 1967a, b, 1968; Bliokh et al. 1968, 1969, 1977, 1980; Galejs 1972; Nickolaenko and Rabinowicz 1974, 2003; Rabinowicz 1979, 1988; Sentman 1987; Nickolaenko and Hayakawa 2002). We only mention that the fields of an anisotropic cavity are expanded into an infinite series of eigen-functions of the perfect cavity, as seen in Eqs. (7.17) and (7.19). By satisfying the boundary conditions , one obtains the infinite algebraic system for the coefficients of this expansion. After finding the coefficients which is a real computational problem, one constructs the field components sought.

The field source is a vertical electric dipole with the current moment $M_C(f) = Ids(f)$ positioned at the equator at the point where the local time is 17 h. The time dependence is exp ($i\omega$ t). The impedance boundary conditions are used at the lower boundary of ionosphere [see equations (7.13) and (7.14)]. When obtaining the components of surface impedance tensor, we use the "hedgehog" model of geomagnetic field: the constant radial magnetic field abruptly changing its direction at the equator (Rabinowicz 1988; Nickolaenko and Hayakawa 2002;

Nickolaenko et al. 2004). The magnetization is therefore antisymmetric between the northern and southern hemispheres, and it is assumed that there is no additional asymmetry in the system (neither day-night nor polar). The components of relevant surface impedance are:

$$z_{\theta\theta} = z_{\varphi\varphi} = z_1 = ikh\left[1 - \frac{\nu(\nu+1)}{(ka)^2}\right] \tag{8.1}$$

$$z_{\varphi\theta} = -z_{\theta\varphi} = z_2 = z_1 \frac{\rho_m \cdot \mathrm{sign}(\cos\theta)}{1+\sqrt{1+\rho_m^2}} \tag{8.2}$$

The line splitting depends on the parameter of anisotropy of the ionosphere plasma $\rho_{\mathrm{m}} = \omega_{\mathrm{r}} / \nu_{\mathrm{e}}$ where ω_{r} is the radial electron gyro-frequency, ν_{e} is the electron effective collision frequency, and $\nu(f)$ is the propagation constant of ELF radio waves.

It is convenient to use Table 8.1 when discussing the field behavior in the vicinity of the first SR mode. The eigen-functions are listed in this table having the zonal quantum number n = 1 and the azimuthal numbers m = −1, m = 0, and m = +1. The observer coordinates have index O, and the source coordinates have index S. The first column of Table 8.1 lists the field components and relevant elements of the modal frequency characteristics. The second column contains the quantum numbers m; it means what a particular index corresponds to the wave. The last column contains the contracted expressions for individual eigen-functions and their frequency characteristics. Since the first resonance mode n = 1 splits into three sub-modes ($m \leq n$), we list the quantities m = −1, m = 0, and m = +1.

Frequency characteristics $R^1_{\mathrm{nm}}(\omega)$ / $R_{\mathrm{nm}}(\omega)$ and 1 / $R_{\mathrm{nm}}(\omega)$ corresponding to different numbers m reach their maxima at various frequencies due to the splitting of eigen-values when $\rho_{\mathrm{m}} \neq 0$. Computations show that conditions $|R_{11}(\omega)| < |R_{10}(\omega)|$ and $|R_{11}(\omega)| < |R_{1-1}(\omega)|$ are held at frequencies above 8 Hz. Here, the function dominates with m = 1, which corresponds to the wave traveling from east to the west. The function with n = 1 and m = −1 has the highest amplitude at frequencies $f \leq 8$ Hz, and here the opposite conditions are valid: $|R_{11}(\omega)| > |R_{10}(\omega)|$ and $|R_{11}(\omega)| > |R_{1-1}(\omega)|$. This is the wave propagating from west to the east.

Diagrams in Fig. 8.1 illustrate expected effects of the line splitting. The top pictogram shows the angular field distribution for the three sub-levels of the first SR mode and the direction of wave propagation. The second diagram shows the relative phase shift of the vertical electric field component when observed at two longitudinally separated observatories (see below). The third plot depicts expected polarization of the field, an effect also arising from the line splitting. The lowest plot in Fig. 8.1 schematically depicts the idealized amplitude characteristics of the split individual sub-levels.

Consider "high" frequencies where the n = 1 and m = 1 eigen-function dominates. This wave is traveling from east to west and reaches the highest

Table 8.1 Eigen-functions corresponding to the first SR mode n = 1

Field component	*Azimuthal index*	*Eigen-function*
E_r	Standing wave, m = 0	$\frac{R^1_{10}}{R_{10}} \cos\theta_S \cos\theta_0$
	Wave traveling from east to west, m = 1	$\frac{R^1_{11}}{2R_{11}} \sin\theta_S \sin\theta_0 e^{i(\varphi_0-\varphi_S)}$
	Wave traveling from west to east, m = −1	$\frac{R^1_{1-1}}{2R_{1-1}} \sin\theta_S \sin\theta_0 e^{-i(\varphi_0-\varphi_S)}$
H_φ	Standing wave, m = 0	$\frac{-1}{R_{10}} \cos\theta_S \sin\theta_0$
	Wave traveling from east to west, m = 1	$\frac{1}{2R_{11}} \sin\theta_S \cos\theta_0 e^{i(\varphi_0-\varphi_S)}$
	Wave traveling from west to east, m = −1	$\frac{1}{2R_{1-1}} \sin\theta_S \cos\theta_0 e^{-i(\varphi_0-\varphi_S)}$
H_θ	Standing wave, m = 0	0
	Wave traveling from east to west, m = 1	$\frac{-i}{2R_{11}} \sin\theta_S e^{i(\varphi_0-\varphi_S)}$
	Wave traveling from west to east, m = −1	$\frac{+i}{2R_{1-1}} \sin\theta_S e^{-i(\varphi_0-\varphi_S)}$
Resonance factors	*Azimuthal index*	*Explicit impressions for resonance factors*
R_{nm}	Standing wave, m = 0	$R_{1,0} = \omega_1^2 - \omega^2 + ic\omega\frac{z_1}{h}$
	Wave traveling from east to west, m = 1	$R_{1,1} = \omega_1^2 - \omega^2 + ic\omega\frac{z_1}{h} + \frac{3}{4}c\omega\frac{z_2}{h}$
	Wave traveling from west to east, m = −1	$R_{1,-1} = \omega_1^2 - \omega^2 + ic\omega\frac{z_1}{h} - \frac{3}{4}c\omega\frac{z_2}{h}$
R^1_{nm}	Standing wave, m = 0	$R^1_{1,0} = \omega - ic\frac{z_1}{h}$
	Wave traveling from east to west, m = 1	$R^1_{1,1} = \omega - ic\frac{z_1}{h} - \frac{3}{4}c\frac{z_2}{h}$
	Wave traveling from west to east, m = −1	$R^1_{1,-1} = \omega - ic\frac{z_1}{h} + \frac{3}{4}c\frac{z_2}{h}$

amplitude above 8 Hz. In contrast, the eigen-function with n = 1 and m = −1 travels from west to east and dominates at the lower frequencies. The separation of frequency characteristics being essentially the Stark effect might be explained as follows.

The outer geomagnetic field forces the ionosphere electrons to turn around the geomagnetic field line with the gyro-frequency ω_e. The waves m = −1 and m = +1 circle the globe in opposite directions and interact with the ionosphere electrons in different ways. The electric vector of m = +1 wave "rotates in the direction of electron gyro-motion", and this provides a greater damping in comparison with the m = 0 standing wave or with the m = −1 wave traveling in the opposite direction. Since the geomagnetic field modifies not only the losses, but also the phase velocity, these become functions of propagation direction, and there appears the splitting of resonance frequencies. The bottom diagram of Fig. 8.1 shows the splitting: here three distinct maxima are seen, and each of them corresponds to a separate index m.

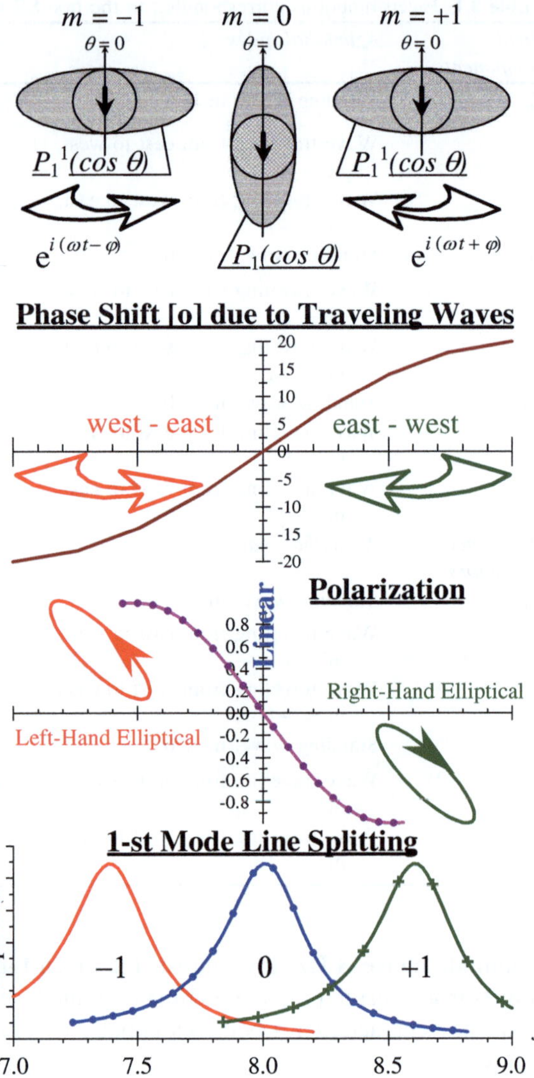

Fig. 8.1 Expected properties of the fields when the degeneracy of resonance oscillations is lifted

8.2 Measurements of Vertical Electric Field Component

The historically first attempt of detecting the line splitting exploited the synchronous and coherent records of the vertical electric field at two sites widely separated along the longitude. Observations were conducted at the base Kharkov (50° N and 37° E)—Ulan-Ude (50° N and 105° E) (Bliokh et al. 1971; Bormotov et al. 1971). The SR signal was translated from Ulan-Ude to Kharkov by a telephone line with the help of phase modulation of the 1.8 kHz carrier frequency.

Signals from two observation sites were logged on the same magnetic tape by an ordinary tape recorder together with the 500 Hz pilot signal from the local generator. The record lasted for 10 min, and 10 recording sessions were held per day uniformly distributed in time. The measurement campaign was of 1 month duration. After completion of the measurement campaign, the time delay was found pertinent to the phone line. A special "radar" was used for the purpose that sent pulses from Kharkov to the short-cut line end at Ulan-Ude. The retardation of this pulse provided the doubled time delay in the line during measurements.

Afterwards, the Kharkov and Ulan-Ude records were played back to recover their coherence and re-recorded again on the tape loop of the special recorder. The synchronization of records was recovered during the playback by appropriate displacement of the magnetic heads compensating the delay in the phone line. The coherence was maintained by the 500 Hz pilot signal. The dual synchronous and coherent record thus obtained was played back with the ten-fold speed, and the special five-channel (five SR modes) analog spectral analyzer performed the spectral processing. The estimates were obtained of power and cross-spectra (coherent power and the mutual phase shift as functions of frequency) averaged over 10 min. Each channel contained 25 discrete frequencies, so that five SR modes were obtained in detail. The equipment and processing were described by Bliokh et al. (1971, 1977, 1980) and Bormotov et al. (1971).

The bottom diagram of Fig. 8.1 explains the idea of measurements: here the triplet is shown into which the first resonance mode splits. We depicted three distinct peaks, while the realistic frequency characteristics are wider and considerably overlap. The left peak corresponds to the $n = 1$ and $m = -1$ eigen-function, which is the wave traveling from west to east, as the wide arrow indicates in the upper diagram. The central peak corresponds to $n = 1$ and $m = 0$ (standing wave), and the right peak is $n = 1$ and $m = +1$ (the east west traveling wave). The values of m index are shown in the figure. The second and third diagrams depict the spatial phase shifts of traveling waves when detected at longitudinally separated sites and the polarization of the magnetic field vector.

The idea of experiments was as follows. The standing wave ($m = 0$) is dominant at the central frequency of 8 Hz, so that the mutual phase shift must be zero here, provided that the signal is detected at two longitudinally separated observatories. The eigen-function with $m = +1$ dominates at higher frequencies, and the westward traveling wave prevails in this band. Consequently, the phase must lead at the eastern site (Ulan-Ude) relative to the western observatory (Kharkov). Similarly, the sub-function with $m = -1$ determines the field at lower frequencies where the wave travels eastward, so that the Ulan-Ude signal is delayed in phase. This is illustrated by the second diagram of Fig. 8.1. These simple speculations indicate that the splitting of eigen-values causes the phase shift increasing with frequency: from negative to positive values in the vicinity of the first resonance mode. Thus, the line splitting must reveal itself as a regular frequency increase of the phase of the complex cross-spectrum $\arg\{< E_1(\omega) \cdot E_2^*(\omega) >\}$. The angular brackets denote averaging over the ensemble of realizations, $E_1(\omega)$ and $E_2^*(\omega)$ are the complex spectra of the signals at Kharkov and Ulan-Ude respectively, and the

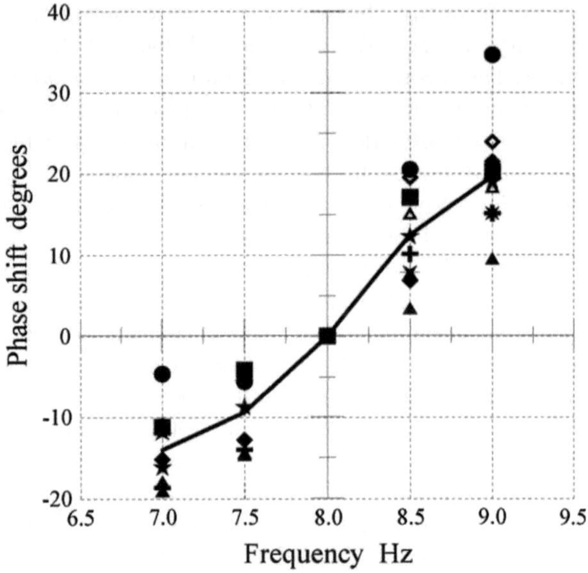

Fig. 8.2 Frequency dependence of the phase difference measured at the Kharkov—Ulan-Ude base in 1969

asterisk denotes complex conjugate. The frequency trend of the cross-spectrum phase must be absent when the line splitting is missing (Bormotov et al. 1971; Bliokh et al. 1971).

Figure 8.2 demonstrates typical experimental results obtained in the measurements. The frequency varying from 6.5 to 9.5 Hz is plotted along the abscissa on linear scale. The phase of cross-spectrum is shown in degrees along the ordinate. The points designate particular phase shifts between two records. The equivalent bandwidth of filter was 0.25 Hz and the duration of the record was 10 min. Since the standing wave must prevail at the 8 Hz frequency, the phase shift was set to zero at this frequency. The solid line in Fig. 8.2 depicts the average frequency dependence of mutual phase shift. It was concluded that the line splitting was successfully detected since the experiment showed expected behavior of phase (Bormotov et al. 1971; Bliokh et al. 1971). Measurements at the meridional base Severomorsk—Kara-Dag did not show such an evident and stable frequency variation of the phase shift, and this also validated the conclusion on a successful detection of line splitting.

No cumbersome model computations were available of the cross-spectra in the Earth–ionosphere cavity with the anisotropic upper boundary at that time. Later, the phase variations similar to those shown in Fig. 8.2 were obtained in the isotropic and uniform cavity model. It was acknowledged that traveling waves always exist in such a cavity due to losses. Indeed, direct and antipodal waves arrive at an observer (we will discuss these waves when treating the pulsed ELF radio signals in Chap. 9). These waves have different amplitudes owing to the different attenuation, because relevant propagation paths have different lengths. There appears "uncompensated remnant" or the wave propagating to the source antipode since the direct wave arrives along the shorter path and therefore has a higher amplitude. The residual traveling wave provides a phase shift between two signals, with the shift being sometimes similar to that observed experimentally. By varying the source position in respect to the observers at Kharkov (50°N, 37°E) and Ulan-Ude (50°N, 105°E), one obtains different frequency variations of the mutual phase. The particular frequency dependence depends on the propagation direction of the residual wave. So, in order to prove the detection of SR splitting, one has to know the actual location of the field source. Experimental results of Fig. 8.2, so similar to the expectation, were not an explicit proof of the line splitting. The necessary measurements are feasible nowadays when the source position might be found from the synchronous and coherent records of Q-bursts at longitudinally separated observatories.

In fact, geometry of the above experiment relative to the real position of the global thunderstorm centers in South-East Asia and Africa should provide in the uniform isotropic cavity the mutual phase shift decreasing with frequency instead of its growth observed experimentally. Thus, the data of Fig. 8.2 agree with the concept of the line splitting, still, the direct comparison with the model data was absent and the unknown source position during records remained a real obstacle.

The results obtained later for the gyrotropic cavity model (Nickolaenko and Rabinowicz 1974; Bezrodny et al. 1977; Bliokh et al. 1977, 1980; Rabinowicz 1979; Nickolaenko and Hayakawa 2002) showed the prevalence of the single wave traveling from west to east. The $m = -1$ sideband has the highest amplitude, and the $\exp[i(\omega t - \varphi)]$ wave determines the field behavior almost everywhere in the cavity when the frequency is close to the first SR mode.

8.3 Measurements of Horizontal Magnetic Field

The detection of line splitting by using the polarization of horizontal magnetic fields was addressed in the works by Sentman (1987, 1989) and Labendz (1998). Preliminary considerations were based on the $\pm 90°$ phase shift (see Table 8.1) between the sidebands of H_φ and H_θ field components, which has to appear at the first SR frequency when the line splitting takes place. As a result, the complete horizontal magnetic vector rotates in opposite directions when we vary the frequency around the first resonance peak (see Fig. 8.1). The vector tip draws a circle when the amplitudes of two orthogonal field components are equal and the phase difference is 90°. The trajectory is a straight line when the phase shift is zero. When the phase shift is not zero, but smaller than 90°, the complete horizontal vector outlines an ellipse, see Fig. 8.1.

Ellipticity appears owing to the factor (im) present in the H_θ field (see Table 8.1). As the splitting separates resonance characteristics of different m, the polarization must depend on frequency. Let the eigen-function with $n = 1$ and $m = -1$ have the highest amplitude at frequencies below 8 Hz. In this case, the component $H_X = H_\varphi$ leads in phase relative to $H_Y = H_\theta$, and the complete horizontal magnetic field vector rotates counter-clockwise, so that we have the LHP. The rotation becomes clockwise at frequencies above 8 Hz, where the subfunction n = 1, m = +1 is of the highest amplitude, and the mutual phase shift changes its sign. Polarization becomes linear at intermediate frequencies where contributions from m = −1 and m = +1 waves are equal, or where the subfunction n = 1, m = 0 dominates. These arguments lead to the idea that a line splitting may be established by using frequency variations of the polarization sense around the first SR frequency (Sentman 1987). Elementary considerations predict a rapid frequency variation of the polarization sense from positive to negative values shown in the third plot of Fig. 8.1.

Experiments were carried out in the frequency band from 3 to 60 Hz by Sentman (1987, 1989) at Table Mountain, California (34°N and 117°W). They showed that the signal ellipticity repeats the SR pattern in the band from 5 to 34 Hz. The focus in that report was on diurnal properties of the ellipticity $e_1(f)$, and it was shown that the elliptical polarization reached its maximum after local sunset or local sunrise. It was positive several hours after local sunrise reaching the value of 0.7 near the first SR resonance. The ellipticity decreased with frequency reaching negative values (RHP) at frequencies between the resonance peaks. It became maximum negative near the 8 Hz resonance immediately after local sunset (04:00–05:00 UT) reaching the −0.8 level. The limited dynamic range used to plot these data did not permit us to show the expected rapid variations of polarization sense when going across the resonance. However, unpublished ellipticity spectra plotted with greater dynamic range than presented in Sentman's (1987) report, as well as data subsequently obtained in California and in Alaska, quite clearly show the expected switch in polarization exactly as described.

In his second work, Sentman (1989) presented observations of the time domain signature of the polarization of discrete ELF events (Q-bursts). The same field site and the same equipment were used. Transient pulses from the rare powerful strokes also demonstrated RHP and LHP over different frequency bands. No fast frequency variations in the ellipticity were evident owing to the limited frequency resolution of the short-length events (<1 s). Signal processing also showed substantial deviations (in excess of 1 Hz) of the peak frequencies of different field components. This fact was also elucidated to the ionosphere anisotropy without any model support though. Much later, such deviations were confirmed in the model computations for the anisotropic cavity (Nickolaenko and Rabinowicz 2003).

We must note that dynamics of ellipticity might indicate the cosmic origin of the effects. At any rate, diurnal variations of the orientation of the major axis of polarization ellipse are unexpectedly close to the behavior of geomagnetic pulsation Pc1 which are associated with the magneto-hydro-dynamic waves propagating along the field lines in the Earth's magnetosphere (Kleimenova 1965; Zybin and Kleimenova 1965).

There are two more details. The first one is that the theory does not predict fast variations of the polarization sense neither in time of the day nor with frequency. The second, the computed deviations of the peak frequencies might reach 0.5 Hz between the H_X and H_Y components (SR background signal) in the uniform isotropic cavity, provided that two thunderstorm centers serve as the field sources having the distinct distances from the observer and separated by approximately right angle. Then, the H_X channel will be driven by the strokes of one center, while the H_Y channel detects the lightning from the other center. On the other hand, computations showed that peak frequencies might deviate by 1 Hz or so of the of H_φ and H_θ fields in the anisotropic cavity model with a single point equatorial source circling the globe (Nickolaenko and Rabinowicz 2003).

The experiment by Labendz (1998) was carried out in Germany (51°N and 9.5°E) in the frequency band from 0.1 to 20 Hz. The H_X and H_Y field components were detected by induction coil antennas; the vertical electric rod antenna was used together with two orthogonal electric grounded antennas. The results were in general correspondence to the data by Sentman (1987, 1989). The ellipticity was a function of time. Negative sense was detected all through the day with a short period of positive sense during the late evening. Special processing was made to separate components of different polarizations. The following split frequencies were announced: $f(m = -1) = 7.82$ Hz, $f(m = 0) = 8.05$ Hz, and $f(m = +1) = 7.98$ Hz. The "central" frequency 8.05 Hz corresponded to the most frequent linearly polarized events. The sidebands were connected with the LHP and the RHP respectively. Unfortunately, we cannot consider these results as robust. The reason is not only in applications of Z-transform, but also the processed signal fragments were extremely short, about a few tenths of a second. Such brief parts cannot provide the frequency resolution of 1/100 Hz mentioned in the paper. One might suspect that results of signal processing were volatile, and the super-resolution was an artifact.

No computational data were presented demonstrating that the above experiments indicate the line splitting. Interpretation was based on speculations, and observations are found to contradict in part to the expectations based on approximate preliminary considerations.

8.4 What the Line Splitting Should Look Like?

Polarization of a magnetic field vector (Fowler et al. 1967) is formally described in a manner similar to that of conventional optical polarization, i.e., by introducing the coherence matrix (e.g., Born and Wolf 1964), but instead of electric field, the magnetic component is used. Details of obtaining polarization were described by Sentman (1987, 1989), Nickolaenko et al. (2002, 2004), and Nickolaenko and Hayakawa (2002) and addressed in Chap. 7. Our model exploits the "hedgehog" geomagnetic field and an equatorial point vertical electric dipole source placed at 17 h local time. For $n = 1$ mode (vicinity of 8 Hz), such a source excites only the "side" waves with $m = \pm 1$ traveling from west to east and from east to west (see Table 8.1). The absence of standing wave $m = 0$ guarantees the greatest possible frequency separation of individual sub-peaks in the power spectrum. The source current moment is a constant. Solution of the problem is found as an expansion over eigen-modes of the perfect Earth–ionosphere cavity. An infinite system of algebraic equations is obtained for the expansion coefficients. Procedure of the solution in the SR band is complicated, and it was addressed by Nickolaenko et al. (2002, 2004) and Nickolaenko and Hayakawa (2002), so that we confine ourselves to presenting only the basic results.

Figure 8.3 shows the spectral data of model computations. The abscissa depicts the frequency in Hz, and the ordinate shows the characteristics of individual sub-modes on a logarithmic scale. Marked thin lines depict the frequency response of individual sub-modes: red curve with diamonds shows the $|R_{1-1}(f)|^{-2}$ function, the blue line with dots corresponds to $|R_{10}(f)|^{-2}$, and the green line with pluses shows the function $|R_{11}(f)|^{-2}$. It is clear that the frequency response is the highest for the eigen-function $|R_{1-1}(f)|^{-2}$ with $m = -1$. We must note that the highest amplitude (and the lowest attenuation) of the curve $m = -1$ corresponds to the lowest resonance frequency. This paradox comes from the way in what the non-diagonal tensor component of the surface impedance enters the resonance characteristics in Table 8.1. As a result, the smallest imaginary part in the eigen-value is combined with its highest real part, which is impossible in an isotropic cavity.

We must emphasize a property of the cavity with anisotropic ionosphere: at resonance frequencies, the wave traveling from west to east noticeably exceeds in amplitude that traveling in the opposite direction and the standing wave. Therefore, the solution with $m = -1$ plays a dominant role at the first SR, and the field behavior is described by the wave $\exp[i(\omega t - \varphi)]$ almost everywhere in the cavity. This property is overlooked in approximate considerations of line splitting (e.g. Bezrodny 2005), while playing the decisive role in forming spatial distribution of fields in the anisotropic cavity.

The thick black line in Fig. 8.3 shows the resulting power spectrum of the complete horizontal east–west magnetic field. We show the spectrum for the closest distance between the equatorial source and the observer at 50° N and 0° E when the propagation path coincides with the meridian. Despite the largest separation of individual sub-modes, no fine structure appears in the resulting resonance curve as the natural line width substantially exceeds the splitting. Moreover, the amplitudes of individual frequency characteristics are completely different, and a "competition" between the sub-levels is impossible. This contradicts with expectations based on the elementary considerations, and so the polarization pattern depicted in Fig. 8.1 appears to be unrealistic. A single sub-mode is dominant around the resonance peak, and this mode determines an invariable polarization sense.

The formal description of elliptic polarization was already given in Chap. 7, but we reproduce the important relations here. To find the field polarization, we use the coherence matrix:

$$J = \begin{Vmatrix} j_{XX} & j_{XY} \\ j_{YX} & j_{YY} \end{Vmatrix} = \begin{Vmatrix} \langle H_X H_X^* \rangle & \langle H_X H_Y^* \rangle \\ \langle H_Y H_X^* \rangle & \langle H_Y H_Y^* \rangle \end{Vmatrix} \tag{8.3}$$

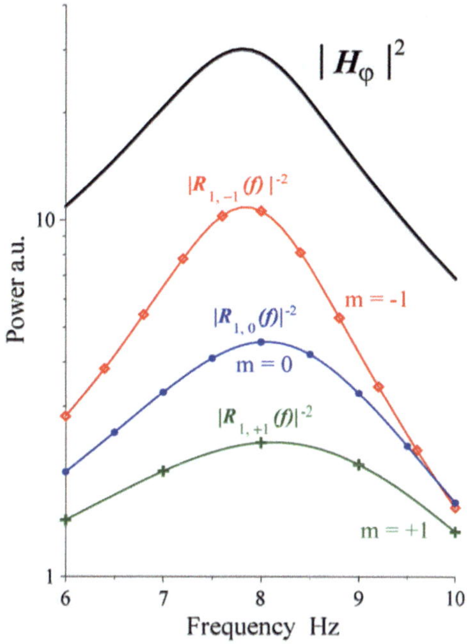

Fig. 8.3 Power spectra of the field component H_φ (f) computed for different sub-levels in the vicinity of the first resonance frequency. The "hedgehog" geomagnetic field model is used

The angular brackets denote averaging over the ensemble. The following relations are valid $H_X \equiv H_\varphi \equiv H_{WE}$ and $H_Y \equiv -H_\theta \equiv H_{SN}$. The field components are:

$$H_X(\omega) = A_1 exp[i(\omega t + \eta_1)] \tag{8.4}$$

$$H_Y(\omega) = A_2 exp[i(\omega t + \eta_2)] \tag{8.5}$$

The polarization sense (the direction of field rotation) is expressed though the angle β, which is found from the equation:

$$\sin(2\beta) = \frac{2\,Im(j_{XY})}{\sqrt{(j_{XX} - j_{YY})^2 + 4 j_{XY} j_{YX}}} = \frac{2A_1A_2}{A_1^2 + A_2^2}\sin(\eta_2 - \eta_1) \tag{8.6}$$

The counter-clockwise rotation corresponds to $\beta > 0$, which is called the left-hand polarization (LHP). The magnetic field vector rotates in the opposite, clockwise direction when the angle $\beta < 0$, and this is the right-hand polarization (RHP). The ellipticity of polarization is found from:

$$e_l = \tan(\beta) \tag{8.7}$$

This is positive $e_l > 0$ for the LHP waves, $e_l < 0$ for the RHP waves, and $e_l = 0$ for linear polarization (or depolarized field). Thus, after computing the H_φ and H_θ

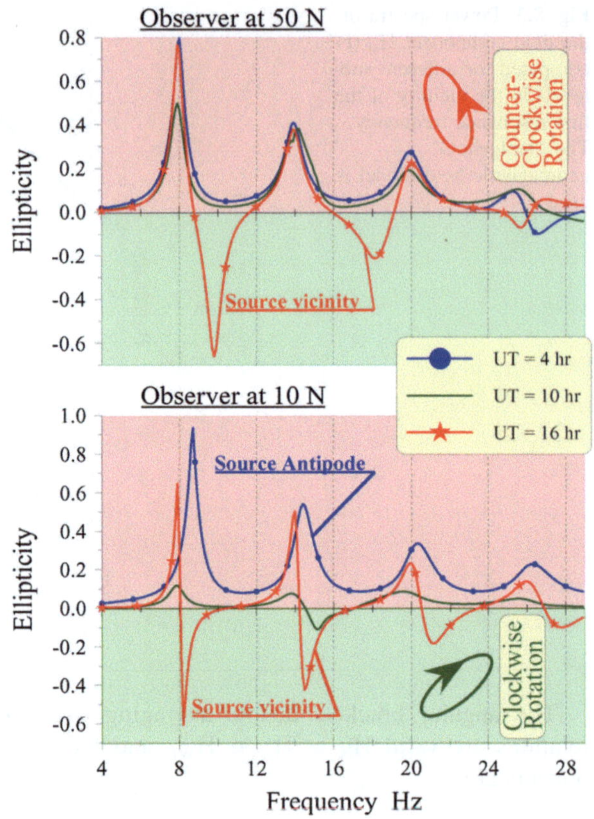

Fig. 8.4 Computed spectra of ellipticity in the "hedgehog" geomagnetic field for the equatorial source at 17 h LT. The observer is located at zero longitude and 50° N latitude (*top*), and at zero longitude and 10° N latitude (*bottom*)

field components as functions of frequency and coordinates of an observer, we can find the wave polarization.

Figure 8.4 shows the ellipticity spectra computed for the parameter of gyrotropy $\rho_m = 0.5$. Particular curves correspond to 0400 UT (blue line with dots), 1000 UT (the smooth green curve), and 1,600 UT (red line with stars). In the first case, the observer is close to the source antipode; it is in the source vicinity in the last one, as the equatorial source is positioned at 17 h LT. The time 1000 UT corresponds to an intermediate distance. When the source–observer distance is great, the horizontal magnetic field vector rotates counter-clockwise (LHP) at the peak frequencies of SR (see Fig. 8.4). The result is conditioned by domination of the eigen-faction with m = −1 in a major part of the cavity. The LHP is found when the observer is in the Northern hemisphere, and it turns into the RHP in the Southern hemisphere.

The top plot in Fig. 8.4 shows data for the middle latitude observer (50° N and 0° E), and the bottom plot corresponds to the tropical position of the site (10° N and 0° E). One may see that the spectrum of ellipticity reminds us of SR power spectrum when the source–observer distance is large. Such behavior was noted in the experiment by Sentman (1987), which is explained by the traveling wave regime. The standing wave regime occurs only in a close vicinity of equatorial source, and here, the sense of polarization may rapidly vary with frequency around the Schumann resonance peaks (Nickolaenko et al. 2002, 2004). This behavior was expected from the elementary consideration shown in Fig. 8.1 when two traveling wave have comparable amplitudes, and different peak frequencies. The negative polarization appears just above the resonance, provided that the observer is close to the equatorial source. The bottom plot in Fig. 8.4 shows the relevant "derivative" pattern appearing at short source–observer distances.

Polarization spectrum becomes less pronounced at the intermediate source–observer distances, but it tends to repeat the resonance pattern. The model data have much in common with observations of 80s with an exclusion that the wave ellipticity never becomes negative at all frequencies. We will demonstrate below that stability of the direction of the field rotation obtained in the major part of the Earth-ionosphere cavity is in accord with the modern observations.

8.5 Comparison of Model and Experimental Data

We demonstrated in Fig. 7.5 the experimental results based on SR monitoring at Lehta, Karelia, Kola Peninsula (64°N and 34°E) and at Karymshino, Kamchatka (53°N and 157°E). This figure showed that waves observed were of LHP at resonance frequencies. Figure 8.5 demonstrates another example in more detail. It compares the results of two days of Lehta record with the model computations for $\rho_m = 1$. The point vertical electric dipole was positioned on the equator at the point with 17 h local time, and the relevant computational dynamic pattern of ellipticity is depicted in the upper frame of Fig. 8.5. The UT time is shown on the abscissa in hours of the 48 h interval. The ordinate shows the frequency from 4 to 24 Hz. The ellipticity is shown by color, and the color scales are placed shown at the right border of plots.

The model predicts that the sense of polarization is positive at the resonance frequencies and does not vary during the day. This is seen especially clearly in the vicinity of the first and the second modes. The negative polarization might appear only in the valleys between the resonance peaks.

Two lower plots present experimental data collected at Lehta during 48 h of continuous record from 18 to 20 September 2000. The lower plot depicts the 'raw' field ellipticity averaged over the interval of 1 h. The frequency resolution of this plot is better than 0.1 Hz; it is conditioned by the 12 s length of signal segments used in FFT procedures. Such a resolution is redundant, and the dynamic spectrum became fragmented with respect to frequencies, so that the general structure is lost in details. The middle frame in Fig. 8.5 depicts the same experimental graph after its smoothing over the frequencies, and so the frequency resolution was reduced to 0.5 Hz. A horizontal brown-orange strip emerges around the 8 Hz frequency corresponding to the positive polarization (LHP) of the field. Similarly, but with a

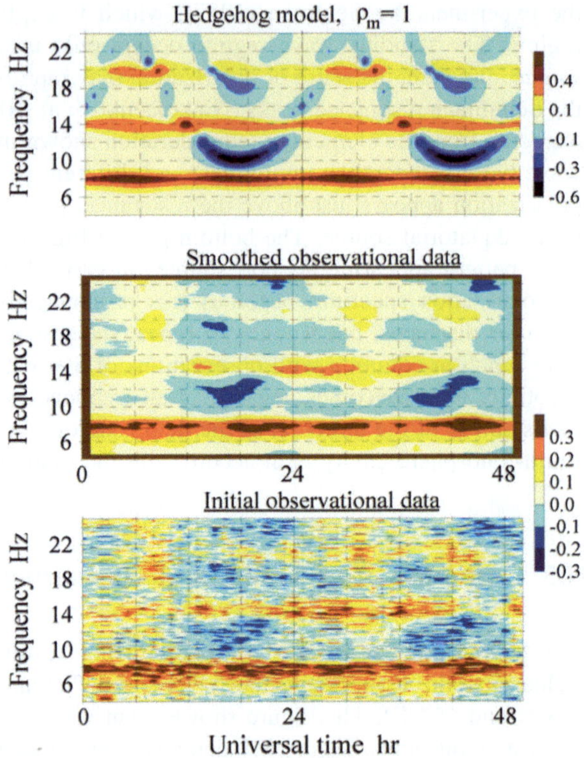

Fig. 8.5 Detailed comparison of dynamic spectra of ellipticity at Lehta: model and experiment

yellow permanent trek, it is seen around the second mode frequency: the ellipticity is smaller there, but its tendency to persist is obvious. The blue crescent areas of negative ellipticity are seen between resonance peaks relevant to short source-observer distances. We see that the model computations successfully predict the ellipticity pattern. An optimist will find some more coincident details in the smoothed experimental and the model dynamic spectra, especially around the first and the second mode frequencies.

The simplest global thunderstorm distribution was used in the computation: a point equatorial source of constant intensity moving around the globe. However, the correspondence between computations and the experiment is high. The reciprocity is probably explained by the independence of polarization of the signal amplitude. The ellipticity depends rather on the phase relations in the cavity, which are connected with the ionosphere properties and the propagation geometry.

We have demonstrated that only a difference in the spatial properties of individual eigen-functions facilitates the detection of the SR line splitting. Splitting separates individual peaks in the spectral line and causes noticeable modifications of the field behavior, which depends on frequency. Consequently, the line splitting not unambiguously detectable in the single-station power spectra may be discerned by exploiting spatial characteristics of the fields, which are different for different modes.

The lifting of resonance degeneracy has been sought using two kinds of experiments. Measurements of the vertical electric field were made of the mutual phase shift at two longitudinally separated points, and the expected frequency dependence was found. It was later found that there is an impact of the source–observer geometry on the observed phase, which needs to be taken into account.

Polarization measurements of the horizontal magnetic field vector confirmed that there occurs the rotation of field vector, and the expected elliptical polarization was found. Modeling of the Earth-ionosphere cavity with anisotropic upper boundary showed that waves travel from the west to east in the major part of the cavity at the SR frequencies. Detailed comparison with modern observations showed similarity of dynamic spectra. Thus, the SR line splitting has been finally resolved. A preliminary estimate for the effective parameter of gyrotropy gives $0.5 < \rho_{\mathrm{m}} < 1$. Further detailed research may permit us to obtain an improved value of this parameter.

References

Bezrodny VG, Nickolaenko AP, Sinitsin VG (1977) Waves in natural waveguides. J Atmos Terr Phys 39:661–688

Bezrodny VG (2005) Asymptotic theory of Schumann resonance fields in gyrotropic Earth—ionosphere cavity. In: Litvinenko LN, Yamplosky YM (eds) Electromagnetic manifestations of geophysical effects at Antarctica. Kharkov, Radioastronomical Institute of NASU, pp 105–121 (in Russian)

Bliokh PV, Nickolaenko AP, Filippov YF (1968) Diurnal variations of eigen frequencies of the Earth-ionosphere resonator due to eccentricity of the geomagnetic field. Geomagn Aeron 8:250–260 (Geomagn Aeron Engl Transl 8:198–206)

Bliokh PV, Nickolaenko AP, Filippov YF (1969) Excitation of the Earth-ionosphere wavegiude by the lightning discharges and the influence of geomagnetic field on noise spectrum. Alta Freq 38(3–4):189–192

Bliokh PV, Bormotov VN, Kontorovich VM, Nickolaenko AP, Sapogova NA, Shulga VF, Filippov YF (1971) On the degeneracy lift in the spherical earth-ionosphere cavity. Preprint 10:58 (Institute of Radiophysics and Electronics of the Ukrainian Academy of Sciences, Kharkov, Ukraine, in Russian)

Bliokh PV, Nickolaenko AP, Filippov YF (1977) Global electromagnetic resonances in the earth-ionosphere cavity. Naukova Dumka, Kiev, p 199 (in Russian)

Bliokh PV, Nickolaenko AP, Filippov YF (1980) Schumann resonances in the earth-ionosphere cavity. In: Jones DL (ed) The Smithsonian/NASA astrophysics data system. Peter Peregrinus, London, p 168

Bormotov VN, Lazebny BV, Nickolaenko AP, Sapogova NA, Shulga VF (1971) Removal of the resonance degeneration in the spherical cavity between the Earth and the ionosphere. Paper presented at URSI symposium on wave propagation, union radio science international, Tbilisi, Georgia

Born M, Wolf E (1964) Principles of optics. Elsevier, New York 854

Fowler RA, Kotick BJ, Elliott RD (1967) Polarization analysis of natural and artificially induced geomagnetic micropulsation. J Geophys Res 72:2871–2883

Galejs J (1972) Terrestrial propagation of long electromagnetic waves. Elsevier, New York, p 362

Kleimenova NG (1965) Some observations of the natural electromagnetic field in the 1–20 Hz band at the Tiksi and Lovosero polar stations. Fizika Zemli 3:97–102 (in Russian)
Labendz D (1998) Investigation of Schumann resonance polarization parameters. J Atmos Solar-Terr Phys 60:1779–1789
Large DB, Wait JR (1966) Cavity resonator modes in a cylindrically stratified magnetoplasma. Radio Sci. 1:655–660
Large DB, Wait JR (1967a) Resonances in a cylindrical earth-ionosphere cavity model with a dipolar magnetic field. J Geophys Res 72:5395–5400
Large DB, Wait JR (1967b) Resonances of the thin shell model of the Earth-ionosphere cavity with dipolar geomagnetic field. Radio Sci 2:695–700
Large DB, Wait JR (1968) Influence of radial magnetic field on the resonances of a spherical plasma cavity. Radio Sci 3:633–637
Marple SL Jr (1987) Digital spectral analyses with applications. Prentice-Hall, Upper Saddle River, NJ, p 584
Nickolaenko AP, Rabinowicz LM (1974) Electromagnetic field in the spherical Earth-ionosphere cavity with gyrotropic upper boundary. Preprint 36:33 (Institute of Radiophysics and Electronics of the Ukrainian Academy of Sciences, Kharkov, Ukraine)
Nickolaenko AP, Hayakawa M (2002) Resonance in the Earth-ionosphere cavity. Springer, New York, p 380
Nickolaenko AP, Rabinowicz LM, Shvets AV, Schekotov AY (2002) Detection of Schumann resonance line splitting. Radiofiz Electron 7:498–508 (in Russian)
Nickolaenko AP, Rabinowicz LM (2003) Diurnal variations of peak frequencies in the gyrotropic Earth-ionosphere cavity. Radiofiz Electron 8:357–365 (in Russian)
Nickolaenko AP, Rabinowicz LM, Shvets AV, Schekotov AY (2004) Polarization characteristics of low frequency resonances. Izv VUZov Radiofizika 47(4):267–291 (in Russian)
Nickolaenko AP, Sentman DD (2007) Line splitting in the Schumann resonance oscillations. Radio Sci 42:RS2–RS13. doi:10.1029/2006RS003473
Rabinowicz LM (1979) Computations of electromagnetic field in gyrotropic spherical Earth-ionosphere cavity with the source placed at middle latitudes. Preprint 141:20 (Institute of Radiophysics and Electronics of the Ukrainian Academy of Sciences, Kharkov, Ukraine)
Rabinowicz LM (1988) Global electromagnetic resonances in non-uniform and anisotropic Earth-ionosphere cavity. Ph.D. thesis, Kharkov (in Russian)
Sentman DD (1987) Magnetic elliptical polarization of Schumann resonances. Radio Sci 22:595–606
Sentman DD (1989) Detection of elliptical polarization and mode splitting in discrete Schumann resonance excitations. J Atmos Terr Phys 51:507–519
Tanahashi S (1976) Detection of line splitting of Schumann resonance from ordinary data. J Atmos Terr Phys 38:135–142
Wait JR (1962) Electromagnetic waves in stratified media. Elsevier, New York, p 372
Wait JR (1965) Cavity resonances for a spherical Earth with a concentric anisotropic shell. J Atmos Terr Phys 27:81–89
Zybin KY, Kleimenova NG (1965) Amplitude spectrum of micropulsations in the 1-20 Hz band. Geomag Aeron 5:1125–1126 (in Russian)

Chapter 9
Transient Events

This chapter is concerned with the complete processing procedure of an ELF transient recoded by a typical SR receiver. Waveforms of the Poynting vector clearly indicate the direct and antipodal waves arriving at the observer. Amplitude spectra are obtained of vertical electric and two horizontal magnetic field components together with the complex spectra of the wave impedance and of the Poynting vector. The period of the phase oscillations of the Poynting vector (or the wave impedance) is used for establishing the source distance. After finding the source bearing, we obtain the coordinates of a lightning source. By comparing the observed spectra with the model data, we estimate the source current moment as a function of frequency. We also demonstrate that the field polarization of the recorded Q-burst is close to that of continuous field component obtained in the previous chapter. Afterwards, the wide-band records of the Q-bursts with exceptional quality are compared with the model waveforms showing exclusive reciprocity of data. The propagation velocity is evaluated of the pulsed ELF signal, and results of observations are explained. Finally, the worldwide data are presented on the parametric ELF pulse generated by a galactic gamma-ray flare of Dec. 27, 2004. A comparison of simultaneous records at six observatories allowed us to establish the source polarity and to locate the parametric source. The peculiarities are addressed of this unusual source in comparison with the ordinary terrestrial lightning strokes.

9.1 Samples of Q-Bursts Recorded by Typical SR Receiver

The term Q-bursts was introduced by Ogawa et al. (1966a, b): it refers to discrete natural radio pulses that have a duration of 0.3–1.5 s and are detected all over the globe (Ogawa et al. 1967). These pulses arrive from exclusively powerful lightning discharges, and their amplitude may exceed that of regular SR background by a factor of ten. We present here a typical ELF radio pulse (Q-burst), discuss its characteristics and demonstrate how to locate its source and to find its source spectrum.

The technique of locating the causative source was suggested by Jones (1970a, b), was experimentally applied for the first time by Jones and Kemp (1970), and

A. Nickolaenko and M. Hayakawa, *Schumann Resonance for Tyros*,
Springer Geophysics, DOI: 10.1007/978-4-431-54358-9_9,

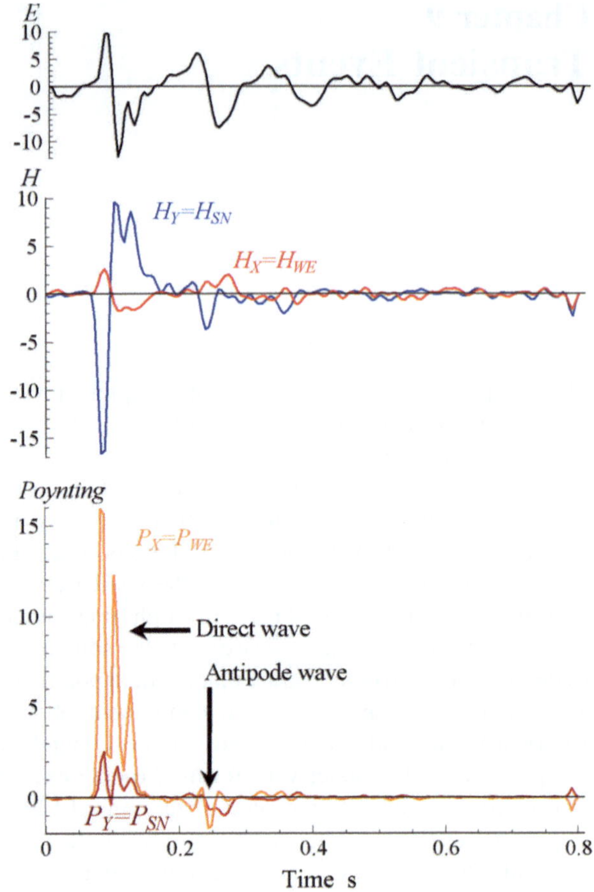

Fig. 9.1 Waveforms of a typical ELF transient signal (Q-burst)

was elaborated by Nickolaenko and Kudintseva (1994). Source location was performed in many works (e.g. Lazebny and Nickolaenko 1976a, b; Burke and Jones 1995; Boccippio et al. 1995; Sato et al. 2003; Greenberg and Price 2004, 2007; Nakamura et al. 2010). Special interest in the Q-bursts and in the slow tail atmospherics appeared again after establishing their association with the lightning strokes that are the source of transient luminous events (TLEs) in the mesosphere (Boccippio et al. 1995; Sukhorukov and Stubbe 1997; Füllekrug and Reising 1998; Füllekrug et al. 1998; Marshall et al. 1998; Füllekrug 1999; Füllekrug et al. 2000; Huang et al. 1999; Füllekrug Constable 2000; Hobara et al. 2001, 2006; Price et al. 2002, 2004; Sato et al. 2003; Marshall et al. 2005; Neubert et al. 2005; Shalimov and Bösinger 2006; Greenberg et al. 2007; Hayakawa et al. 2011, 2012; Whitley et al. 2011; Surkov and Hayakawa 2012).

The pulse shown in Fig. 9.1 was recorded at the Lehta observatory, Karelia, Russia, (64° N and 34° E) on February 21, 2000, around 21 h 20 min UT. The two upper plots demonstrate waveforms of the field components E, H_{WE} and H_{SN}

recorded with a sampling frequency of 204.8 Hz. Time is shown along the abscissa in seconds, and arbitrary units are used along the ordinate. A short pre-history of the pulse is also shown. Pulses of this outline are regarded as Q-bursts (quiet bursts), because they have a regular waveform with attenuating sinusoid of 8 Hz frequency. This wave is especially clear in the E–field record.

The third plot in Fig. 9.1 depicts temporal variations of the Poynting vector $\vec{P} = \vec{E} \times \vec{H}$ with the components $P_X = P_{WE} = -E \cdot H_Y = -E \cdot H_{SN}$ and $P_Y = P_{SN} = E \cdot H_X = E \cdot H_{WE}$. Direct and antipodal waves can be clearly seen in the Poynting vector: These arrive at an observer from opposite directions and therefore have opposite signs. Amplitudes of direct and antipodal waves (the power flux) are noticeably different due to losses in the Earth–ionosphere cavity.

Polarization of the horizontal magnetic field pertinent to a Q–burst usually deviates from the linear one. This is demonstrated in the left panel of Fig. 9.2. Here we depict the Lissajous figure (phasor) of the horizontal magnetic field and of the Umov–Poynting vector. In the left panel of Fig. 9.2, the horizontal axis shows temporal variations of the H_X field component (the west–east field), while the vertical axis depicts the H_Y field (the south–north component). The axes scales are equal, so that the tip of circularly polarized wave will form a circle. The arrow in the left frame shows the rotation direction.

Though we have chosen one of the best realizations of ELF transients, the Lissajous figure deviates from both a straight line and from an elliptic curve. The right panel in Fig. 9.2 shows evolution of the Poynting vector. An optimist can resolve here the three excursions of the vector from the origin, which were clearly seen in the lower plot of Fig. 9.1. Variations occur in a narrow sector directed away from the source. Plots of Fig. 9.2 show visibly that the wave arrival angles

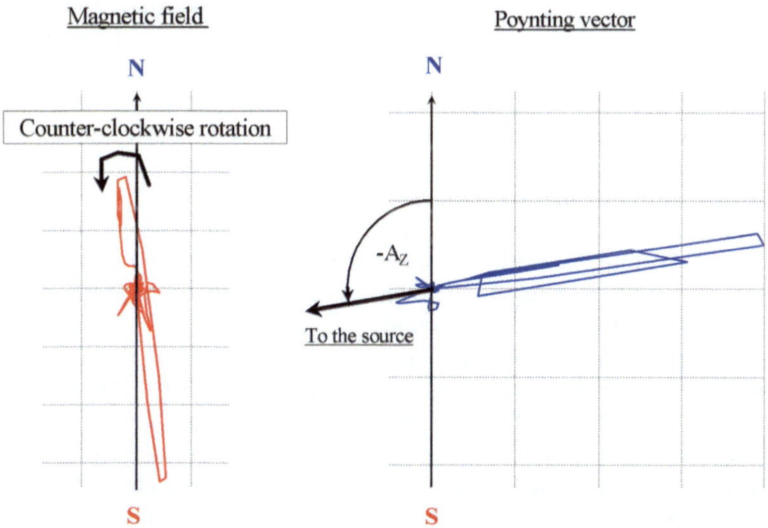

Fig. 9.2 Lissajous figures of the horizontal magnetic field and Umov–Poynting vector for the Q-burst shown in Fig. 9.1

mutually agree when seen from the orientation of magnetic field and from direction of Poynting vector, and they point to the source positioned in the third quadrant (source bearing about 260°).

Rotation of the horizontal magnetic field vector was conditioned most probably by the anisotropy of lower ionosphere because we know from the calibration that matching of the frequency characteristics of the receiving channels was very high. In particular, the mutual phase shifts were smaller than 3°, and magnetic antennas had no influence on each other. The anisotripy causes the SR line splitting addressed in Chap. 8. The non-linear polarization of the wide band radio signal seen in Fig. 9.2 is a "remnant" of SR line splitting. Unfortunately, the polarization of a model ELF pulse has not been computed yet in the time domain. This problem is very interesting, but it presents a real challenge since one has to construct the wide-band complex spectra of the fields in the anisotropic cavity, and afterwards the fields should "pass" through a realistic frequency characteristic of the receiver and then be transformed into the time domain pulse. The major difficulties occur in the first stage.

Results of spectral processing of ELF transient are shown in Fig. 9.3. The signal frequency is plotted on the linear scale along the abscissa in Hz. The ordinates show the spectral amplitudes of separate field components on linear scale. Signal processing was made with the help of standard FFT procedure that uses 2,048 signal samples. Since the original record contained 165 samples, it was initially centered and zero padded afterwards up to 2,048 points. Curves in Fig. 9.3 present the amplitude spectra of vertical electric ($|E(f)|$) and the complete horizontal magnetic field component ($|H_\varphi(f)|$) accompanied by the amplitude spectrum of the wave impedance ($|W(f)| = |E(f)|/|H_\varphi(f)|$). Spectrum of the vertical electric field demonstrates a perfect SR pattern. This is practically absent in the horizontal magnetic field. Two lower plots in Fig. 9.2 demonstrate spectrum of the real part of Umov–Poynting vector being the power flux in the Earth–ionosphere cavity and the spectrum of its argument which coincides with that of the wave impedance. Amplitude and phase spectra of the wave impedance were smoothed by moving averaging. The smoothed spectra are shown by thick lines and the thin lines depict the initial spectra.

Fast oscillations of the wave impedance at higher frequencies are conditioned by the noise component prevailing due to reduction of magnetic field with the frequency. Therefore, we must treat the spectra in 30–35 Hz segment as "spoiled" by the stochastic SR background: here the spectral density of the Q-bursts becomes comparable with that of the background.

To evaluate the particular source–observer distance, we use the first four oscillations of the wave impedance (e.g. Bliokh et al. 1977, 1980; Nickolaenko and Kudintseva 1994; Burke and Jones 1995, 1996; Nickolaenko and Hayakawa 2002; Nakamura et al. 2010). For this purpose, we must measure the median period of these oscillations, and this is equal to $< \Delta f > = 7$ Hz when evaluated from the modulus of the wave impedance $|W(f)|$and$< \Delta f > = 7.66$ Hz when found from the phase variations $\mathrm{Arg}\{W(f)\} = \mathrm{Arg}\{\mathrm{P}(f)\}$. Relevant values are shown in Fig. 9.3.

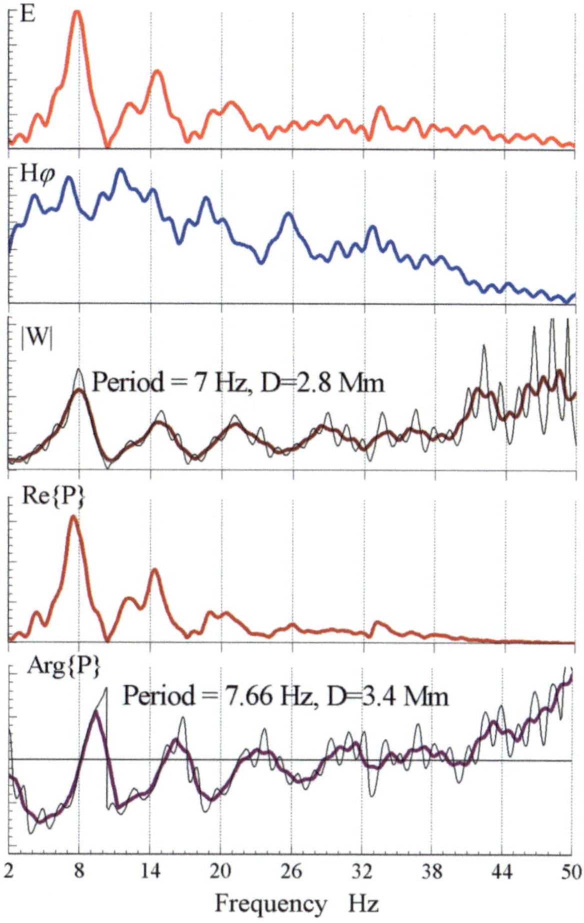

Fig. 9.3 Spectra of the fields, wave impedance, and Umov–Poynting vector

To derive the source–observer distance, we use the simple formula (Burke and Jones 1995, 1996; Nickolaenko and Hayakawa 2002):

$$D = 20 - \frac{120}{<\Delta f>} \tag{9.1}$$

Here the source–observer distance is measured in Mm (1 Mm = 1,000 km) and the average period of oscillations $<\Delta f>$ is measured in Hz.

By substituting the above periods into Eq. (9.1), we obtain two distances: 2.85 and 4.33 Mm. In combination with the source bearing of 261°, the distances provide the coordinates: (52° N and 9° W)—this is a point in the Atlantic Ocean west of England and south of Ireland, and (41° N and 21° W)—not far away from Azores Islands. Since the standard deviation of the source bearing was about ±6°, the particular discharge should be positioned inside a "rectangular" zone ±360 km wide and ±740 km long. The data corresponds to many independent

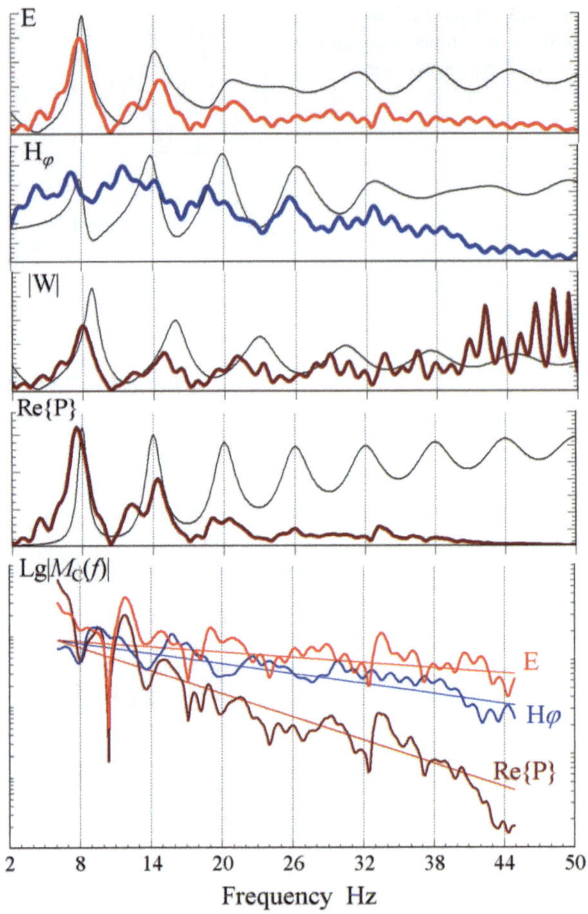

Fig. 9.4 Comparison of observed and the model spectra for the source–observed distance of 3.6 Mm. The lowest frame shows fitting of the amplitude spectra of the source current moment derived from different field components

estimates of the accuracy of the stroke location by using the ELF transients being about 1 Mm (e.g. Bliokh et al. 1977, 1980; Burke and Jones 1995, 1996; Boccippio et al. 1995).

After obtaining the median source–observer distance of 3.6 Mm, we can compute the expected spectra of all field components for the delta-source and then derive the spectrum of the source amplitude. Experimental and the model spectra are compared in the upper plots of Fig. 9.4: the vertical electric field, complete horizontal magnetic field, wave impedance and the Poynting vector. The thin curves depict the computed amplitude spectra corresponding to the 3.6 Mm delta-source distance. The thick lines repeat the experimental spectra that we have shown in Fig. 9.3. One may observe a good correspondence between the model and experiment, although the spectra of observed magnetic field deviate from the model. The lowest frame in Fig. 9.4 contains three spectra of the source current moment evaluated from different experimental spectra. Here the logarithmic scale is used along the ordinate. Every curve is supplied with the label indicating the

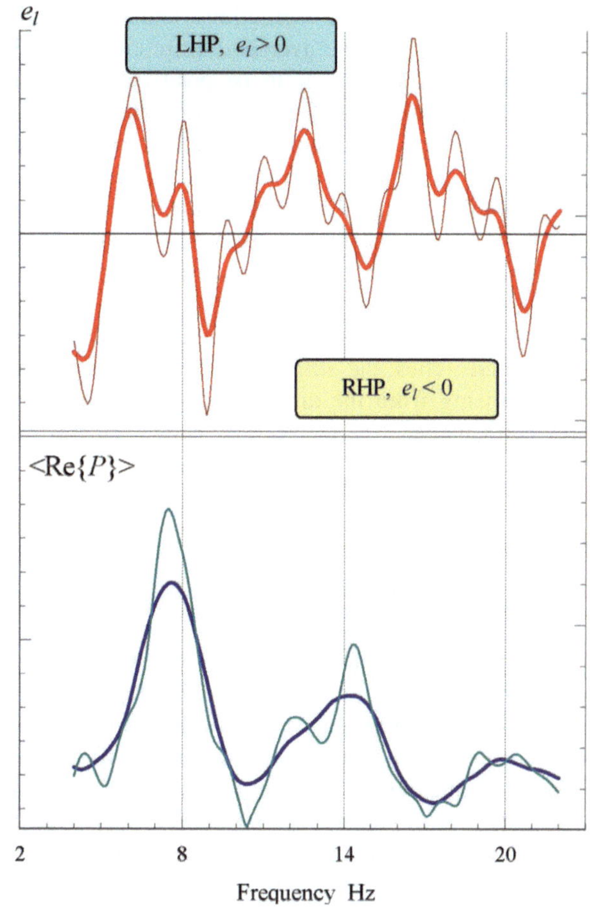

Fig. 9.5 Polarization of horizontal magnetic field pertinent to the Q-burst and relevant resonance spectra of the real part of the Umov–Poynting vector

field component from which it was obtained. It is obvious that the decrease of the source amplitude spectrum is present in all estimates despite their mutual deviations. The "red" spectrum of the parent lightning discharge is a typical feature observed experimentally. Every spectrum of the source current moment was fitted with the exponential frequency dependence seen in the figure as a straight line. The following scale frequencies f_S were obtained for the spectra of source current moment $M_C(f) = M_0 \cdot \exp\{-f/f_S\}$ derived from the vertical electric, horizontal magnetic, and the Poynting vector correspondingly: 34.9, 18.2, 7.9 Hz.

To conclude this section, let us turn to the polarization of the ELF transient. Elements of the coherence matrix usually contain the averaged values, while we treat the single event. To stabilize the spectral estimates, we averaged them by using the moving average (or spectral window), i.e., the value at a given frequency was an average of the estimate at this frequency and at ±5 adjacent frequencies.

The upper plot in Fig. 9.5 shows the smoothed (thick lines) and initial (thin lines) spectra of the ellipticity e_l that was derived by Eq. (8.7). We observe that smoothing removed fast oscillations, and the behavior of ellipticity pertinent to a Q-burst is similar to the averaged ellipticity of SR background addressed in two previous chapters. It tends to be positive at the resonance frequencies and just below them, while the "negative rotation" is met in the valleys just above the resonance modes. To demonstrate this clearly, we plot the relevant spectrum of the real part of the Umov–Poynting vector in the lower frame of Fig. 9.5. Spectra of the power flux indicate in addition that phenomenon of the global electromagnetic resonance facilitates radio propagation around the Earth's surface. This is a distinction of SR from the typical resonance associated with the "locked in" or "trapped" waves.

Obviously the areas of positive ellipticity are found at the SR peaks and just below them. Thus, the ellipticity of ELF transients also confirms the line splitting of SR arising from the anisotropy of the ionosphere that was addressed in the previous Chapter.

9.2 Comparison of Observed and Model Waveforms of Q-bursts

Experimental data of the previous section represented typical quasi-sinusoidal waveforms observed in both the field components. Such an outline of Q-bursts is usual for the records performed by the standard SR receivers with the band-pass width from a few hertz to approximately 40 Hz. An ELF transient contains the oscillation attenuating in time of a more or less complicated outline. Therefore, when the first model waveforms of Q-bursts were computed (Nickolaenko et al. 1999), their waveforms seemed extraordinary as these contained only the pairs of pulses instead of an attenuating sinusoid. We collected the model waveforms in Fig. 9.6. The left frames in this figure demonstrate the temporal evolutions of E- and H-fields (in mV/m and μA/m). The time is plotted on the abscissa in milliseconds. Pulses originate from the positive lightning stroke, so that the vertical electric field has a negative onset. The current moment of the vertical dipole source was equal to 10^8 A·m, and its spectral density was constant ("white" source). Computations correspond to the "standard" propagation constant: $\nu(f) = (f - 2)/6 - i \cdot f/100$ in the isotropic cavity. The time domain solution, parameters of the models, and relevant characteristic results are presented in Chap. 14 specially devoted to SR modeling.

The left and lower plots of Fig. 9.6 show a superposition of pulsed waveforms when the source–observer distance varies with a step of 1 Mm. Pulses from the 5, 10, 15, 19, and 20 Mm distances are marked by color. The upper right plot in Fig. 9.6 is the "calibration" curve relevant to the above field source. This is the distance dependence of the initial negative excursion of the $E(t)$ pulse measured in dB against 1 mV/m. In fact, this dependence might be recognized in the outline of superposition of the $E(t)$ pulses.

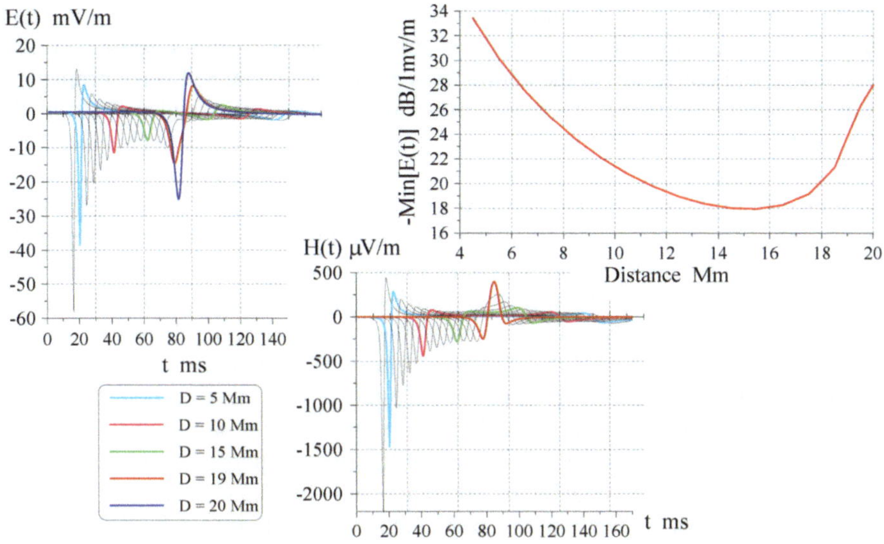

Fig. 9.6 Model waveforms of ELF transient and distance variations of computed pulse onset in the vertical electric field component

Many features of ELF pulses are demonstrated in Fig. 9.6. For instance, one may observe an increase of pulse amplitude at the antipodal distances. Pulses merge at the source antipode in the electric field and they compensate each other in the horizontal magnetic field. The plots demonstrate how the pulsed amplitude is reduced when the circular wave front extends from the point vertical source. The front starts to converge to the source antipode after passing the 10 Mm distance, but geometric convergence of the wave is combined with its attenuation in the real cavity. This is why the pulse amplitude starts to increase from distances somewhat greater than 10 Mm. Since the energy losses increase with signal frequency, the electromagnetic energy at higher frequencies decreases faster and the signal initially "white" acquires the "reddish" spectrum. This is observed in the time domain as a gradual widening of the pulse. Geometric focusing at the source antipode cannot compensate the losses, so that the antipodal field maximum is of finite amplitude. All these features are seen in Fig. 9.6. The left frame in Fig. 9.6 shows in particular that the electric pulse acquires the minimum amplitude at distances close to 15.5 Mm. The signal focusing at the source antipode increases amplitude by $\sim$10 dB from its minimum level. Antipodal "amplification" facilitates the detection of ELF transients arriving from greater distances, and we will see that this is actually so in the observations.

The model waveform contains only two pulses: the direct and antipodal waves. This pair multiply returns to an observer after the "round-the-world travel". Such a feature seems odd for the spherical cavity, where "all propagation paths are possible". However, the elementary consideration demonstrates that there is nothing odd. Consider the point source occupying a pole and emitting a single pulse. The expanding circular wave front initially moves from the source, crosses the equator and converges at the antipode. Afterwards, the pulsed wave propagates from the source antipode, it crosses the equator once again and merges at the source point. This process continues until the amplitude becomes negligible due to wave attenuation. Imagine that there is an observer positioned at some meridian at a distance from the source. This observer will detect the wave expanding from the source first. Then, the wave from the source antipode will arrive forming the second pulse of the above mentioned pair. Thus, an observer in the spherical cavity detects only two pulses: the direct and antipodal as if propagation takes place along the great circle only. Of course this does not concern the distance variation of the pulsed amplitude, and this function reflects the sphericity of the cavity.

From the physical point of view, only paired pulses must be detected in the observation arriving along the complementary great circle arcs. These are transformed into a quasi-sinusoidal waveform by the time response function of the SR receiver. To validate the model data, one must convolve the time domain solutions with the time response function of a particular receiver. Fortunately, we know the unique experimental data allowing for the direct comparison of model and observations (Ogawa and Komatsu 2007).

We compare below the model and experimental waveforms recorded in the extraordinary wide frequency band. The experimental data were published by Ogawa and Komatsu (2007). Records practically did not contain the industrial 50 Hz interference, so that notch filtering was not necessary. The waveforms were recorded in an unusually wide frequency band with the high time resolution. The receiver bandwidth was 1 Hz–11 kHz, which included the whole ELF band (3–3,000 Hz) and a substantial part of the VLF band (3–30 kHz). The sampling frequency was equal to 22 kHz. The vertical electric field component was recorded at Kochi, Japan (33.3° N and 133.4° E) during the years of 2003–2005. Such measurements are impossible at the majority of field sites. Individual pulses were selected visually from the records performed in the fair weather conditions with the help of classical ball antenna.

The source–observer distance was evaluated directly from the time domain record by using the arrival times of direct, antipodal, and round-the-world waves (Ogawa and Komatsu 2007). Let us denote the length of the Earth circumference by L_E ($L_E = 40$ Mm), the source–observer distance by D, and the velocity of pulse propagation by V_P. Then, by starting the time from the moment of a lightning discharge, we have the time delay $t_P = D/V_P$ of the first pulse (direct wave) at the observatory. The second pulse (antipode wave) will have $t_A = (L_E - D)/V_P$, and the first round-the-world wave arrives at $t_R = (L_E + D)/V_P$. We have in mind that the round-the-world propagation path is the great circle arc: source–observer–source antipode–source–observer. The following system of equations is readily obtained:

$$\begin{cases} 2D = V_P(t_R - t_A) \\ L_E = V_P(t_R - t_P) \end{cases} \tag{9.2}$$

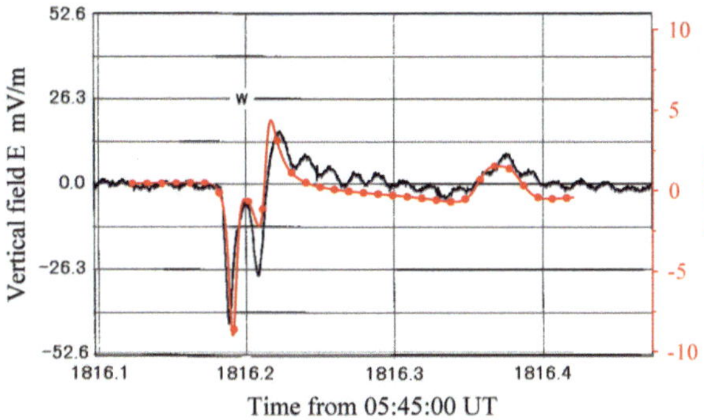

Fig. 9.7 Comparison of the model and observational bursts of W type, $D = 17.4$ Mm

By resolving (9.2) we have: $V_P = L_E/(t_R - t_P)$ and $D = (L_E/2) \times (t_R - t_A)/(t_R - t_P)$.

In the case when we know the occurrence time of the stroke, so that the t_P delay is known, other relations might be used: $V_P = L_E/(t_A + t_P)$ and $D = L_E/2 - (V_P/2) \times (t_A - t_P)$. Such an approach was used by Ogawa and Komatsu (2007), who exploited the differences t_A–t_P and t_R–t_P for evaluating the source–observer distances. We apply these "experimental" values when computing relevant waveforms (Nickolaenko et al. 2008). Afterwards, we adjust the distances to improve correspondence of the model waveforms to those observed.

A comparison of data is made in Figs. 9.7, 9.8, 9.9, and 9.10. Computed waveforms are shown by red curves with dots, and experimental waveforms are shown by the black lines. The similarity of curves is apparent indicating the correctness of the distance estimation in the time domain for all types of ELF transients: W, V, and V–V. Thus, the source–observer distance might be evaluated without computing amplitude spectra, provided that the bandwidth of the receiver covers the whole ELF band at least.

The red dotted model curves are superimposed on the images of experimental records (the numerical data of observations was not available unfortunately). Curves in Fig. 9.7 demonstrate the W-type waveforms in which the direct and antipodal waves interfere, and the source distance was evaluated as 17.4 Mm. The abscissa depicts the time in seconds from the 5 h, 45 min moment UT. The left ordinate shows the observed vertical electric field in mV/m, and the right ordinate corresponds to the model *E*(*t*) filed measured in the same units. The current moment of the field source was 10^8 A·m.

If we superpose the middle sub-pulses of the waveforms, the rest of the curves become rather close. This confirms the 17.4 M distance found by Ogawa and Komatsu (2007) from the observational data. It is interesting to note that the model pulse is smaller than that observed, by a factor of about 5. Further comparison indicates that the current moment of 10^8 A·m systematically underestimates the source amplitude. Deviations of the waveforms are present in many plots. This is the industrial 50 Hz interference, which was weak but still present in the records. Atmospherics are also seen in the record.

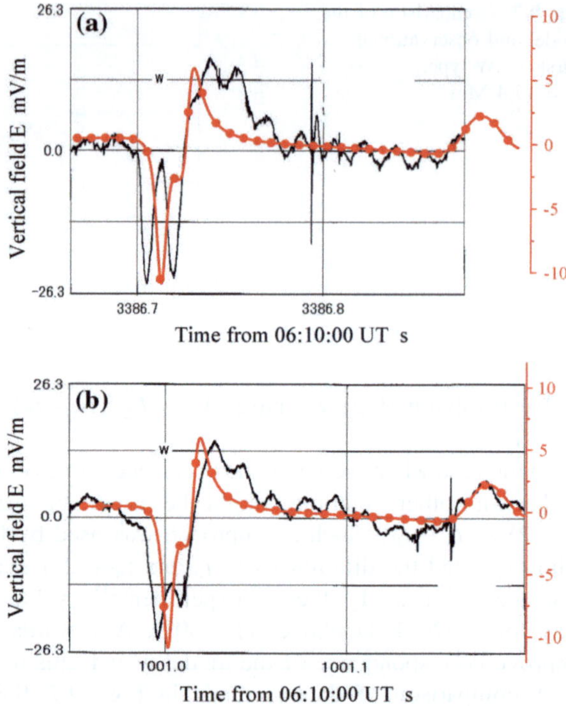

Fig. 9.8 Experimental (*black*) and model (*red*) waveforms of ELF transients: **a** $D = 18.0$ Mm and **b** $D = 18.3$ Mm

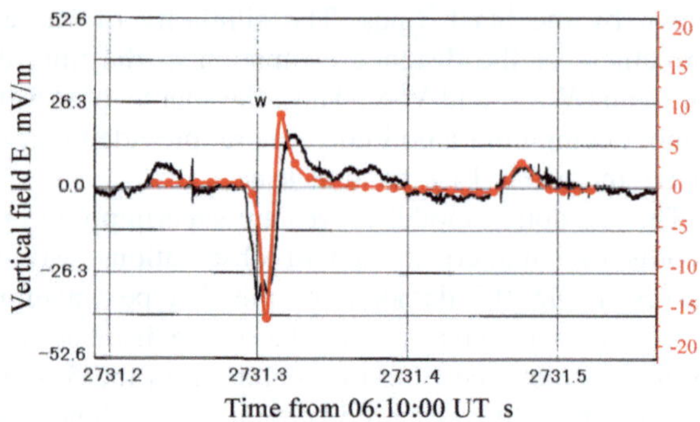

Fig. 9.9 Experimental (*black*) and model (*red*) ELF pulses at 19.2 Mm distance from the source

In Fig. 9.8 we compare observed and the model pulses computed for two distances of 18.3 (a) and 18.0 Mm (b). Again, the reciprocity of model and experiment is obvious. By comparing the plots, we note that the median sub-peak in the pulses of the W-type is more pronounced in the experimental data than in the model. Time separation of the computed peaks of the direct and antipodal waves is noticeably smaller in Fig. 9.8 than that observed experimentally. These departures indicate that the real source–observer distance was somewhat greater than the range evaluated directly from the records, and we will correct this departure below.

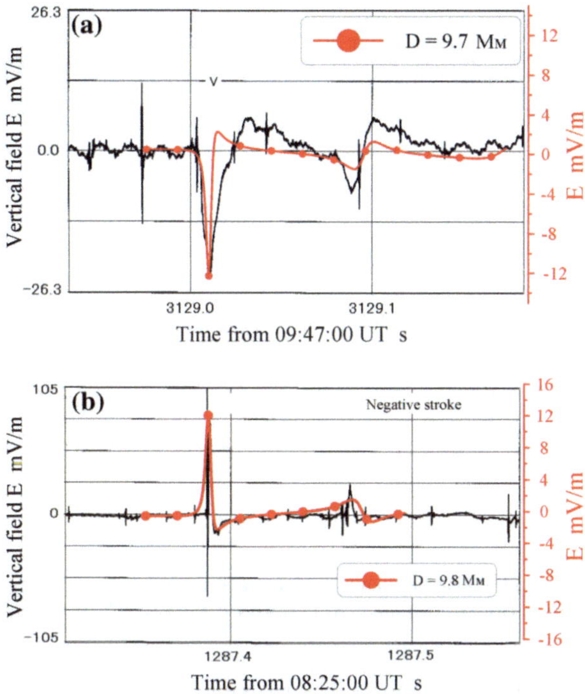

Fig. 9.10 Experimental and model pulses for equatorial source distances: **a** distance from the positive lightning stroke is 9.7 Mm, **b** the negative stroke distance is 9.8 Mm

Both of the frames of Fig. 9.8 demonstrate that ELF background signal is indeed a superposition of many pulses arriving from discharges all over the globe. The powerful atmospherics are clearly seen including those with the so-called "slow" tails. These latter are pulses arriving from distances of a few Mm and having two spectral peaks: one is in a few kHz band, and the other in 10–100 Hz. This is why the slow tail atmospherics have a compact high frequency "head" followed by a slow half-wave or "tail". This kind of signal was actively investigated in the middle of XX century (e.g. Al'pert 1973, 1990).

We continue a comparison of theory and observations in Figs. 9.9 and 9.10. The source distance is close to the antipodal (19.2 Mm) in Fig. 9.9, and it is an "equatorial" one (6.7 and 9.8 Mm) in Fig. 9.10. In the first case shown in Fig. 9.9, the middle sub-pulse is hardly visible in the experimental W-type waveform. It is absent in the computed pulse: the severe interference of direct and antipodal waves has killed the model sub-pulse. In a case of equatorial distance to the source, the direct and antipodal waves are well separated in time, so that the observed pulse corresponds to the V–V type.

Waveforms presented above suggest that there is a high similarity between the observation and computation for all types of waveforms. All records have the "correctly positioned" direct and antipodal waves combined with appropriate arrival times of the round-the-world waves. Nevertheless, a detailed comparison reveals some discrepancy, which might arise from the following reasons:

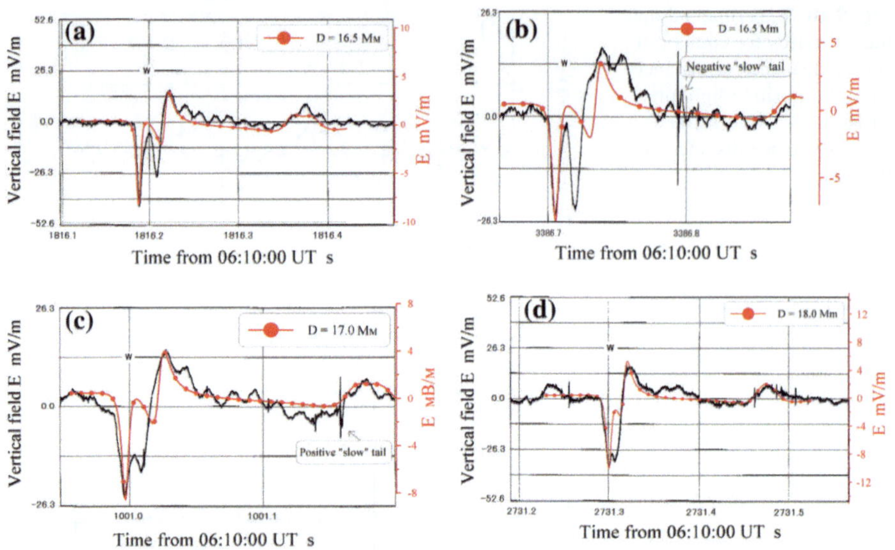

Fig. 9.11 Observed and model waveforms after matching the source—observer distances: frames (**a**), (**b**), (**c**), and (**d**) correspond to Figs. 9.7, 9.8a and b, and 9.9

- Distances evaluated experimentally are not exact, so that deviations will be reduced when an improved source–observer distance is used in the computation.
- The propagation model *v*(*f*) was oversimplified.
- The source current moment was a function of frequency, which was not included into the computation.

We will fit the observational data below and see to what extent such steps would be efficient.

First of all, we must note great source distances pertinent to the majority of recorded pulses: visually selected pulses arrived from the antipodal distances. This singularity is explained by the signal focusing in the spherical cavity shown in Fig. 9.6. Simultaneously with increasing amplitude, the long distance pulses are characterized by closer amplitudes of direct and antipodal waves. These two features might play a role in the preliminary selection. We demonstrate in Fig. 9.11 that the reciprocity increases between the model and experimental data when we slightly modify the source distance. Such "tuning" allowed for matching elements of the waveforms. In particular, the intermediate model sub-pulses became more pronounced, while the peaks of direct and antipodal waves became practically coincident.

One may observe that the following changes were necessary to improve the correspondence of waveforms: *a*—17.4 → 16.5 Mm ($\delta D = -0.9$ Mm); *b*—18.3 → 17.0 Mm ($\delta D = -0.7$ Mm); *c*—18.0 → 17.0 Mm ($\delta D = -1.0$ Mm); and *d*—19.2 → 18.0 Mm ($\delta D = -0.8$ Mm). The distance corrections are smaller than 1 Mm, and this agrees with the commonly accepted accuracy of the source location with the help of ELF radiation. One may conclude that the time domain location of the parent discharges of ELF transients has the same standard deviation of ~ 1 Mm as the technique in the frequency domain.

Waveforms shown in Fig. 9.10 also might be fitted, however this will demand a much more complicated procedure than the distance variation. As we see in Fig. 9.10, positions of model peaks coincide with the experiment, however, the computed waveforms have narrower peaks in comparison with observations. This detail indicates a "red" source spectrum that should be used instead of the "white" one in the computation. The signal from the "red" source will have wider peaks better corresponding to the records. The relevant fitting procedure might be organized in the frequency domain. Similarly to the previous section, we could evaluate the source spectrum and find the corresponding waveforms. Unfortunately, only graphical information was saved in the experiments, so that finding the source spectra was cumbersome.

The other model $\nu(f)$ dependence might be necessary in the detailed comparison with similar observations. For instance, a more sophisticated dependence than a linear $\nu(f)$ might be used. One could desire to account for the global non-uniformities of the ionosphere or its anisotropy. As we see from the data presented above, the correspondence of the model and experiment is exceptionally high especially after matching of the source distance. This is the reason why the conclusion must be made that further modifications including those in the propagation constant would be redundant. Deviations between the theory and experiment are compatible with the accuracy of measurements (STD – standard deviation about 10 %). The latter depends on the ratio of the discrete pulse amplitude to the concurrent SR background. Besides, we must keep in mind irregularities arising from the sporadic nature of lightning strokes.

We used a rare opportunity of comparing the model computations with the unique records of ELF transients in the exceptionally wide frequency band with high sampling frequency and without notch filters. Similarity of the model and observational waveforms validated the uniform model of the Earth–ionosphere cavity bounded by an isotropic ionosphere and having the propagation constant linearly varying with frequency. It also appeared that the time domain heuristic technique for the source distance estimation is as accurate as the commonly used spectral methods.

9.3 Parametric ELF Burst Caused by the Galactic Gamma Flare

We considered in Chap. 6 the impact of a cosmic gamma flare on the spectrum of the SR background. In this section we address the ELF transient associated with the gamma flare. Gamma rays arrived from the magnetar *SGR* 1806–20 on December 27, 2004 and abruptly modified the lower ionosphere, so that sudden amplitude and phase changes were recorded in the very low frequency (VLF; 3–30 kHz) radio transmissions over the Pacific Ocean (Inan et al. 2007). Prior to modifications of the man-made radio signals, an increase was noted in the level of natural VLF noise. The illuminated hemisphere practically coincided with the dayside of the globe: the sub-flare point (20.4° S and 146.2° W) was only 450 km away from the center of the day hemisphere (Hurley et al. 2005; Mereghetti et al. 2005; Palmer et al. 2005; Terasawa et al. 2005; Frederiks et al. 2007). The onset of the flare was reported by GEOTAIL satellite on 21:30.26.35 UT (e.g. Terasawa et al. 2005), but the onset time was re-calculated at the center of the Earth being 21.30:26.65 UT (Tanaka et al. 2011).

A parametric ELF electromagnetic pulse was addressed by Nickolaenko and Hayakawa (2010), Nickolaenko et al. (2010, 2012), and Nickolaenko (2011), which was detected at different SR observatories worldwide (Tanaka et al. 2011). We present all available experimental records of this pulse. The field sites were: Moshiri, Japan (44.37° N; 142.24° E); Onagawa, Japan (38.43° N; 141.48° E); Esrange, Sweden (67.83° N; 21.1° E); Karimshino, Russia (52.83° N; 158.13° E); Nagycenk, Hungary (47.6° N; 16.7° E); and Hornsund, Spitzbergen (77.0°N; 15.5°E). The Esrange, Onagawa and Karimshino sites recorded two orthogonal horizontal magnetic field components, and two magnetic components and the vertical electric field were recorded at Moshiri (Hobara et al. 2000; Ando et al. 2005), Nagycenk and Hornsund. All records were supplied with the GPS time stamps, so that the time axis of individual records might deviate by a few milliseconds utmost. Magnetic antennas at all observatories are parallel to the geography cardinal directions, except at the Karimshino observatory where they are oriented along the geomagnetic coordinates. Owing to the small −6.5° declination, the component D at Karimshino practically coincides with the field H_{WE} (H_X) and the component H is the H_{SN} field (H_Y). The vertical electric field is helpful for establishing the source polarity and the unambiguous source bearing. However, none of observatories presented reliable simultaneous records of all three field components.

Preliminary model computations have yielded the following characteristic properties of the 'parametric' pulse:

1. It arrives around the time of gamma flare.
2. The signal contains only the lowest SR mode.
3. The source polarity must be positive, which means that the onset of the pulse must be negative in the vertical electric field component.
4. Polarization of the pulsed magnetic field must be linear, provided that modification of the global circuit was symmetric.
5. The wave arrival azimuth must point to the center of modification (20.4° S and 146.2° W).
6. Amplitude of the pulse must noticeably exceed the SR background.

We show in Fig. 9.12 fragments of the records observed at six observatories. The time (UT) is shown along the abscissa in the ±1.5 s vicinity of 21:30:26.5 UT. The upper plot shows the vertical electric field component at the Nagycenk

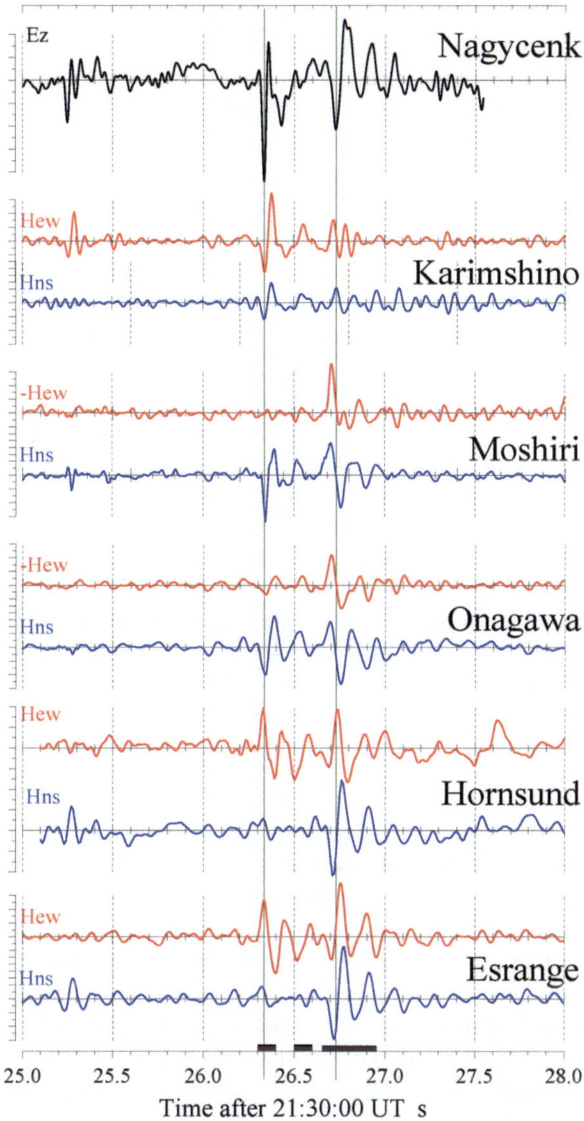

Fig. 9.12 ELF records at different SR observatories. The universal time is shown on the abscissa in the vicinity of the onset of galactic gamma flare

observatory. Magnetic field records at other sites are assembled in pairs. Field amplitudes are plotted in arbitrary units. We changed the sign of the H_{WE} component at Onagawa and Moshiri stations since deviations of sign were noted in previous works when atmospherics were analyzed arriving from the known sources (Schekotov et al. 2007, 2008, 2011). Prior to plotting data in Fig. 9.12, we processed the "raw" records with the help of the low-pass filters and the Singular Spectral Analysis (SSA) or the Caterpillar procedure (Danilov and Zhiglyavsky

1997; Troyan and Hayakawa 2002). This removed remnants of 50 Hz interference and the high frequency noise.

Figure 9.12 demonstrates an outstanding similarity of data collected worldwide, although we notice some departures in the patterns. Deviations originate from different propagation conditions and from the local noise. Of course, differences in particular frequency characteristics of receiving equipments might play some role, however, we use these records as they were.

We place three thick black horizontal bars at the time axis of Fig. 9.12. Each marker denotes a section containing a distinctive pulsed fragment: 26.3–26.4; 26.5–26.6; and 26.7–26.95 s. The wave arrival angle (source azimuth A_Z) and the source triangulation were performed in all these sections by using the waveforms of horizontal magnetic fields. The azimuth A_Z is the angle between the local south–north direction (Y-axis) and the major axis of the wave polarization ellipse. The azimuth is counted clockwise. Polarization characteristics were computed through the elements j_{YY}, j_{XX}, and j_{XY} of the coherence matrix, See (7.5) and (8.3), which were derived from the spectra of the two field components (Fowler et al. 1967).

We applied a special wavelet technique elaborated by Schekotov et al. (2007, 2008) for finding the source bearing. Instead of using the Fourier transform, they applied the expansion upon the complex Gauss wavelet of the third order. The auxiliary angle θ between the X-axis (directed from west to east) and the direction of the major axis of the polarization ellipse was found by using components of the coherence matrix (see Chap. 7):

$$\tan(2\theta) = \frac{2\mathrm{Re}\{H_X \cdot H_Y^*\}}{|H_Y|^2 - |H_X|^2} \tag{9.3}$$

The ensemble of azimuths was estimated for every frequency (the inverse wavelet scale). The median source bearing was estimated by averaging over the frequencies in the 3–20 Hz band. Concurrently, the STDs were found of the source azimuth. The averaged arrival angles were used for constructing the azimuthal histograms with the 10° resolution. The results of every observatory were also presented in a form of the lobes superimposed on the global map, see below. The signal ellipticity e_l, was found as we have done it in previous chapters. The average ellipticity and its standard deviation were derived similarly to the source azimuth: for each sample the averaging was performed over the frequencies of 3–20 Hz.

Figure 9.13 depicts the results of processing in the ±0.5 s vicinity of the arrival time moment 21:30:26.5 UT. The upper frame depicts the Lissajous figures (phasors) of the horizontal magnetic field. One may observe that records carry a typical problem of the Q-bursts: the pulsed signals are elliptically polarized. Thus, establishing the source bearing was a complicated task that is fulfilled by using the above-mentioned source bearing technique. The plots below show (from top to bottom) the recorded waveforms of horizontal magnetic field components, statistics of the source bearings, the field ellipticity, and the wavelet spectra of the complete magnetic field amplitude.

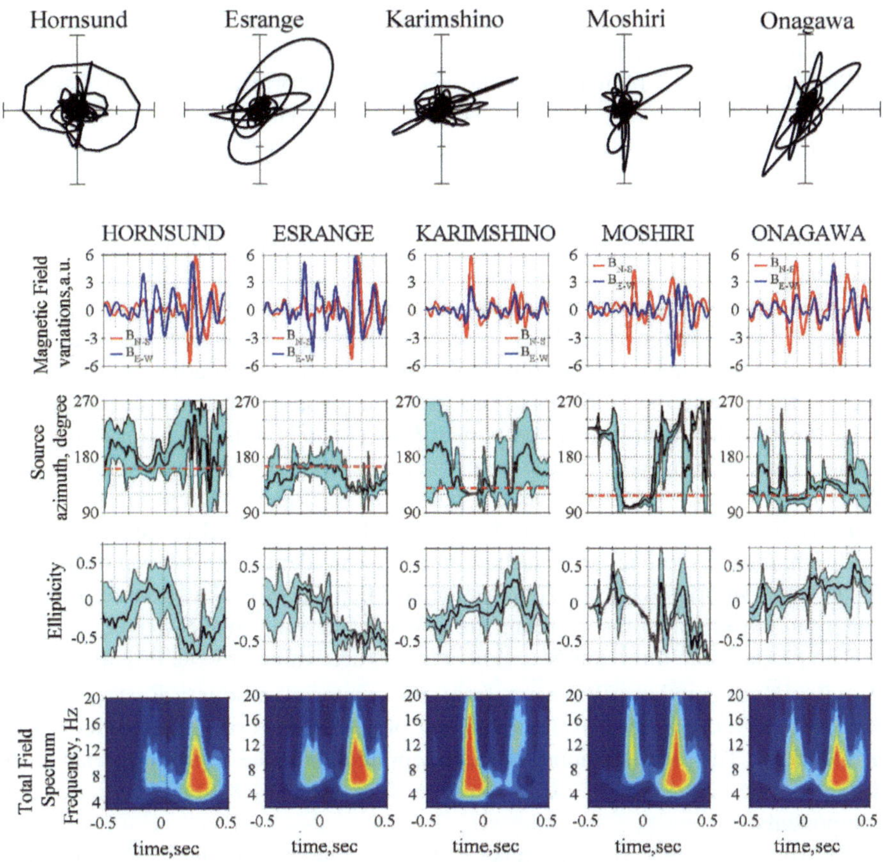

Fig. 9.13 Results of the signal processing in the ±0.5 s vicinity of the gamma flare, $t = 0$ corresponds to 21:30:26.5 s UT

The second line of plots in Fig. 9.13 repeats temporal variations of magnetic fields. The plots demonstrate the reciprocity of dynamic characteristics to specific fragments of the record. The vertical dotted lines correspond to $t = 0$ (the 26.5 s time moment). The third line of plots depicts statistics of the wave arrival azimuth. The thick black line shows the evolution of the average source azimuth. The horizontal red dash-dot lines show the direction from a particular observatory to the sub-flare sources:point. The blue area around the average azimuth indicates the ±STD zone. One may observe that regardless the complicated Lissajous patterns, the experimental data contain intervals of rather stable source bearings associated with the particular pulsed waveforms, and these tend to indicate the sub-flare point.

The fourth row of plots in Fig. 9.13 shows variations of the field ellipticity: the median values are presented by thick black lines and the blue area outlines the standard deviation. The lower frames demonstrate wavelet spectra of the complete magnetic field (in arbitrary units). The frequency band (the inverse wavelet scale) is shown ranging from 2 to 20 Hz.

Plots of Fig. 9.13 indicate that the wavelet transform and relevant statistical processing allow for establishing the intervals of stable wave arrival angles pointing toward the sub-flare point. Within these intervals of stability, the horizontal magnetic field vectors tend to be linearly polarized. Dynamic spectra of the horizontal magnetic field contain two distinct pulses in the time intervals 1 and 3. The field intensity of both these pulses is concentrated around the first SR frequency.

When determining the wave arrival angle by using two orthogonal components of the horizontal magnetic field (goniometer method) (Hayakawa and Ohta 2006; Hayakawa 2009), one obtains the dual source bearings deviating by 180°. The correct direction to the field source can be found by accounting for the polarity of pulse onset in the electric field component $E(t)$. Actually, this is what is done when we use the Poynting vector. The components $H_X(t)$ and $H_Y(t)$ were registered at many observatories, while the electric pulse was recorded at Nagycenk having the negative half-wave at its initiation. Since the polarity of the pulsed vertical electric field remains the same all around the globe, we can identify the source bearing for all the observatories with horizontal magnetic field records. To choose the source direction, we establish the quadrant from which the radio wave arrived.

The propagation direction of a radio wave is determined by the Poynting vector $\vec{P} = \vec{E} \times \vec{H}$ orientation. We have from the vectorial product: $P_X(t) = -E(t) \cdot H_Y(t)$ and $P_Y(t) = E(t) \cdot H_X(t)$, and the list of polarities is readily obtained shown in Table 9.1. This table demonstrates correspondence of the source quadrant to the different combinations of pulse polarities in the horizontal magnetic field components.

By using Table 9.1, one can readily find the source quadrant observed from a particular field-site. The observed polarity of the pulse onsets in magnetic field records must be used for the purpose. The relevant results are summarized in Table 9.2. The first columns of this table lists the observatories, and the second and third column indicate the polarities of pulse onsets in the $H_X(t)$ and $H_Y(t)$ field components. One may see that polarity of the pulse in $H_X(t)$ record was too small at Moshiri to derive its polarity. For the rest of observatories, the fourth column shows the source quadrant found from the signs of pulse onsets, and the fifth column lists the azimuths of the sub-flare sources:point taken at the particular sites.

Table 9.1 Dependence of the source position on the polarity of the pulse onsets in the $H_X(t)$ and $H_Y(t)$ components

Polarity of the pulse onset $E(t)$	–	–	–	–
Polarity of the pulse onset $H_X(t)$	+	+	–	–
Polarity of the pulse onset $H_Y(t)$	–	+	–	+
Polarity of the pulse onset $P_X(t)$	–	+	+	–
Polarity of the pulse onset $P_Y(t)$	–	–	+	+
Quadrant where the source is found	**I**	**II**	**III**	**IV**

Table 9.2 The source quadrant derived for the first parametric pulse

Observatory	Polarity of pulse onset		Source	Geometrical
	$H_X(t)$	$H_Y(t)$	Quadrant	SRC azimuth
Esrange	+	+	II	−16°
Hornsund	+	+	II	−33°
Karimshino	–	+	III	129°
Moshiri	?	–	?	117°
Onagawa	–	–	III	117°

Comparison of the last two columns indicates that the ELF records at distant observatories correspond to the source centered at the sub-flare point. This is why we will consider the Pacific Ocean in what follows.

A more sophisticated analysis is possible based on the cross-Poynting vector, which implies the $E(t)$ realization from the Nagycenk observatory and the $H_X(t)$ and $H_Y(t)$ records from all other observatories. The cross-Poynting vector cannot be regarded as a rigorous field characteristic. However, its temporal variations, especially, the relevant Lissajous figures are exceptionally informative when analyzing the mutual correspondence of experimental data. We collected the relevant plots for every observatory in Fig. 9.14.

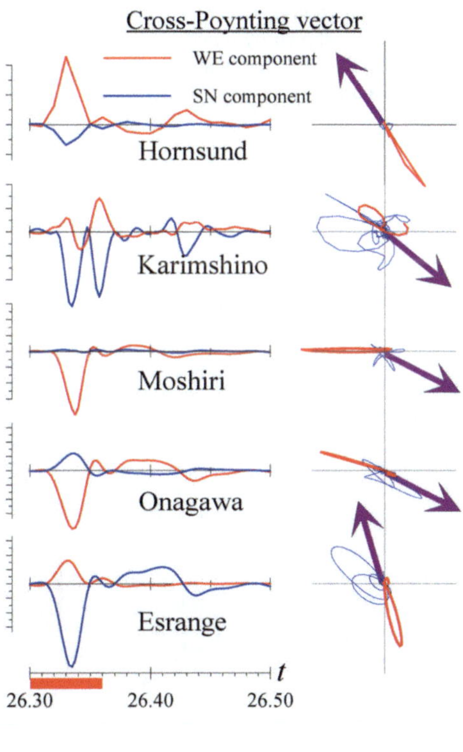

Fig. 9.14 Temporal variations and relevant Lissajous plots of the cross-Poynting vectors derived for all observatories

The right plots in Fig. 9.14 depict the temporal dependence of orthogonal components of the cross-Poynting vector. The $P_X(t) = -E(t) \cdot H_Y(t)$ component is shown in red line and the $P_Y(t) = E(t) \cdot H_X(t)$ projection is shown in blue line. The time interval was chosen corresponding to the arrival of the first, "early" pulse. The left plots show the Lissajous figures of the cross-Poynting vector. These are shown in thin blue lines in the whole interval of 0.2 s. The thick red loops show variations pertinent to the time interval 26.3–26.36 s. The relevant interval is marked at the time axis by the horizontal red bar, and this is the time when the fist half-wave of the pulsed ELF radio signal arrived. The violet arrows depict the bearings of the sub-flare point from every observatory. Since the Poynting vector is directed away from the field source, the red loops must be opposite to the violet arrows, which we really observe. Thus, the field records confirm that the first pulse actually has arrived from the sub-flare point positioned in the Pacific. One has to admit that all these features agree with the model predictions for the parametric ELF pulse.

After obtaining the set of source bearings and the area where the source was positioned, we can perform the source triangulation. Three numbered panels in Fig. 9.15 depict angular histograms of the source bearings obtained at every

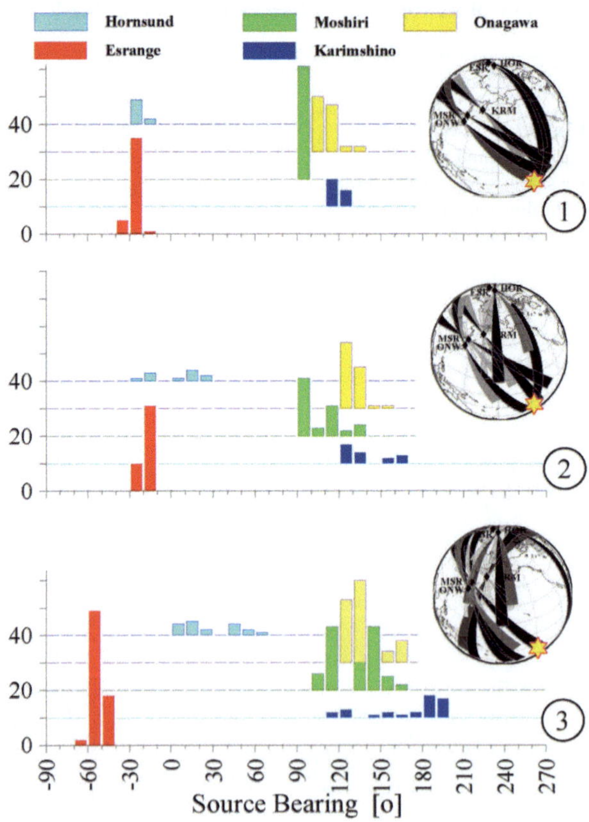

Fig. 9.15 Angular histograms of the source bearings. *Circular* insets show the globe and the *source triangulation* from the observatories shown by *black diamonds*. The *star* marks the sub-flare point

observatory. The abscissa depicts the source azimuth in degrees varying with a step of 10°. The ordinate shows the number of samples when the source bearing was found within the particular sector. The histograms from the sites differ in color and the circled numbers denote the time intervals [−0.2; −0.1] s, [0.0; 0.1] s, and [0.2; 0.45] s correspondingly. We compare the results of source triangulation with the sub-flare point in the circular diagrams of Fig. 9.15 for the three characteristic time intervals.

Each circular inset in Fig. 9.15 illustrates the spherical projections of triangulation based on the relevant angular histograms. The length of the lobes and their darkness reflect the number of source bearings falling into the particular sector. Positions of the field-sites are shown by the black diamonds, and the star marks the sub-flare point. The abbreviations ESR, KRM, MSR, NGK, HRN, and ONG correspond to the observatories Esrange, Karimshino, Moshiri, Nagycenk, Hornsund, and Onagawa. One may observe that the lobes of diagrams tend to intersect in a close vicinity of the sub-flare point for the first and in the second time intervals (the direct and antipodal waves of the first pulse). Serious scatter is present in the third time interval (the second pulse). To clarify geometry of the source triangulation, we overlaid the individual histograms in Fig. 9.16.

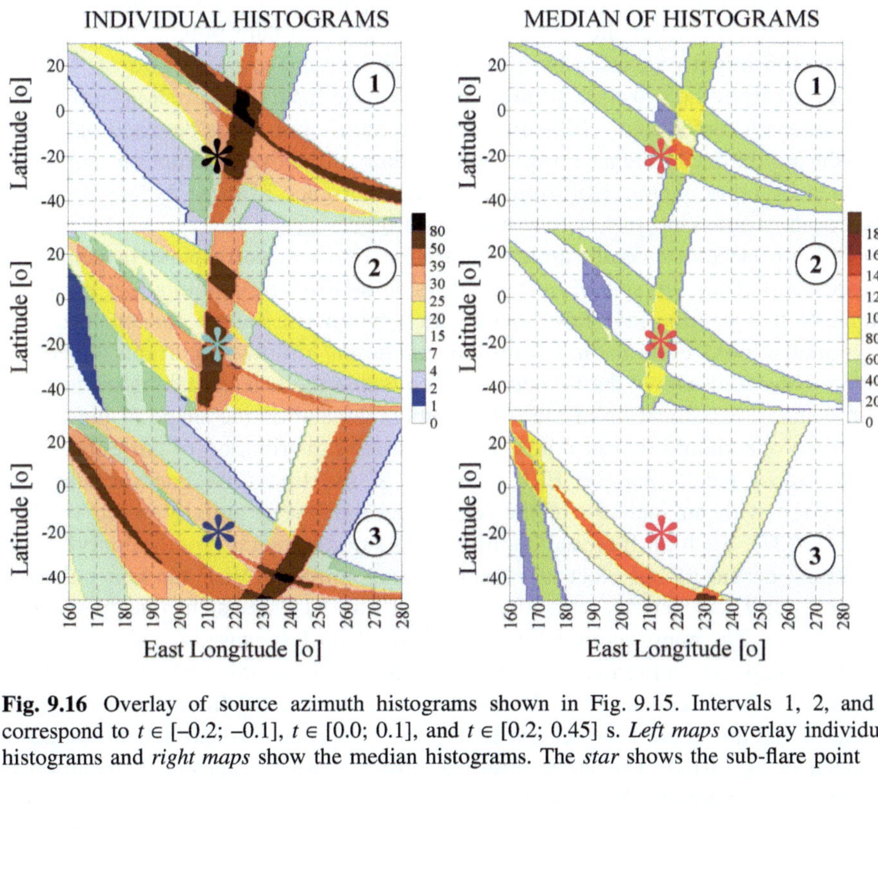

Fig. 9.16 Overlay of source azimuth histograms shown in Fig. 9.15. Intervals 1, 2, and 3 correspond to $t \in [-0.2; -0.1]$, $t \in [0.0; 0.1]$, and $t \in [0.2; 0.45]$ s. *Left maps* overlay individual histograms and *right maps* show the median histograms. The *star* shows the sub-flare point

Figure 9.16 illustrates the global maps in the Cartesian projection. The east longitude is shown along the abscissa in degrees ranging from 160 to 280° E (280° E corresponds to 80° W). The latitude is plotted along the ordinate in degrees, with the positive northern latitude. The star in every frame marks the sub-flare point of 20° S and 146° W. The left maps contain the overlay of individual histograms of Fig. 9.15, and the right maps imply only the median, "centroids" of histograms.

To obtain the maps of Fig. 9.16, we moved along the globe with the 1° step and computed the bearings from every observatory. This allowed for summation of the relevant numbers of individual source bearing histograms (or of their "centroids"). Thus, moving step by step, we obtained the intersections of histograms and constructed the 2D maps (for the fixed time interval). The color indicates how often a given point of the globe is met in the observed source bearings. In a sense, maps of Fig. 9.16 reflect the "spatial distribution" of the source activity. One may expect in an ideal case that the profile of the map reflects spatial distribution of the source currents. The word "ideal case" means that the ratio of the pulse amplitude to the background noise is large. As one may observe from Fig. 9.12, the pulse amplitudes exceeded the SR background by a factor of 4–5, which is not very high.

Nevertheless, individual histograms tend to intersect in the close vicinity of the sub-flare point in the first and the second time intervals, while they are scattered in the third interval. One may suggest that the parametric "source activity" was fractured and its center did not coincide with the sub-flare point. The structured distribution might be caused by a relatively low signal to noise ratio of ~ 5, and deviations from the sub-flare point might also arise from the natural asymmetry of the source current system. Still, since the parametric pulse was detected only once and during the most intense cosmic gamma flare, all above assumptions should be verified by other events. Unfortunately, the occurrence of these events is beyond the reach of mankind power (Or, maybe, fortunately?).

ELF observations demonstrate that source bearings correspond in general to expectation. They are spread over the dayside hemisphere and this fact indirectly suggests that gamma rays have caused the asymmetric source currents. Such an asymmetry looks quite natural from the standpoint of upward currents "feeding" the global circuit above the thunderstorm cells and the leakage taking place in the fair weather. Modification of the leakage must depend on the shape of the ground relief, so one should expect a non-uniform spatial distribution of currents.

We must recall that the ELF transient of the first time interval might be associated with an increase in the VLF electromagnetic activity noted by Inan et al. (2007). Such an increase could be explained by some "preliminary" modifications of the global electric circuit causing multiple VLF random pulses or "noise", and this activity was integrated into an ELF transient due to its huge wavelength. The major gamma flux that arrived later has caused the abrupt change in the ionosphere and relevant "trimpi" events in the records of VLF transmitters (Inan et al. 2007; Tanaka et al. 2008). It also modulated SR frequencies treated in Chap. 6 and has caused the second ELF pulse (Nickolaenko and Schekotov 2011; Tanaka et al. 2011). The latter originates from a novel transient distribution of parametric currents in the atmosphere.

It is interesting to compare the model and observed pulses. We apply the simplest pattern of parametric currents caused by the ionizing radiation uniform in space (Nickolaenko 2011). Air conductivity between the ionosphere and the ground suddenly increases giving rise to the leakage current. We assume that the cosmic radiation suddenly increases the leakage current from its regular density of 1–2 pA/m^2. Modification lasts for 1/60 of a second, and the current returns to its regular value afterwards. The elementary current moment of the ‘switch on–switch off’ type has the constant spectrum in the frequency band of SR. According to VLF observations, gamma rays covered at least a ‘cap’ of 60° radius, and we accept this size in the computation. The elementary current moment (relevant to the air column of 1 m^2 cross-section) is a product of the ionosphere height (60 km) and the surplus current density (1 pA/m^2). An elementary source provides the elementary field at the observatory, so that the complete field is an integral over all sources. The complex spectra were used in the computation with an account for the delays caused by the Earth’s curvature. We used the uniform cavity model and the expressions for the electromagnetic field components provided by the source of current moment $M_C(\omega) = 1$ (see Chap. 14). The simple heuristic linear dependence of the propagation constant was used: $\nu(f) = (f - 2) / 6 - i{\cdot}f / 100$.

The field produced by distributed currents is a convolution of elementary spectra with the spatial distribution of the source current moment. The procedure has the clear physical meaning. The parametric source is formed by many small elements, and each of them contributes to the resulting field. Elementary fields arrive at the observatory with individual amplitudes and phases. By summing them, we obtain resulting spectra of a particular field component: the convolution of elementary fields. We must have in mind that owing to a huge wavelength, an elementary source must be small in comparison with λ, but it is large in the common sense.

The parametric source with size comparable with the hemisphere can excite only the lowest SR modes because the waves from its different sections interfere destructively at higher frequencies. Excitation at lower frequencies plays an important role in distinguishing the parametric pulse from other ELF transients.

Computations also show that the great size of source ‘equalizes’ the amplitudes of two orthogonal horizontal magnetic field components. For instance, the H_X and H_Y amplitudes at Karimshino become close while the sub-flare point is positioned at the ~130° azimuth corresponding to the 2:1 ratio. In modeling, we did not assume the uniform frequency response of the electric channels in the receivers. Instead, we accounted for a typical gain of the electric antenna (see Chap. 2):

$$G_E = \frac{i\omega RC_0}{1 + i\omega R(C_0 + C_1)} \tag{9.4}$$

Here $C_0 = 25$ pF is the self-capacitance of the antenna active electrode (a sphere of ~55 cm diameter), $C_1 = 50$ pF is the input capacitance of antenna pre-amplifier, and $R = 10^9\ \Omega$ is the input resistance of antenna pre-amplifier. Such an

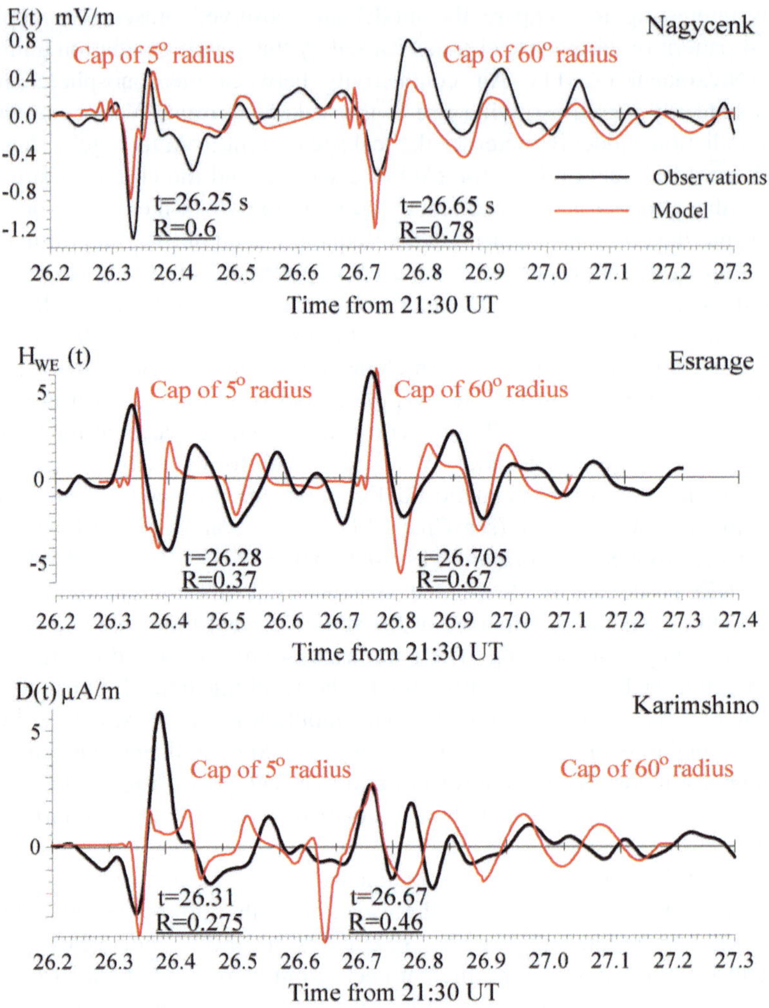

Fig. 9.17 Comparison of observed and model pulsed waveforms

antenna is a high-pass filter, and the observed pulse is slightly deformed, because its ULF component is removed.

A detailed comparison of the observed and model waveforms is made in Fig. 9.17. We show the records and model for Nagycenk (*E* field), Esrange (H_{WE} field), and Karimshino (*D*-field component). The time is shown on the abscissa from 21:30:26.2 to 21:30:27.4 s. The signal amplitude is shown on the ordinate. The vertical electric field component was measured in mV/m, and we used the μA/m for magnetic components. This measure is widely used in radio engineering because the product of the *H*- field in μA/m with the free space wave impedance of

$120\pi\ \Omega$ provides the electric field in μV/m. There is an equality: 1 μA/m = 0.4π pT = 1.256 pT.

Experimental waveforms are shown in Fig. 9.17 in black thick lines, and the model data are depicted in red curves. By moving the model pulses against the time and computing the cross-correlation coefficients, we estimate the arrival time of pulses at every site. These are listed in Fig. 9.17. The cross-correlation coefficients vary in time, and for finding the arrival time we have chosen the highest absolute value of cross-correlation coefficient. These parameters are also listed in the figure. The arrival times obtained match fairly well at all observatories.

The onset period of the first pulse was remarkably shorter in all records. To obtain such a narrow pulse, we had to reduce the radius of modification to 5°. Simultaneously, modification of the leakage current density was increased by an order of magnitude (to 30 pA/m^2) for obtaining the appropriate pulsed amplitude. The second pulses were computed with the modification of the 60° radius and for the transient current of 1 pA/m^2 amplitude.

Correspondence of the model and experimental waveforms is evident. The first pulse arrived 0.3–0.4 s prior to the major gamma flux and the second one was detected almost simultaneously with the characteristic 'trimpi' event in the trans-Pacific VLF radio transmissions. The model and observed ELF waveforms match each other, and deviations present in Fig. 9.6 might be attributed to an impact of the natural background noise.

We demonstrated that the most intense galactic gamma flare of Dec. 27, 2004 was associated with the Q-bursts detected worldwide. Close matching of arrival times, the single mode SR spectra, the source polarity, the wave arrival angle, and the linear polarization allow us to conclude that both pulses are feasible to arise from the modification of the global electric circuit.

References

Al'pert YL (1973) Radio wave propagation and the ionosphere, vols 1 and 2. Consultants Bureau, New York, [1972, Nauka, Moscow (in Russian)]

Al'pert YL (1990) Space plasma, vols 1 and 2. Cambridge University Press, Cambridge, New York, Melbourne, Sydney

Ando Y, Maltsev P, Sukhynyuk A, Goto T, Yamauchi T, Hobara Y, Sekiguchi M, Ikegami Y, Sera M, Korepanov VE, Hayakawa M (2005) New ELF observation system at Moshiri, Japan and assessment of acquired data. J Atmos Electr 25(1):29–39

Bliokh PV, Nickolaenko AP, Filippov YF (1977) Global electromagnetic resonances in the Earth-ionosphere cavity, Naukova Dumka, Kiev (in Russian), P 199

Bliokh PV, Nickolaenko AP, Filippov YF (1980) Schumann resonances in the Earth-ionosphere cavity, New York, London, Paris, Peter Perigrinus, p 168

Boccippio DJ, Williams ER, Heckman SJ, Lyons WA, Baker IT, Boldi R (1995) Sprites, ELF transients and positive ground strokes. Science 269:1088–1091

Burke CP, Jones DLl (1995) Global radiolocation in the lower ELF frequency band. J Geophys Res 100:26263–26271

Burke CP, Jones DLl (1996) On the polarity and continuing currents in unusually large lightning flashes deduced from ELF events. J Atmos Terr Phys 58:531–540
Danilov DL, Zhiglyavsky AA (eds) (1997) Principal component of the time series: the caterpillar method. St.-Petersburg state University, St-Petersburg, Russia (in Russian), p 307
Fowler RA, Kotick BJ, Elliot RD (1967) Polarization analysis of natural and artificially induced geomagnetic micropulsations. J Geophys Res 72:2871–2875
Frederiks DD, Golenetskii SV, Palshin VD, Aptekar RL, Ilyinskii VN, Oleinik FP, Mazets EP, Cline TL (2007) Giant flare in SGR 1806-20 and its Compton reflection from the Moon. Astron Lett 33:1–18. doi:10.1134/S106377370701001X
Füllekrug M (1999) Global lightning triangulation. In: Proceedings of 11th international conference on atmospheric electricity, Guntersville, Alabama, June 7–11, 1999, pp 709–711
Füllekrug M, Constable S (2000) Global triangulation of lightning discharges. Geophys Res Lett 27:333–336
Füllekrug M, Reising SC (1998) Excitations of Earth-ionosphere cavity resonances by sprite-associated lightning flashes. Geophys Res Lett 25:4145–4148
Füllekrug M, Fraser-Smith AC, Reising SC (1998) Ultra-slow tails of sprite associated lightning flashes. Geophys Res Lett 25:3495–3498
Füllekrug M, Constable S, Heinson G, Sato M, Takahashi Y, Price C, Williams E (2000) Global lightning acquisition system installed. EOS Trans AGU 81(30):333
Greenberg E, Price C (2004) A global lightning location algorithm based on the electromagnetic signature in the Schumann resonance band. J Geophys Res 109:D21111. doi:10.1029/2004JD004845
Greenberg E, Price C (2007), Diurnal variations of ELF transients and background noise in the Schumann resonance band. Radio Sci 42:RS2S08. doi:10.1029/2006RS003477
Greenberg E, Price C, Yair Y, Ganot M, Bor J, Sátori G (2007) ELF transients associated with sprites and elves in eastern Mediterranean winter thunderstorms. J Atmos Solar-Terr Phys 69(13):1569–1586
Hayakawa M (2009) A review on direction finding of VLF/ELF sferics. J Atmos Electr 29(1):35–52
Hayakawa M, Ohta K (2006) The importance of direction finding technique for the study of VLF/ELF sferics and whistlers. IEEJ Trans Fundam Mater 126(2):65–70
Hayakawa M, Nickolaenko AP, Shvets AV, Hobara Y (2011) Recent studies of Schumann resonance and ELF transients. In: Wood MD (ed) Lightning: properties, formation and types, Chap. 3, Nova Sci Pub, pp 39–71
Hayakawa M, Hobara Y, Suzuki T (2012) Lightning effects in the mesosphere and ionosphere. In: Cooray V (ed) Lightning electromagnetics, Chap. 16, Inst. Engineering and Technology, pp 611–646
Hobara Y, Iwasaki N, Hayashida T, Tsuchiya N, Williams ER, Sera M, Ikegami Y, Hayakawa M (2000) New ELF observation site in Moshiri, Hokkaido, Japan and the results of preliminary data analysis. J Atmos Electr 20:99–109
Hobara Y, Iwasaki N, Hayashida T, Hayakawa M, Ohta K, Fukunish H (2001) Interrelation between ELF transients and ionospheric disturbances in association with sprites and elves. Geophys Res Lett 28:935–938
Hobara Y, Hayakawa M, Williams E, Boldi R, Downes E (2006), Location and electrical properties of sprite-producing lightning from a single ELF site. In: Füllekrug M, Mareev EA, Rycroft MJ (eds) Sprites, elves and intense lightning discharges, vol 225. NATO Sci. Ser., Ser. II, Springer, Dordrecht, Netherlands, pp 211–235
Huang E, Williams E, Boldi R, Heckman S, Lyons W, Taylor M, Nelson T, Wong C (1999) Criteria for sprites and elves based on Schumann resonance observations. J Geophys Res 104:16943–16964
Hurley K, Boggs SE, Smith DM, Duncan RC, Lin R, Zoglauer A, Krucker S, Hurford G, Hudson H, Wigger C, Hajdas W, Thompson C, Mitrofanov I, Sanin A, Boynton W, Fellows C, von Kienlin A, Lichti G, Rau A, Cline T (2005) An exceptionally bright flare from SGR 1806–20 and the origins of short-duration big gamma-ray bursts. Nature 434:1098–1103

Inan US, Lehtinen NG, Moore RC, Hurley K, Boggs S, Smith DM, Fishman GJ (2007) Massive disturbance of the daytime lower ionosphere by the giant g-ray flare from magnetar SGR 1806–20. Geophys Res Lett 34:L08103. doi:10.1029/2006GL029145

Jones D Ll (1970a) Numerical computations of terrestrial ELF electromagnetic wave fields in the frequency domain. Radio Sci 5:803–809

Jones DLL (1970b) Propagation of ELF pulses in the Earth-ionosphere cavity and application to slow tail sferics. Radio Sci 5:1153–1163

Jones DLL, Kemp DT (1970) Experimental and theoretical observations of Schumann resonances. J Atmos Terr Phys 32:1095–1108

Lazebny BV, Nickolaenko AP (1976a) Synchronous observations of the ELF bursts in the frequency range of the Schumann resonances. Geomag Aeron 16:121–126 (in Russian)

Lazebny BV, Nickolaenko AP (1976b) Diurnal variations of the ELF bursts number observed synchronously at Kharkov and Ulan-Ude. Geomagn Aeron 16:372–373 (in Russian)

Marshall IH, Hale LC, Croskey CL, Lyons WA (1998) Electromagnetics of sprite and elve-associated sferics. J Atmos Solar-Terr Phys 60:771–786

Marshall RA, Inan US, Neubert T, Hughes A, Sátori G, Bor J, Collier A, Allin TH (2005) Optical observations geomagnetically conjugate to sprite-producing lightning discharges. Ann Geophys 23(6):2231–2237

Mereghetti S, Götz D, vonKienlin A, Rau A, Lichti G, Weidenspointner G, Jean P (2005) The first giant flare from SGR 1806-20: observations using the anticoincidence shield of the spectrometer on INTEGRAL. Astrophys J Lett 624:L105–L108. doi:10.1086/430669

Nakamura T, Sekiguchi M, Hobara Y, Hayakawa M (2010) A comparison of different source location methods for ELF transients by using the parent lightning discharges with known positions. J Geophys Res 115:A00E39. doi:10.1029/2009JA014992

Neubert T, Allin TH, Blanc E, Farges T, Haldoupis C, Mika A, Soula S, Knutsson L, van der Velde O, Marshall RA, Inan U, Sartori G, Bor J, Hughes A, Collier A, Laursen S, Rasmussen IL (2005) Co-ordinated observations of transient luminous events during the EuroSprite2003 campaign. J Atmos Solar-Terr Phys 67:807–820. doi:10.1016/j.jastp.2005.02.004

Nickolaenko AP (2011) Source models for "parametric" Q–burst. J Atmos Electr 31(2):95–110

Nickolaenko AP, Hayakawa M (2002) Resonances in the Earth-ionosphere cavity. Kluwer Academic Publishers, Dordrecht-Boston-London

Nickolaenko AP, Hayakawa M (2010) Model disturbance of Schumann resonance by the SGR 1806–20 γ –ray flare on December 27, 2004. J Atmos Electr 30(1):1–11

Nickolaenko AP, Kudintseva IG (1994) A modified technique to locate the sources of ELF transient events. J Atmos Solar-Terr Phys 56:1493–1498

Nickolaenko AP, Schekotov AYu (2011) Experimental detection of an ELF radio pulse associated with the gamma-ray burst of December 27, 2004. Radiophys Quantum Electron 54(1):2011. doi:10.1007/s11141-011-9268-6

Nickolaenko AP, Hayakawa M, Kudintseva IG, Myand SV, Rabinowicz LM (1999) ELF sub-ionospheric pulse in time domain. Geophys Res Lett 26(999–1):002

Nickolaenko AP, Hayakawa M, Ogawa T, Komatsu M (2008), Q-bursts: a comparison of experimental and computed ELF waveforms. Radio Sci 43:RS4014. doi:10.1029/2008RS003838

Nickolaenko AP, Kudintseva IG, Pechonaya O, Hayakawa M, Nakamura T, Hobara Y, Tanaka Y (2010) Impact of a gamma-ray burst on the Schumann resonance. Radiophys Quantum Electron 53(9–10):542–556. doi:10.1007/3

Nickolaenko AP, Kudintseva IG, Pechony O, Hayakawa M, Hobara Y, Tanaka YT (2012) The effect of a gamma ray flare on Schumann resonances. Ann Geophys 30:1321–1329. doi:10.5194/angeo-30-1321-2012

Ogawa T, Komatsu M (2007) Analysis of Q-burst waveforms. Radio Sci 42:RS2S18. doi:10.1029/2006RS003493

Ogawa T, Tanaka Y, Miura T (1966a) On the frequency response of the ball antenna for measuring ELF noise signals. Spec Contr Geophys Inst Kyoto Univ 6:9–12

Ogawa T, Tanaka Y, Miura T, Yasuhara M (1966b) Observations of natural ELF and VLF electromagnetic noises by using ball antennas. J Geomagn Geoelectr 18:443–454

Ogawa T, Tanaka Y, Fraser-Smith AC, Gendrin R (1967) Worldwide simultaneity of a Q-burst in the Schumann resonance frequency range. J Geomagn Geoelectr 19:377–384

Palmer DM, Barthelmy S, Gehrels N, Kippen RM, Cayton T, Kouveliotou C, Eichler D, Wijers RAMJ, Woods PM, Granot J, Lyubarsky YE, Ramirez-Ruiz E, Barbier L, Chester M, Cummings J, Fenimore EE, Finger MH, Gaensler BM, Hullinger D, Krimm H, Markwardt CB, Nousek JA, Parsons A, Patel S, Sakamoto T, Sato G, Suzuki M, Tueller J (2005) A giant γ-ray flare from the magnetar SGR 1806-20. Nature 434:1107–1109

Price, C., M. Asfur, W. Lyons and T. Nelson (2002), An improved ELF/VLF method for globally geolocating sprite-produced lightning, *Geophys. Res. Lett.*, **29** **(3)**, doi: 10.1029/2001GL013519

Price C, Greenberg E, Yair Y, Sàtori G, Bor J, Fukunish H, Sato M, Israelevich P, Moalem M, Devir A, Levin Z, Joseph JH, Mayo I, Ziv B, Sternlieb A (2004) Ground-based detection of TLE-producing intense lightning during the MEIDEX mission on board the space shuttle Columbia. Geophys Res Lett 31(20):L20107. doi:10.1029/2004GL020711

Sato M, Fukunishi H, Kikuchi M, Yamagishi H, Lyons WA (2003) Validation of sprite-inducing cloud-to-ground lightning based on ELF observations at Syowa station in Antarctica. J Atmos Solar-Terr Phys 65:607–614

Schekotov AY, Molchanov OA, Hayakawa M, Fedorov EN, Chebrov VN, Sinitsin VI, Gordeev EE, Belyaev GG, Yagova NV (2007) ULF/ELF magnetic field variations from atmosphere induced by seismicity. Radio Sci 42:RS6S90. doi: 10.1029/2005RS003441

Schekotov AY, Molchanov OA, Hayakawa M, Fedorov EN, Chebrov VN, Sinitsin VI, Gordeev EE, Andreevsky SE, Belyaev GG, Yagova NV, Gladishev VA, Baransky LN (2008) About possibility to locate an EQ epicenter using parameters of ELF/ULF preseismic emission. Nat Hazards Earth Syst Sci 8:1237–1242

Schekotov AY, Pilipenko V, Shiokawa K, Fedorov E (2011) ULF impulsive magnetic response at mid-latitudes to lightning activity. Earth Planets Space 63(2):119–128. doi:10.5047/eps.2010.12.009

Shalimov SL, Bösinger T (2006) An alternative explanation for the ultra-slow tail of sprite-associated lightning discharges. J Atmos Solar-Terr Phys 68:814–820. doi:10.1016/j.jastp.2005.12.001

Sukhorukov AI, Stubbe P (1997) On ELF pulses from remote lightning triggering sprites. Geophys Res Lett 24(13):1639–1642

Surkov VV, Hayakawa M (2012) Underlying mechanisms of transient luminous events: a review. Ann Geophysicae 30:1185–1212

Tanaka YT, Terasawa T, Yoshida M, Horie T, Hayakawa M (2008) Ionospheric disturbances caused by SGR 1900 + 14 giant gamma ray flare in 1998: constraints on the energy spectrum of the flare. J Geophys Res 113:A07307. doi:10.1029/2008JA013119

Tanaka YT, Hayakawa M, Hobara Y, Nickolaenko AP, Yamashita K, Sato M, Takahashi Y, Terasawa T, Takahashi T (2011) Detection of transient ELF emission caused by the extremely intense cosmic gamma-ray flare of 27 December 2004. Geophys Res Lett 38:L08805. doi:10.1029/2011GL047008

Terasawa T, Tanaka YT, Takei Y, Kawai N, Yoshida A, Nomoto K, Yoshikawa I, Saito Y, Kasaba Y, Takashima T, Mukai T, Noda H, Murakami T, Watanabe K, Muraki Y, Yokoyama T, Hoshino M (2005) Repeated injections of energy in the first 600 ms of the giant flare of SGR 1806–20. Nature 434:1110–1111. doi:10.1038/nature03573

Troyan V, Hayakawa M (2002) Inverse geophysical problems. Terrapub, Tokyo, p 289

Whitley T, Fullekrug M, Rycroft M, Bennett A, Wyatt F, Elliott D, Heinson G, Hitchman A, Lewis A, Sefako R, Fourie P, Dyers J, Thomson A, Flower S (2011), Worldwide extremely low frequency magnetic field sensor network for sprite studies. Radio Sci 46:RS4007. doi: 10.1029/2010RS004523

Chapter 10
Inverse Problem of SR

This chapter describes the inverse electromagnetic problems in the ELF band. In a usual, direct problem, the fields are found for the particular source (or source distribution) in the Earth–ionosphere cavity with definite boundary conditions. Inverse problems exploit the observed electromagnetic fields as initial information and derive the temporal variations of the source intensity, the unknown source distribution, or propagation conditions. We already considered the prevalent inverse problem in Chap. 5. We also treated the inverse problem linked to an ELF transient in Chap. 9. Here, we concentrate on obtaining accurate temporal and spatial properties of lightning sources responsible for the background signal. Initially, we discuss the shortcomings of traditional measurements of the source dynamics and apply a more progressive technique that separates the universal and local time factors involved in the SR records. For this purpose, we use the simultaneous records at three widely separated points. This allowed us to obtain estimates of the diurnal-seasonal variations of the global thunderstorm intensity on the three-year span. The simultaneously obtained local time factors characterize the movement of sources around the globe. The formally rigorous approach is addressed then, which exploits the Tikhonov technique of tomographic reconstruction. Monitoring data of three globally allocated SR observatories are used as the distance projections, and relevant space–time source distributions are obtained. Details of this procedure are discussed, and the output data are compared with the optical observations from space.

10.1 Universal Time and Local Time Variations

We know that the global electromagnetic or SR is maintained by radiation of the planetary thunderstorms. Therefore, the intensity of resonance must reflect the present state of atmospheric electric activity worldwide. So as to reduce an impact of the modal structure of resonant field, the thunderstorm intensity is evaluated from the cumulative SR intensity, i.e., the field intensity integrated in the frequency band covering the three first modes (Polk 1969; Nickolaenko 1997;

A. Nickolaenko and M. Hayakawa, *Schumann Resonance for Tyros*,
Springer Geophysics, DOI: 10.1007/978-4-431-54358-9_10,

Füllekrug and Fraser-Smith 1997; Nickolaenko et al. 1998, 2008, 2011a, b; Nickolaenko and Hayakawa 2002, 2010a). Integral intensity reflects the thunderstorm activity in a better way than the intensity of individual modes; however, it still contains traces of the field modal structure. An alternative approach was used by Fraser-Smith et al. (1991) who recorded the resonance power in the vicinity of 11 Hz. This frequency lies between the first and the second modes, so that the intensity of signal is less dependent on the source–observer distance, as if the field "does not know" to which mode it belongs. The simplicity is an advantage of this method: one has to use only a band-pass filter and a quadratic detector. The shortcoming of such measurements is a relatively low level of resonance signal, so that impact of different interference might be substantial. Both the methods provide the estimates still depending on the distance from the thunderstorms and containing remnants of the modal structure.

A technique was suggested by Sentman and Fraser (1991) to separate the universal and local time factors involved in the SR, provided that the signals were simultaneously recorded at two widely separated points (California and Northern Australia). This technique was further developed by Pechony and Price (2007) who demonstrated that local time variations reflect the source distance rather than the ionosphere height. It was extended and applied by Nickolaenko and Hayakawa (2007) to the experimental data collected at Lehta and Moshiri sites. A conclusion was made that the local time modulations arise predominantly from the source–receiver geometry, while the universal time modulations are driven by changes in the global thunderstorm intensity. Let us present essentials of the technique.

We will extend the number of observatories to three. The concept is based on the following assumptions:

1. Integral SR intensity $P\ (t_{\mathrm{U}},\ \lambda_i)$ at a site with the east longitude λ_i and at the universal time moment t_{U} is a product of the universal function $U\ (t_{\mathrm{U}})$ that depends only on the universal time and the local modulating function $L\ (t_{\mathrm{L}})$ that varies in the local time:

$$P(t_U, \lambda_i) = U(t_U) \cdot L(t_L) \tag{10.1}$$

2. The local time (hr) is related to the universal time and the longitude λ_i (radians) in the following way:

$$t_L = t_U + 12\lambda_i/\pi \tag{10.2}$$

3. The local time modulating function $L\ (t_U)$ is the complex periodic function of time with the period of 24 h. The following Fourier expansion is valid for it at the particular latitude λ_k:

$$L_k(t_U) = 1 + \sum_{n=1}^{N} A_n \cdot \exp\{in\Omega_E(t_U + l_k)\} \tag{10.3}$$

where $l_k = 12\lambda_k/\pi$ is measured in hr and the angular velocity of Earth rotation $\Omega_E = \pi/12$ is given in 1/hr.

4. We accept that the average of local modulation function is equal to unity, and only the real parts Re $\{L\ (t_U)\}$ have a physical meaning.

The above features were postulated by Sentman and Fraser (1991) with an exception that Eq. (10.3) contained only the first term $n = 1$. The complex amplitude A_1 characterizes the daily 24 h component. The semi-diurnal variation has a period of 12 h and amplitude A_2, etc. Higher order terms will be taken into account up to the fifths. The magnitude and initial phase of modulation function $L\ (t_U)$ depend on the complex amplitudes A_n, which satisfy the condition:

$$|A_n|^2 \ll 1. \tag{10.4}$$

Intensities $P_1(t_U) = P\ (t_U,\ \lambda_1)$, $P_2(t_U) = P\ (t_U,\ \lambda_2)$, and $P_3(t_U) = P\ (t_U,\ \lambda_3)$ were monitored at three observatories positioned at the longitudes λ_1, λ_2, and λ_3. The geographic coordinates of sites were: Lehta (64.4° N, 34° E), Moshiri (44.4° N, 142.2 E), and West Greenwich (41.6° N, 71.6° W). There are three possible pairs: Lehta–Moshiri, Lehta–West Greenwich, and Moshiri–West Greenwich. Each pair provides the following identity (Nickolaenko and Hayakawa 2007):

$$\begin{aligned}
\ln(P_k/P_p) &= \ln(L_k) - \ln(L_p) \approx A_1\{\exp[i\Omega_E(t_U + l_k)] - \exp[i\Omega_E(t_U + l_k)]\} \\
&\quad + A_2\{\exp[i2\Omega_E(t_U + l_k)] - \exp[i2\Omega_E(t_U + l_k)]\} + \ldots \\
&= A_1\exp(it_U\Omega_E)\left[\exp(i\Omega_E l_k) - \exp(i\Omega_E l_p)\right] \\
&\quad + A_2\exp(2i\Omega_E t_U)\left[\exp(2i\Omega_E l_k) - \exp(2i\Omega_E l_p)\right] + \ldots
\end{aligned} \tag{10.5}$$

where the indices k and p correspond to the k-th and p-th observatory.

SR was monitored simultaneously at three field sites, and we obtained the monthly averaged diurnal variations collected from August 1999 to December 2001. For this purpose, the intensity $P = (H_{NS})^2 + (H_{EW})^2$ of the horizontal magnetic field components was derived from the records of H_{NS} and H_{EW} fields and integrated over three first resonance modes. This procedure reduces an impact of distance between the observer and thunderstorms and emphasizes the role of the source intensity (see Chap. 5). Thus, the functions $P_k(t_U)$ was obtained that we use in the analysis. The unknown complex amplitudes A_n were found from Eq. (10.5) for each pair of sites by multiplying its both sides by exp $(-in\Omega t_U)$ and integrating over 24 h:

$$A_n = \frac{1}{24}\frac{\int_0^{24} dt_U \ln(P_k/P_p)\exp(-in\Omega_E t_U)}{\left[\exp(in\Omega_E l_k) - \exp(in\Omega_E l_p)\right]} \tag{10.6}$$

One obtains the following relation for the denominator of Eq. (10.6) after introducing the median longitude $\lambda_{\rm kp} = (\lambda_{\rm k} + \lambda_{\rm p})/2$ and the longitude deviation $\Delta\lambda_{\rm kp} = (\lambda_{\rm k} - \lambda_{\rm p})/2$.

$$\exp(in\Omega_E l_k) - \exp(in\Omega_E l_p) = -2i\sin(n\Omega_E \Delta\lambda_{kp})\exp(-in\Omega_E l_{kp}) \qquad (10.7)$$

We demonstrate below that separation of the universal and local time terms provides reasonable results. The $U\,(t_U)$ functions represent the global thunderstorm intensity. The $L\,(t_L)$ variations usually allow for the source proximity interpretation, provided that spatial distribution of the field sources is relatively simple. Sophisticated thunderstorm distributions in space result in excessively complicated $L\,(t_L)$ functions, while the recovered $U(t_U)$ terms continue to replicate the thunderstorm intensity.

By using a pair of records $P_1(t_U)$ and $P_1(t_U)$, one obtains the A_n coefficients of the expansion (10.6). These are substituted into Eq. (10.3) and provide the local time factor $L(t_U)$ for the particular pair. Functions $L_1\,(t_{\rm L})$ and $L_2\,(t_{\rm L})$ are coincident in the local time, but their pattern depends on the particular couple of sites. Local time factors are used in Eq. (10.1) as functions of UT to obtain the universal diurnal variations $U_1(t_{\rm U})$ and $U_2(t_{\rm U})$. Thus, two functions are obtained for a pair of sites:

$$\begin{aligned} U_1(t_U) &= P_1(t_U)/L_1(t_U) \\ U_2(t_U) &= P_2(t_U)/L_2(t_U) \end{aligned} \qquad (10.8)$$

Three dual sets of SR records provide six $U(t_{\rm U})$ diurnal patterns, and we can derive the average universal pattern $\langle U_k(t_U)\rangle$ and its standard deviation. An alternative and simpler signal processing was suggested by Nickolaenko and Hayakawa (2007) based on the geometric averaging of original records. There are three 'dual' estimates:

$$\begin{aligned} G_{1,2} &= \sqrt{P(t_U,\lambda_1)\cdot P(t_U,\lambda_2)} \propto U_{1,2}(t_U) \\ G_{1,3} &= \sqrt{P(t_U,\lambda_1)\cdot P(t_U,\lambda_3)} \propto U_{1,3}(t_U) \\ G_{3,2} &= \sqrt{P(t_U,\lambda_3)\cdot P(t_U,\lambda_2)} \propto U_{3,2}(t_U) \end{aligned} \qquad (10.9)$$

and one "triple" estimate of the current intensity of the global thunderstorms:

$$G_3(t_U) = \sqrt{P(t_U,\lambda_1)\cdot P(t_U,\lambda_2)\cdot P(t_U,\lambda_3)} \propto U_{1,2,3}(t_U) \qquad (10.10)$$

The physical background of geometric averaging is rather simple: when thunderstorms approach one of the globally separated observatories, they simultaneously retreat from the other. Therefore, the product of resonance intensities is less dependent on the source distance, or on the L functions. We demonstrate below that the geometric averaging is equally efficient for estimating the global thunderstorm activity.

Figure 10.1 depicts the 'initial' data recorded at three field sites. The year/moth information is shown on the abscissa. Daily variations $P(t_U)$ are shown on the

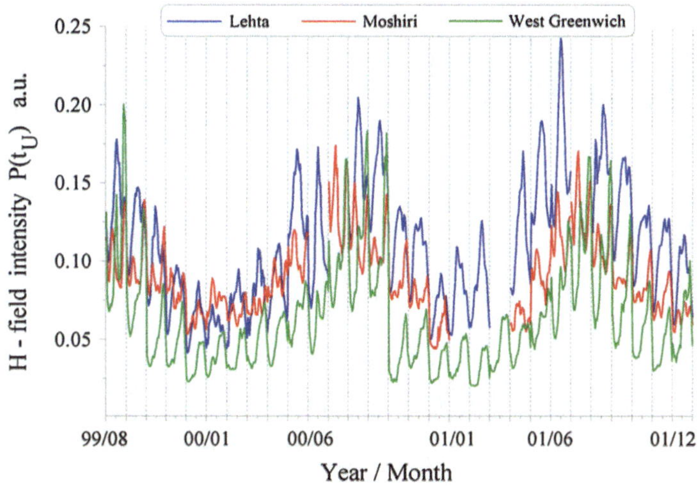

Fig. 10.1 Survey of data collected at three SR observatories

ordinate in arbitrary units. Blue lines show the Lehta data, while the red and green lines correspond to Moshiri and West Greenwich data. The presentation is similar to that already used in Chap. 5: the abscissa contains separate strips corresponding to 29 months. The strips are separated by vertical dotted lines and each of them presents the monthly averaged diurnal variations of cumulative intensity. There are gaps in plots, which mean the absence of particular experimental data.

We note that the integrated SR intensity of Fig. 10.1 reflects the thunderstorm activity: the well-known variations on the annual and diurnal scales are recognized here. In particular, one can recognize the African—American peak in the thunderstorm intensity combined with its decrease over the Pacific (early hours of UT). Frequency integration inhibited the modal structure in the distance dependence, so that initial $P(t_U)$ data tend to represent the global thunderstorm activity in a much better way than the intensities of separate resonance modes.

Pronounced seasonal/diurnal variations in Fig 10.1 occur over a 'podium' present at all three observatories: daily patterns are 'elevated' over the abscissa and never approach the zero level. This vertical shift or the podium originates from the 'out of time' thunderstorm activity causing also the depolarized SR fields arriving from all directions to an observer (Yatsevich et al. 2006, 2008).

As one may observe, the Lehta curves occupy the highest position in Fig 10.1. If we ascribe the 0 dB for the Lehta experimental data, the signals at the Moshiri and West Greenwich observatories will be of –3 and –6 dB correspondingly. Systematic deviations in intensity might be attributed to different latitudes of observatories, hence, to different distance from the global thunderstorm centers. However, particular deviations in Fig. 10.1 were most probably caused by the different calibration. Magnetic antennas at each site are of somewhat different

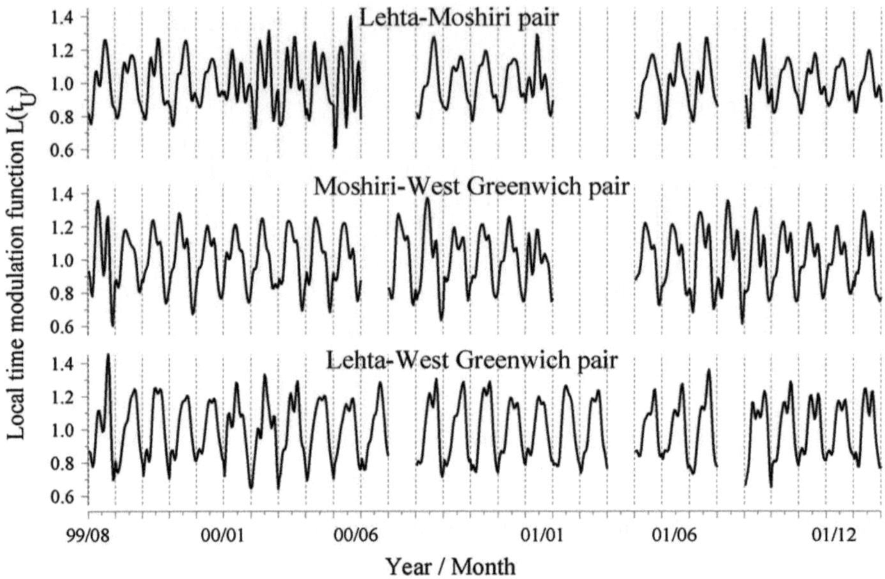

Fig. 10.2 Local time factors $L\ (t_U)$ found for particular pairs of SR records

construction, and they have never been calibrated together. Simultaneously, the systematic deviation in intensity is insignificant for establishing diurnal patterns.

Figure 10.2 shows the local time factors $L\ (t_U)$ derived for each pair of sites by using Eq. (10.3). Data presentation is similar to that of Fig. 10.1: the same 29 vertical strips, each representing the diurnal $L\ (t_U)$ pattern for a particular month. The upper plot depicts the monthly averaged diurnal variations $L\ (t_U)$ for the Lehta–Moshiri pair. The middle and lower plots refer to Moshiri–West Greenwich and Lehta–West Greenwich pairs. Gaps in curves indicate the situation when an observatory was out of work. One may observe that outline (sometimes rather complicated) of the local functions depends on the pair of sites. Thus, expectations of stable similarity of all three functions are not satisfied. For example, the pattern at the Lehta–Moshiri base during the February–May period of 2000 contains two–three diurnal peaks, and these are almost invisible in the plots derived for the Lehta–West Greenwich and Moshiri–West Greenwich pairs. At the same time, the range of daily variations is stable, and the patterns have a similar shape for the same months of different years. Such behavior implies that local time variations correspond to a reproducible physical process.

After obtaining local time modulating factors, we computed the universal time factors from Eq. (10.8). Each pair of sites provides a couple of $U_k\ (t_U) = P_k\ (t_U)\ /\ L\ (t_U)$ estimates, which are collected in Fig. 10.3 in the same fashion: 29 strips containing the 24 h daily variations. We will regard the $U_k(t_U)$ and $U_p(t_U)$ functions as "recovered". These functions are the six estimates for the diurnal variations of the global thunderstorm activity.

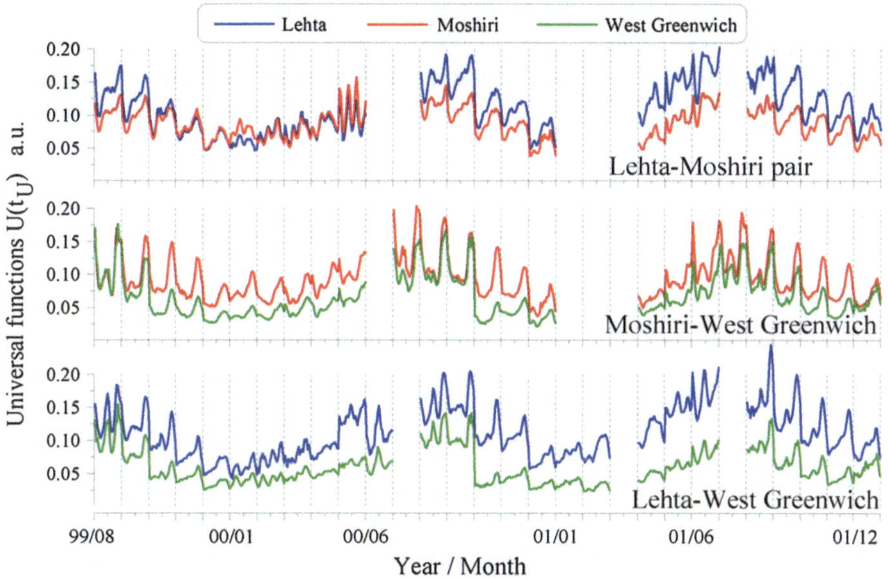

Fig. 10.3 Universal time functions $U(t_U)$ found for particular pairs of SR records

Three vertically separated panels contain paired plots in Fig. 10.3. The upper panel depicts the Lehta–Moshiri results. The red line shows the universal time variations $U_1(t_U)$ found as the $P_1(t_U, \lambda_{Mo})/L(t_U)$ pertinent to the Moshiri–Lehta base. The blue curve depicts the Lehta $U_2(t_U)$ function found similarly from the Lehta $P_2(t_U, \lambda_{Lht})$ record. The middle plot in Fig. 10.3 shows results for the Moshiri–West Greenwich observatories. The red line is the Moshiri universal time variations U_3 (t_U) and the green line demonstrates the West Greenwich U_4 (t_U) function. The lower plot depicts results for the Lehta–West Greenwich data sets. The blue line depicts the Lehta universal time variations U_5 (t_U), and the green line demonstrates the West Greenwich U_6 (t_U) function.

As one may see, all six universal time patterns have much in common, which demonstrate distinct and similar variations on both the diurnal and seasonal scales. Contributions from different global thunderstorm centers might be recognized in the plots. Mutual deviations are also obvious. Again, the green curves (West Greenwich) occupy the lowest position, but such a stable difference does not affect the pattern outline.

Decomposition of the signal into the universal and local time terms retains the podium in the global lightning activity: diurnal patterns are elevated over the abscissa. Moreover, both the "podium" height and the amplitude range of diurnal variations vary in accord: the both simultaneously grow or reduce from month to month. This behavior suggests that both characteristics are driven by the global thunderstorms and only the diurnal changes arise from the motion of a compact center around the globe, while the "podium" is maintained by strokes scattered over the planet (see Chap. 5).

Six 'individual' plots of Fig. 10.3 are the 'independent' estimates for the global thunderstorm intensity. We now can statistically process the $U\ (t_U)$ data and find the average diurnal variation

$$\langle U(t_U)\rangle = \frac{1}{6}\sum_{n=1}^{5} U_n(t_U), \tag{10.11}$$

shown in the black line in Fig. 10.4. Concurrently, we can estimate the relevant STD:

$$\mathrm{STD_U} = \left\{\frac{1}{5}\sum_{n=1}^{5}\left[U_n(t_U) - \langle U(t_U)\rangle\right]^2\right\}^{1/2}, \tag{10.12}$$

The STD data are not shown in Fig. 10.4 for simplicity.

The alternative data processing technique implies the geometric averaging of initial records. We computed the corresponding patterns by using the pairs (the Lehta–Moshiri, Lehta–West Greenwich, and Moshiri–West Greenwich sites) and found the paired geometric averages $G_{1,2}$, $G_{1,3}$, and $G_{2,3}$ defined by Eq. (10.9) together with the 'triple' geometric average $G_{1,2,3}$ (Eq. (10.10)). Then all four functions were averaged giving the $\langle G(t_U)\rangle$ variations (red line in Fig. 10.4).

After processing the experiment, we obtained two sorts of estimates for the diurnal/seasonal variations of the global thunderstorm intensity. The first kind of them is based on the resolving of the local and universal time factors. The $\langle U(t_U)\rangle$ function is shown in Fig. 10.4 in black line. Results of geometric averaging

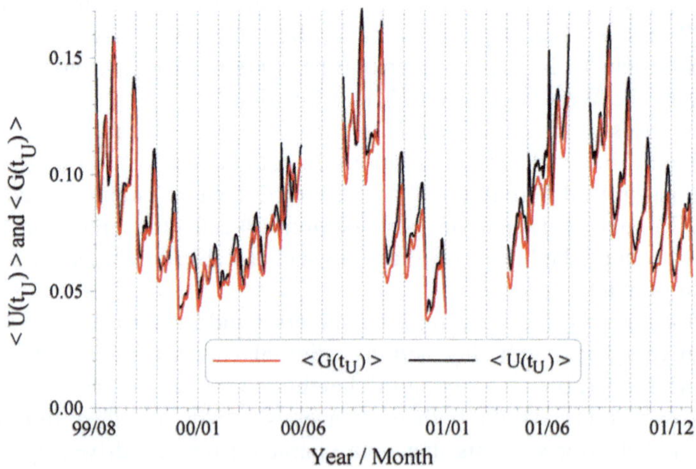

Fig. 10.4 Average diurnal/seasonal variations of the global thunderstorm activity found from $U\ (t_U)$ and $G\ (t_U)$ functions

include four quantities, and their average $\langle G(t_U)\rangle$ is shown in the same figure in red line.

Again the abscissa contains 29 strips with relevant daily variations. The universal time is used in each strip, and the intensity of thunderstorms is plotted on the ordinate in arbitrary units. As one may observe, completely different techniques provide exceptionally consistent data: the positions of minima and maxima practically coincide, and the curves are very close. The two lines come so close together that sometimes they are hardly separate in their plot.

Figure 10.5 surveys the universal diurnal patterns pertinent to four seasons of a year: winter (December, January, February—upper plots), spring (March, April, May—the second row of plots), summer (June, July, August—the third row of

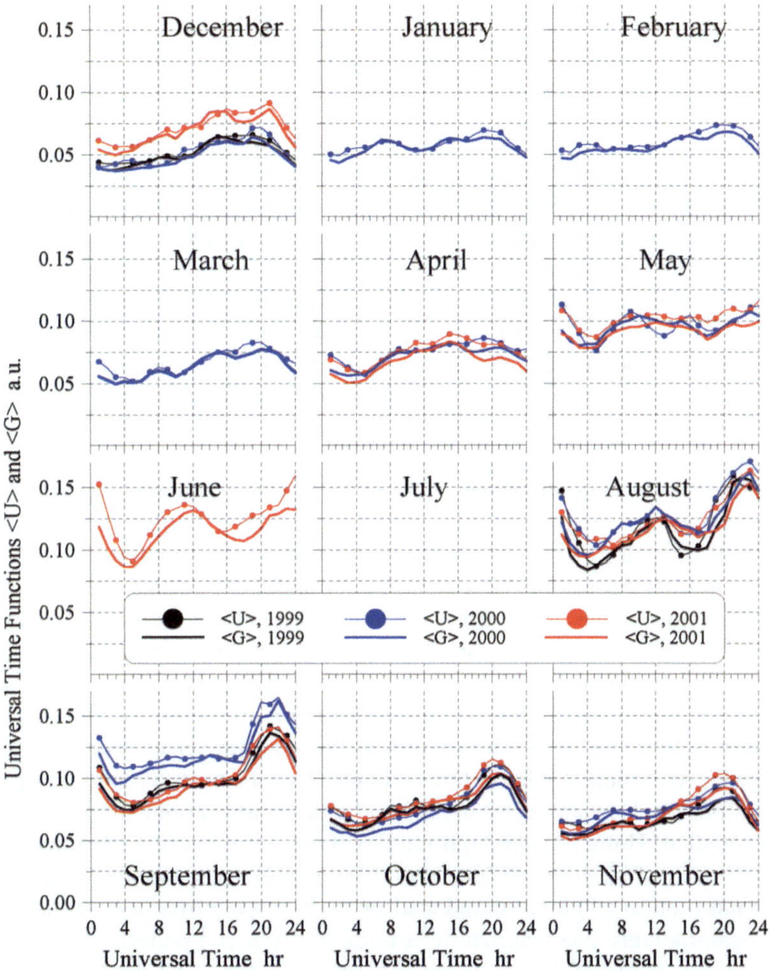

Fig. 10.5 Comparison of two types of output data: survey of diurnal variations of the global thunderstorm activity $\langle U\rangle$ and $\langle G\rangle$ estimated for each month of different years of observations

plots), and autumn (September, October, November—lower plots). The frames of individual months comprise the averaged "recovered" data $\langle U(t_U)\rangle$ (curves with dots) and the $\langle G(t_U)\rangle$ functions (smooth curves). Data for different years are collected in the same monthly panels demonstrating the inter-annual variability in the global thunderstorm intensity. The black lines refer to 1999; the blue curves, 2000; and the red lines, the year of 2001. It is seen that experimental data sets for July were always incomplete, so the relevant frame is empty. One may observe a close correspondence of estimates that we already noted when describing Fig. 10.4.

The processing of SR data results in distinct seasonal variations of thunderstorm intensity, while the diurnal variations are less pronounced. The intensity of global thunderstorms is found to vary by a factor of 2 during the day. The summer (boreal) activity exceeds that of the winter by a factor of about 3. It is interesting to note that daily patterns of a particular month repeat year after year thus indicating that diurnal change and the space–time distributions of global thunderstorms for a given month are annually replicated with minor deviations, as it was noted by Nickolaenko et al. (1998). Inter-annual deviations (if any) are observed as vertical shifts of the "monthly averaged diurnal variations". Such shifts are comparable with the range of the two-fold diurnal variations (e.g. Sátori et al. 2008). Three distinct peaks in the diurnal curve are rarely observed in the $\langle U(t_U)\rangle$ and $\langle G(t_U)\rangle$ functions being associated with the global centers.

It is important to remind that the recovered data $\langle U(t_U)\rangle$ and the geometrical average functions $\langle G(t_U)\rangle$ were obtained directly from the ELF intensity, which is driven by electromagnetic radiation of the worldwide thunderstorm activity. Simultaneous SR records provide robust estimates of thunderstorms; their advantage is that the electromagnetic fields automatically "collect and weight" contributions from all lightning strokes.

Figure 10.6 presents the three different local $L\,(t_L)$ functions. The blue curve corresponds to the Lehta—Moshiri pair, the green line depicts the Lehta—West Greenwich data, and red curve shows the Moshiri—West Greenwich pair. Their comparison tests the validity of assumptions 1 and 2. According to these, the only distinction among the local time modulating functions is the temporal shift accounting for the longitudes of sites. The soundness of this expectation was addressed by Nickolaenko et al. (2011) where the model data were used for evaluating the accuracy of the 'output' information on the source intensity and on its motion around the globe. It is worth noting that universal time functions are recovered even when the local time modulations seem to be very complicated. Expectations of close similarity in the local time factors among all the pairs might not be satisfied. However, the high reciprocity is observed of experimental $L\,(t_L)$ curves shown in Fig. 10.6: all functions have much in common and tend to reflect the diurnal motion of global thunderstorms with respect to any observation site.

To quantitatively characterize departures in the $L\,(t_L)$ patterns, we performed statistical processing of data and depict its results in Fig. 10.7. Here the upper plot shows the average local time variations $\langle L(t_L)\rangle$ and the lower plot depicts the corresponding standard deviation normalized by the average variation (in %).

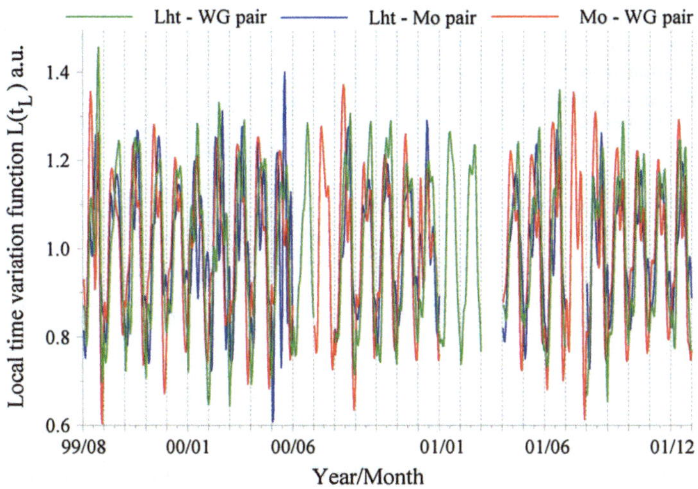

Fig. 10.6 Comparison of local time modulating factors

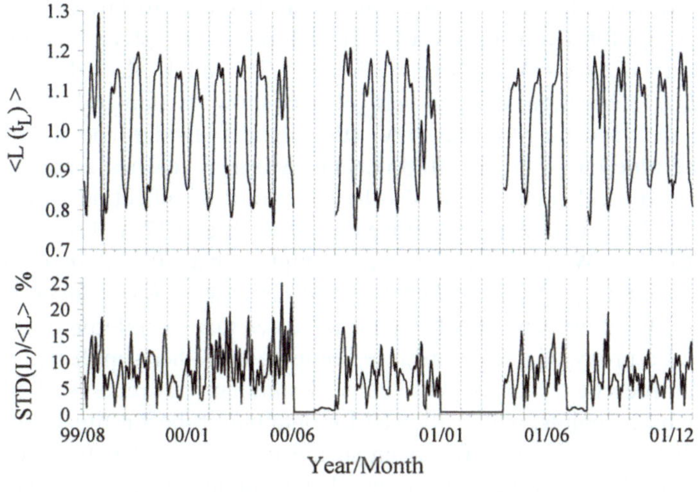

Fig. 10.7 Median local time variations (*upper plot*) and their standard deviations in percents (*lower plot*)

Discrepancies among the local time modulating functions appear to be rather modest, and the characteristic value is about 10 %.

The $L(t_L)$ factors typically have two maxima during the day. We can attribute these maxima to the thunderstorm proximity conditioned by their motion around the globe. Peaks correspond to the times when thunderstorms approach the observatories or come close to their antipodes. Such simple and "universal"

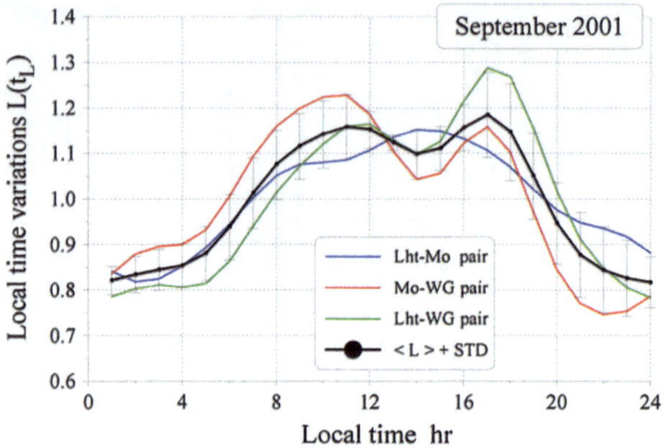

Fig. 10.8 Local time modulating factors and standard deviation from median $L(tL)$ s as function of local time for September 2001

patterns are observed often, but not always. The local factors may have an unequal number of maxima during certain months for the particular pairs.

To demonstrate such a situation, we depict the September 2001 segment in Fig. 10.8. The local time is shown on the abscissa in hr. Three individual $L\ (t_L)$ functions are plotted. The one obtained for the Lehta—Moshiri pair is shown in blue line, $L\ (t_L)$ for Moshiri—West Greenwich pair is in red, and the green graph shows the for Lehta—West Greenwich base. The thick black line depicts the median $\langle L(t_L)\rangle$ variation, and the vertical bars show its standard deviation characterizing the 'accuracy of recovery' (the highest standard deviation is about 10 %). The blue curve in Fig. 10.8 (Lehta—Moshiri) reaches a maximum positioned at 14 h local time, and it also has the point of inflection around 9–10 h. Other curves have two peaks positioned at 10–11 h and 17 h. The morning peak for the Moshiri—West Greenwich pair is higher than the afternoon maximum while the morning peak of the Lehta—West Greenwich $L\ (t_L)$ function is lower than that of the afternoon. This behavior indicates that the interpretation of $L\ (t_L)$ functions as a proxy of source distance seems to be rather simplistic.

The data presented include 29 months of simultaneous records of SR at three widely separated points, which allowed for obtaining robust estimates of intensity of the global thunderstorms. To obtain such estimates, one has to decompose the daily patterns of cumulative intensity into local and universal time variation and to use the latter as the proxy for the global thunderstorm activity. There is an alternative technique that is based on the geometrical averaging of the paired and the triple record. This approach gives an "independent" estimate of the source intensity. The results of these two techniques agree exclusively well. The standard deviation of established intensities does not exceed the 10 % level. All observatories were positioned in the Northern hemisphere. It is obvious that the accuracy of estimates might be improved by locating one of the observatories in the Southern hemisphere.

SR monitoring shows that the global thunderstorm activity is characterized by the two-fold diurnal variation, while the variation with season may reach the factor of three. The "podium" is always present in the activity. Feasibly, it is formed by radiation of "out of time" electric storms that permanently occur all over the globe.

The local time variations are usually bi-modal and contain the morning and the afternoon peaks. Sometimes, the morning peak becomes more pronounced, but sometimes it vanishes. However, two peaks are habitual, at least, for the present data set. In accordance with Eq. (10.7), the form of local time variation is governed by the median longitude of two observatories $\lambda_{kp} = (\lambda_k + \lambda_p)/2$. This is why its position at the time axis can change for different pairs.

Model computations based on OTD (Optical Transient Detector) data indicated that sometimes too many peaks might appear in the "recovered" $L\ (t_L)$ functions. Such a situation usually arises when the spatial distribution of global thunderstorms is complicated. To demonstrate this, we computed the integrated SR intensity in the uniform Earth–ionosphere cavity by using the DMM (Diurnal Monthly Mean) maps of average global distribution of lightning flashes (Pechony and Price 2007; Pechony 2007; Pechony et al. 2007; Yatsevich et al. 2006, 2008). DMM is a special processing of original OTD data files. The raw orbital data collected during five years by OTD were used to obtain the monthly mean diurnal variations with the hourly resolution. Data for a given hour were averaged over all the days a month (for all five years). The procedure provides 24 maps of global lightning distribution one for an hour. The single 'output' map contains averaged information from $\sim$150 raw maps (see Pechony and Price 2007; Pechony 2007 for details). We plot in Fig. 10.9 the results obtained with the DMM–OTD data for the month of September.

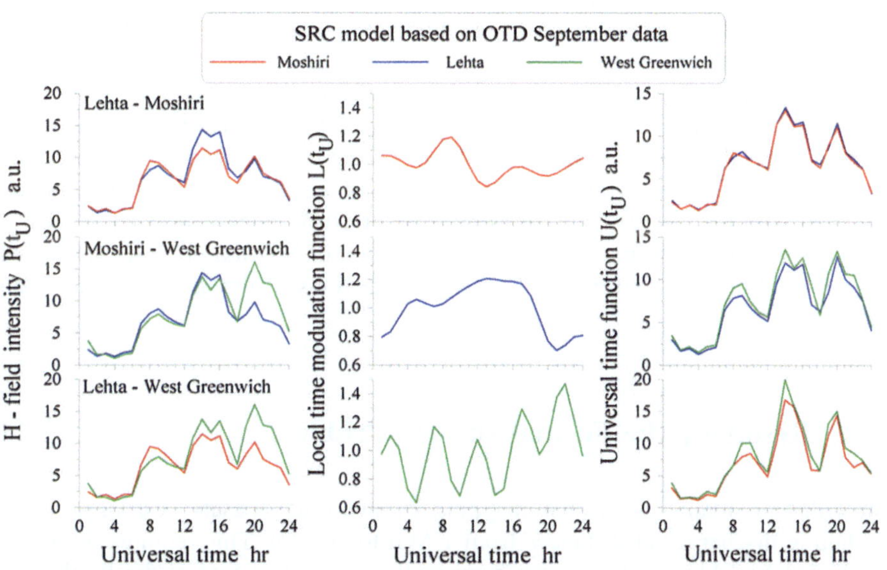

Fig. 10.9 Model data for DMM-OTD distribution of lightning flashes in September

We computed the field components at three observatories for the lightning activity given with the resolution of 2.5° × 2.5°. Cumulative intensity of the complete horizontal magnetic field was found by integrating the power spectra from 5 to 23 Hz.

The model results are presented in Fig. 10.9. The rows of plots (from top to bottom) correspond to the Lehta—Moshiri, Lehta—West Greenwich, and Moshiri—West Greenwich pairs. Again, the blue lines depict Lehta data, the red curves are Moshiri, and the green lines show the West Greenwich. The left plots show the "initial" cumulative intensities $P\ (t_U)$ at different sites. The 'original' plots have much in common, although their similarity has reduced owing to a more complicated source model. It is interesting to note that the SR intensity in the DMM–OTD model distinctly exhibits three peaks. Each of them corresponds to a particular thunderstorm center in South-East Asia ($\sim$9 h), Africa ($\sim$15 h), or America ($\sim$20 h). Experimental SR data for the September do not contain these distinct peaks (see Fig. 10.1).

The middle column of plots in Fig. 10.9 shows the local time factors $L\ (t_U)$ derived for different pairs of observatories. One may note that local modulations become outstandingly different from pair to pair, and the lower $L\ (t_U)$ function (Moshiri—West Greenwich pair) contains five oscillations. When the model implied a single compact source, the $L\ (t_L)$ functions reflected the source daily motion: the higher peak corresponded to the closest distance, and the smaller peak was the 'antipodal' distance. Local variations obtained in the September DMM–OTD model do not allow for such an interpretation. Thus, the local time function as an indicator of source proximity might meet difficulties when the spatial distribution of activity incorporates many distinct thunderstorm centers. Fortunately, such a complicated spatial distribution of sources was never met in particular observations. Therefore, the local time modulating function never indicated many diurnal oscillations.

Right plots in Fig. 10.9 depict the model universal time factors $U\ (t_U)$. These have much in common again. In the September DMM—OTD model, we clearly observe contributions of three global thunderstorm centers. The six 'output' daily patterns $U\ (t_U)$ of Fig. 10.9 were processed to obtain the average diurnal pattern $\langle U(t_U) \rangle$ with the relevant standard deviation of 10 % (Nickolaenko et al. 2011).

Thus we may conclude that the decomposition of SR data provides robust and efficient estimates of diurnal and seasonal variations of the global thunderstorm activity.

10.2 Resolving the Formal Inverse Problem

In order to obtain the global mapping of lightning activity, we must locate lightning discharges at distances from 1 Mm to 20 Mm. This requires the use of ELF (below 3 kHz) radio waves, because the attenuation rate of the Earth-ionosphere cavity is a few parts of dB/Mm at ELF. Because of severe attenuation of

VLF sferics, one can achieve the regional mapping of lightning activity. There were a few groups working on this topic. In Japan Iwai et al. (1969, 1979) established a VLF direction finding network by means of triangulating the bearings measured at three spaced stations in Japan, whose principle was the conventional orthogonal loop antenna system (so-called goniometer) at a particular frequency of 5 kHz. In USA, Krider et al. (1976) made a major improvement in this goniometer direction finding. They used signals in the frequency range below 60 kHz, and only the part of a lightning signal generated by the cloud-to-ground lightning channel roughly 100 m above the ground, so that they eliminated the non-vertical channel effect. The method was incorporated into the National Lightning Detection Network (NLDN). In 1980s the British Meteorological Office developed the VLF arrival-time-difference system (Lee 1986), and its primary goal was to develop an automatic system for accurately locating thunderstorms over ranges of a few Mm. The operating frequency was 2–23 kHz, and their system worked well for the studies of lightning within the European continent. A detailed review on the different kinds of direction finding of VLF/ELF sferics was made by Hayakawa (2009).

Mapping of the regional lightning activity has been recently very active in different continents including Europe (Betz et al. 2009), South America (Pinto et al. 2009) and the NLDN is still working effectively in Northern America for this purpose. About a decade ago Dowden et al. (2002, 2008) developed a new VLF direction finding on the basis of the arrival time difference at a few stations. The method has been extended into a real time global lightning detection system in recent years. Many observing stations are involved in this worldwide lightning location network (WWLLN) (http::\\wwlln.com). Though it works well for estimating the location and occurrence time of sferics, the rate of lightning detection is rather small of the order of 10 %. As for ELF sferics, we addressed these in the previous Chapter. However, little information can be obtained on the background lightning activity from the ELF transients.

The key advantage of inverse solutions based on SR records is that the records automatically incorporate contributions from all lightning strokes and a relatively small number of observatories are necessary. We will use only three stations in the ELF tomography described below. The ELF tomography is inexpensive as compared to those costly multi-station or satellite observations.

Tomography methods are based on the reconstruction of internal structure of an object from the integral quantities measured along various paths (e.g. Troyan and Hayakawa 2002; Korovkin et al. 2007). A tomographic projection is formed by observations along different paths. A set of projections allows for reconstructing a 2D distribution on the basis of the central slice theorem. Another approach implies the resolving of a system of equations relating the integral quantities along different paths and the pixel values, yielding the desired 2D distribution. The second approach is employed here. Consider an integral lightning intensity within annular concentric stripes or belts on the Earth surface with the center at the observation point. Lightning discharges within a particular stripe are "equidistant" from an observation point and form the same distance signature in the measured SR spectra. Shvets (2001) proposed a technique for decomposing the average power SR spectra measured at a single site into the distance profile of lightning intensity. Each point in such a distance profile represents an integral value of lightning intensity within the given stripe. When a few remote observatories are used, one obtains the distance profiles of the same thunderstorm activity seen under various ''angles of view''. The mathematical treatment of a set of such profiles reconstructs the global lightning distribution over the Earth.

The problem of the mapping with the ELF tomography is resolved in two stages. The first one obtains the distance profiles from the power spectra of SR measured at a number of stations distributed over the planet. The second stage is the 2D reconstruction of the global lightning activity from a set of distance profiles. Both of these stages are ill-posed problems with additional limitations arising from the non-negative nature of signal power and lightning intensity. These are resolved by the regularized least squares method.

To obtain the distance profile, one has to invert the observed SR power spectrum by using the distance dependence of its spectral pattern (Shvets 2001; Ando et al. 2005a; Ando and Hayakawa 2007). This dependence was applied for finding the distance from the stroke that caused the Q–burst in Chap. 9. Spectra of electric and magnetic fields are described by the following relations (see also (5.7) and (5.8)) and plotted in Fig. 10.10:

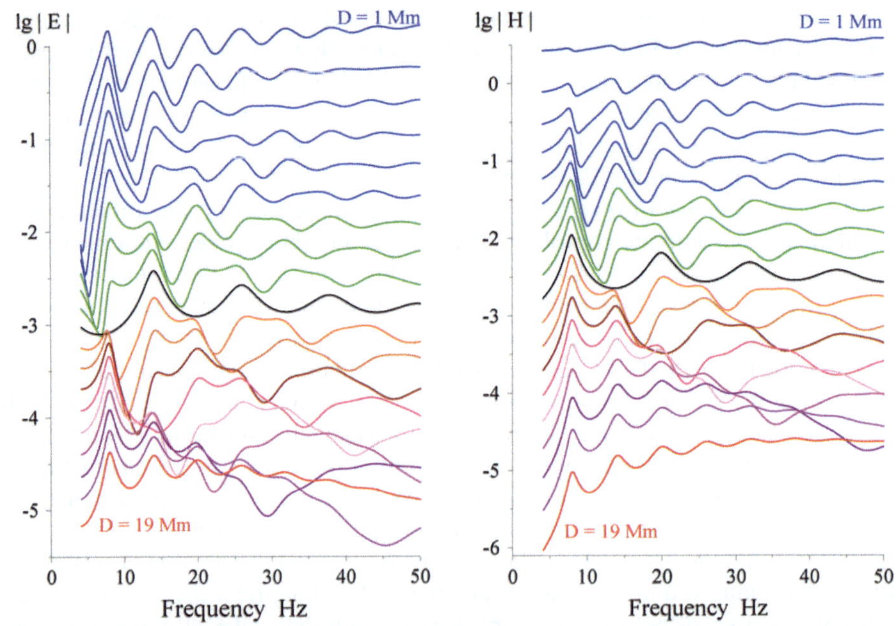

Fig. 10.10 Amplitude spectra of vertical electric and horizontal magnetic field components for the source–observer distances ranging from 1 to 19 Mm

$$E_r(\omega) = \frac{M_C(\omega)}{4ha^2\varepsilon_0} \frac{i\nu(\nu+1)}{\omega} \frac{P_\nu[\cos(\pi - \theta_H)]}{\sin \pi\nu} = M_C(\omega) e(\omega, \theta_H) \qquad (10.13)$$

$$H_\varphi(\omega) = -\frac{M_C(\omega)}{4\,ha} \frac{P_\nu^1[\cos(\pi - \theta_H)]}{\sin \pi\nu} = M_C(\omega) h(\omega, \theta_H) \qquad (10.14)$$

Here a is the Earth's radius, h is the effective height of ionosphere, ε_0 is the free space permeability, $M_C(\omega)$ is the source current moment, θ_H is the source—observer angular distance, $\nu(\omega)$ is the propagation constant, and ω is the circular frequency. The source is positioned at the pole $\theta = 0$ of the spherical coordinate system $\{r, \theta, \varphi\}$ with the origin at the center of the Earth.

Spectra of electric and magnetic fields are shown in Fig. 10.10 for distances ranging from 1 Mm to 19 Mm and varying with the step of 1 Mm. Horizontal axes show the frequency in Hz, and vertical axes depict the spectra on logarithmic scale. To separate different spectra, we shifted each of them vertically by –0.25. The current moment was chosen of $|M_C(\omega)| = 10^8$ A m $(\mathrm{Hz})^{1/2}$ and the electric and magnetic spectral densities were measured in mV/m and μA/m correspondingly. The dispersion dependence ν (f) was specified in the following way (see Chap. 14). The propagation constant ν is found from the relation $\nu\ (\nu + 1) = (kaS_\nu)^2$, where S_ν is the complex refractive index of the TEM wave: $S_\nu = c/V - 5.49i\alpha_D/\mathrm{f}$, k is the free space wave number, c is the velocity of light in vacuum, V is the wave phase velocity, and α_D is the wave attenuation. The particular dependence was derived from the SR parameters Ishaq and Jones (1977): $c/V = 1.64 - 0.1759{\cdot}\ln(f) + 0.01791[\ln(f)]^2$ and $\alpha_D = 0.063{\cdot}f^{\,0.64}$.

One may note that an interference pattern is superimposed on the regular succession of SR peaks in Fig. 10.10. The interference arises from the "beating" of direct and antipodal waves. This feature was discovered by Jones (1969, 1970a, b) who has used it for finding the distance from the observatory to the strokes causing ELF transients. It is important for resolving the inverse problem that every distance has a unique spectral outline. For example, the odd peaks are absent in the $|E\ (f)|$ spectra and the even peaks are absent in the $|H\ (f)|$ spectra at the 10 Mm distance. This distance–pattern link helps in establishing the range of discrete strokes; this was discussed in Chap. 9. It also allows us to obtain the distance distribution of lighting discharges when one uses the SR background spectra (Shvets 2001). After obtaining individual distance profiles of the source intensity, one can combine them for recovering the relevant spatial source distribution (Shvets et al. 2009, 2010).

The particular procedure implies simultaneous records of three SR observatories that monitored the vertical electric field E and two orthogonal components of horizontal magnetic fields H_{WE} and H_{SN} in 1999–2000. Absolute field measurements of SR background were performed at Lehta, Russia, Moshiri, Japan, and West Greenwich, USA. The West Greenwich data contain the 12-minute average power spectra of all three-field components in the frequency band from 5 to 55 Hz. Each resonance mode was recorded in a form of eight Lorentzian curves $S_n(f) = A_n \cdot [1 - (f/f_n)^2 + i \cdot f/(Q_n \cdot f_n)]^{-1}$ (see Chap.1). In other words, the peak amplitudes A_n, frequencies f_n, and the quality factors Q_n were found and stored in the computer memory every 12 min: 24 spectral parameters as a whole. SR background spectra $\left(\left\langle |E(f))|^2\right\rangle, \left\langle |H_{WE}(f)|^2\right\rangle \text{ and } \left\langle |H_{SN}(f)|^2\right\rangle\right)$ were monitored every 10 min at other stations with the frequency resolution 0.1 and 0.25 Hz in the frequency bands 4–40 and 8–870 Hz respectively (Lehta and Moshiri). The description of equipments and preliminary processing might be found in the papers by Ando et al. (2005b), Belyaev et al. (1999), Heckman et al. (1998), and Hobara et al. (2000, 2006).

A preliminary selection of data was necessary prior to the final processing. The data "spoiled" by local interference were rejected with the help of a special automated procedure. The frequency band 5–35 Hz was chosen that included 5 resonance modes. Initially, for the peak frequencies $f_n \in [8; 14{,}3; 20{,}5; 26{,}5; 32]$ Hz and the quality factors $Q_n \in [6; 6; 6{,}5; 7; 8]$ the resonance curves $S_n(f)$ were computed of the unit amplitude in the $\Delta f = \pm 3$ Hz vicinity of the peaks. Afterwards, for every mode, the angle was found between the model vector $S_n(f)$ and the observed $L_n(f)$ vector. The angle was computed as the integral over the discrete frequencies of the following ratio of scalar product: $\chi_n = \cos^{-1}\left\{\frac{(L_n(f), S_n(f))}{|L_n(f)|\cdot|S_n(f)|}\right\}$. The resolving rule demanded that the angle χ_n must not exceed 40° for any individual mode, and their average value should be lower than 22°.

Thus, the set of the three power spectra was selected averaged over the month and relevant to a given hour. Each set corresponds to a particular observatory. By using the spectra of horizontal magnetic field components, we obtained the cumulative SR intensities (also used as initial data in the previous section). When solving the inverse problem, we divide the distances from 0 to 20 Mm into $N = 40$ intervals of 500 km (4.5°) width. The observer is placed at the pole $\theta = 0$. The patterns of the unit frequency responses $h\ (\omega, \theta_k)$ and $e\ (\omega, \theta_k)$ depend on the source–observer distance as shown in Fig. 10.10. A Poisson sub-succession of pulses arrives at the observer from the given distance interval (strip), and the intensities of individual pulses are summed.

$$\left\langle |E_r|^2\right\rangle = L\cdot\sum_{k=1}^{N}\langle A_k^2\rangle\left\langle |M_k(\omega)|^2\right\rangle\left\langle |e(\omega,\theta_k)|^2\right\rangle \tag{10.15}$$

and

$$\left\{\begin{matrix}\left\langle |H_{NS}|^2\right\rangle \\ \left\langle |H_{EW}|^2\right\rangle\end{matrix}\right\} = L\cdot\sum_{k=1}^{N}\langle A_k^2\rangle\left\langle |M_k(\omega)|^2\right\rangle\left\langle |h(\omega,\theta_k)|^2\right\rangle\left\{\begin{matrix}\langle\cos^2(B_k)\rangle \\ \langle\sin^2(B_k)\rangle\end{matrix}\right. \tag{10.16}$$

Here θ_k is the distance to the middle of k-th strip, angular brackets denote averaging over the ensemble, L is the density of pulse flux (events per second), A_k is the thunderstorm intensity in the k-th strip, $M_k(\omega)$ is the current moment of the source, and B_k is the source bearing in the k-th strip.

Obviously the thunderstorm intensity forms a vector $\mathbf{M}(\omega) = \mathbf{L}\langle A_k^2\rangle\left\langle |M_k(\omega)|^2\right\rangle$ whose projections are the different spectral components. This vector has the dimension of [(A·m)^2·c^{-1}], which presents the power spectral density of the strokes of the average current moments $\langle A_k^2\rangle\left\langle |M_k(\omega)|^2\right\rangle$ arriving with the rate **L** from all distance intervals. The angular patterns of magnetic antennas results in the "partial" vectors $\mathbf{M}_{SN}(\omega) = \mathbf{M}(\omega)\langle\cos^2(B_k)\rangle$ and $\mathbf{M}_{WE}(\omega) = \mathbf{M}(f)\langle\sin^2(B_k)\rangle$, and these depend on the thunderstorms azimuths B_k relevant to a particular observatory.

The accuracy of reconstruction is sensitive to the propagation constant $v(f)$ and to the spectrum of current moment. This is why the most accurate description is desirable of the sources and the Earth–ionosphere cavity. It was assumed that the lightning discharge has the channel length $ds = 5$ km, its peak current is $I_P = 20$ kA, and the time constant of current reduction is $\tau = 0.15$ ms (Shvets and Hayakawa 2011). These parameters provide the constant spectral density of the current moment in the whole SR band: $\langle A_k^2\rangle\langle |M_k(\omega)|^2\rangle = \left(I_p\tau ds\right)^2$. Then, the current moment is equal to $M_k(\omega) = L_k\left\langle (Qds)_k^2\right\rangle$ being the squared charge moment relevant to the charge transposed by all lightning strokes of the k-th stripe in a second. The above relation follows from the obvious identity: $Q = \tau \cdot I_P$ (Heckman et al. 1998).

One can present the links of the power spectra of the field components in a form of system of equations or in a form of matrix relation:

$$\mathbf{b} = \mathbf{A} \cdot \mathbf{L}. \tag{10.17}$$

Here **b** denotes the vector of measured power spectra, and **L** is the sought vector of the stroke rate at different distances. Elements of **A** matrix contain the model spectra of the Earth–ionosphere cavity. These latter are computed for the set of distances from the vertical electric dipole source: $A_{ij} = |I_P\tau\ ds|^2 \cdot |e(\omega, \theta_i)|^2$ or $A_{ij} = |I_P\tau\ ds|^2 \cdot |h(\omega, \theta_i)|^2$.

The system (10.17) is solved by the minimization of the squared deviations of the following functional (Tikhonov 1963; Tikhonov and Arsenin 1977):

$$F(\mathbf{x}, \alpha) = \|\mathbf{AL} - \mathbf{b}\|^2 + \alpha_R\|\mathbf{L}\|^2, \qquad \mathbf{L} \geq 0, \tag{10.18}$$

where α_R is the regularization parameter.

The solution is constructed with the help of algorithm of the non-negative least squares (Lawson and Hanson 1974). Detailed description of the procedure and of the impact of different factors on the solution might be found in the publications by Shvets et al. (2009, 2010).

Results of such a processing are demonstrated in Fig. 10.11. Three global maps in this figure depict the lightning rate (events per second) derived from the SR background spectra recorded at Moshiri, West Greenwich, and Lehta at 15 h UT on Jan. 1, 2000. The rates are shown in the equal distance strips. In the rectangular projection we use, the line of equal distance is a sinusoid, so that the distributions

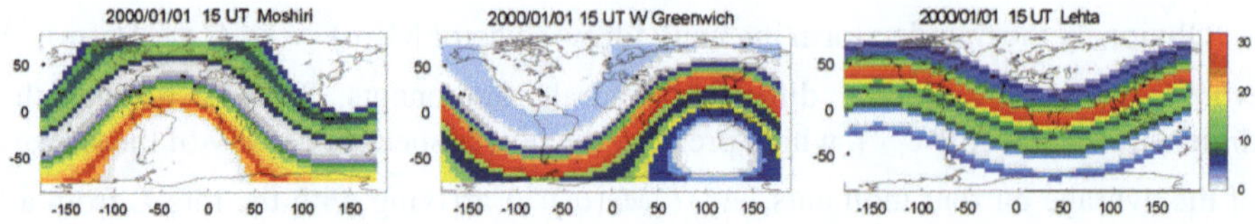

Fig. 10.11 Distance distributions of thunderstorm activity relative observatories Moshiri, West Greenwich, and Lehta that were obtained from power spectra of SR recorded at 15 h UT on Jan. 1, 2000

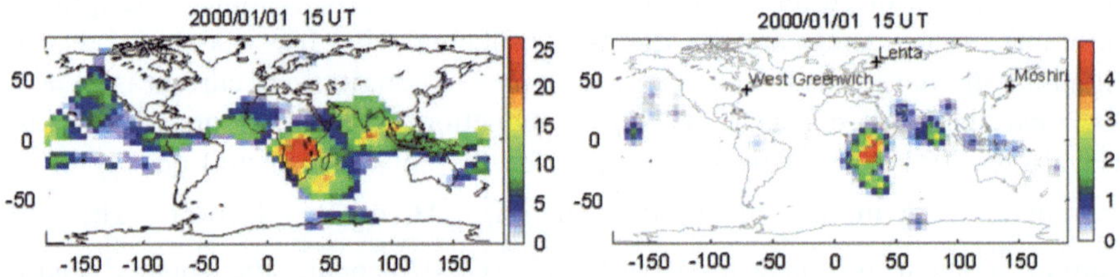

Fig. 10.12 Global thunderstorm distributions recovered from the distance profiles shown in Fig. 10.11, which were obtained at three sites from the SR power spectra for 15 h UT, Jan. 1, 2000. The *left frame* shows reconstruction via geometric average, and the *right frame* depicts the tomographic reconstruction

are seen as different sets of concentric curved color bands. The color scale of the maps is shown on the right.

After obtaining individual distance profiles for every observatory, we can turn to the second stage of solution and construct the spatial distribution of lightning strokes over the globe. The simplest way to do this is computation of the geometric average shown in the left frame of Fig. 10.12. It clearly indicates that global thunderstorms were mainly concentrated at Africa at the particular time moment of 15 h UT on the first of January 2000. The areas of increased activity are also apparent around the Ceylon Island and south of Madagascar. Hopefully, an increase of the number of observatories will increase the number of distance profiles and thus raise the accuracy of lightning location and the resolution in space.

Another, formally rigorous approach is based on the tomographic reconstruction of the spatial distribution from a set of distance projections relevant to different observatories. Each of these detects the thunderstorm radiation at an individual "angle of vision". Results of tomographic reconstruction are shown in the right frame of Fig. 10.12. Here, the crosses indicate the positions of observatories, and the color inking corresponds to the lightning rate. As a whole, geometric averaging and tomographic reconstruction give similar geophysical results. However, the rigorous procedure of tomography allowed for the substantial "cleaning" of global thunderstorm distribution by removing "insignificant" and probably false objects. The most pronounced and compact areas were preserved in the global map. Data obtained are in accord with the climatological expectations and agree with the optical orbital observations by OTD and LIS satellites.

Prior to processing the real data, the detailed modeling was undertaken (Shvets 2001). It showed that global thunderstorm distribution of the tropical continental regions is recovered fairly well when seven stations are working in the Northern hemisphere. Nevertheless, application of only three SR observatories (Moshiri, West Greenwich, and Lehta) allows for monitoring the diurnal motion of thunderstorms and the seasonal drift of activity (Shvets et al. 2009). Three observatories allow us to recover relatively simple distributions when the number of compact active areas is not large. A sophisticated spatial distribution might result in an emergence of some false centers at places where the postulated model activity was absent. The modeling showed that angular patterns of horizontal magnetic antennas might be helpful in the case. Azimuthal information helps in resolving the inverse problem. However, we will not go deep in this issue, but we will only outline the basic idea.

Let the network of SR observatories include K stations distributed all over the globe. Each observatory monitors the orthogonal horizontal magnetic field components. Let us use in reconstructions the "equidistant" cells of the 5°× 5° size. Then, the k-th point has the N-vectors $\mathbf{L}_k^{WE}$ and $\mathbf{L}_k^{SN}$ (partial profiles), each corresponding to the distance profile of lightning rate that was obtained from the inversion of SR power spectra of the field components H_{WE} and H_{SN}. These distance profiles of all stations compose the vector-column of the partial profiles $\mathbf{L} = \left[\mathbf{L}_1^{EW}, \mathbf{L}_1^{NS}, \mathbf{L}_2^{EW}, \mathbf{L}_2^{NS}, \ldots, \mathbf{L}_K^{EW}, \mathbf{L}_K^{NS}\right]^T$ (the upper script T denotes transposition). The following system of equations might be constructed for this vector, which is established by the link between the spatial distribution of source s and the L projections:

$$\mathbf{L} = \mathbf{W} \cdot \mathbf{s} \tag{10.19}$$

The system matrix W has L columns and $2KN$ lines. The model field components in this matrix correspond to the cells of the global map with an account for the wave arrival angles. The system (10.19) is resolved with the algorithm of the non-negative least squares (Lawson and Hanson 1974) by minimization of the functional $F(\mathrm{s}) = \|\mathrm{W} \cdot \mathrm{s} - \mathrm{L}\|^2$. The dimension of the system should be reduced by removing the trivial distance profiles that are equal to zero. This procedure allows us to reduce the rank of matrix W. Besides, the sub-polar regions are excluded, having latitudes higher than 75°.

We demonstrate the model results of application of partial distance profiles in Fig. 10.13. The initial (postulated) model distribution of thunderstorms corresponds to median optical OTD/LIS observations in October 1995–2002 (e.g. http://thunder.msfc.nasa.gov/). It is shown in the upper left frame of Fig. 10.13. The global map depicts the lightning flash rate (arbitrary units) in the cells of 5° × 5° dimension. The distance distributions of lightning strokes were computed by using these maps which are shown in the upper right part of the figure. Three frames here show individual profiles of cumulative intensity (red curve), that of the H_{WE}

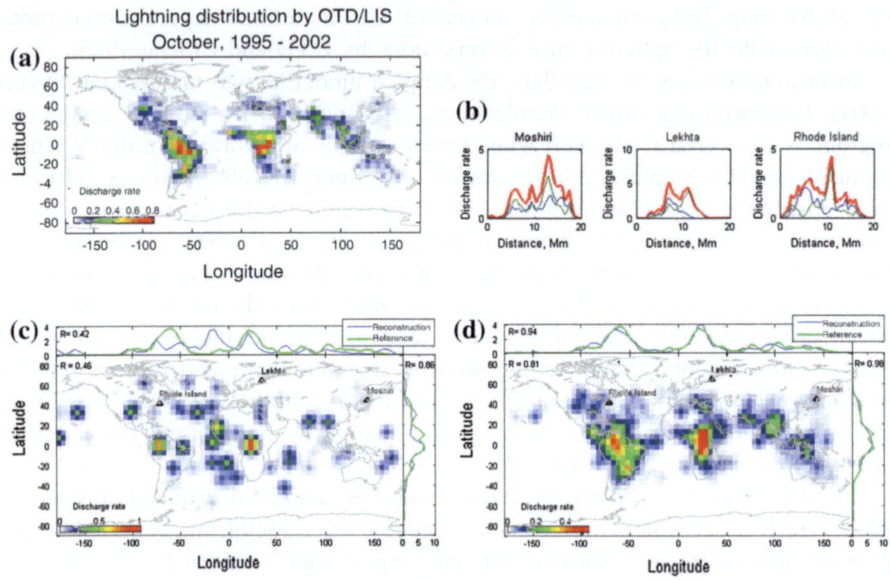

Fig. 10.13 Model reconstructions: **a** median OTD/LIS distribution of flashes in October used as initial information; **b** distance profiles at different sites: *red* cumulative, *blue* H_{WE}, and *green* H_{SN} field; **c** tomographic reconstruction based on profiles of cumulative intensity; **d** reconstruction applying the partial profiles

component (blue curve), and of the H_{SN} field (green line). These profiles were used in the tomographic reconstruction based on the cumulative intensity (the lower left map) and based on partial distance profiles, i.e., with an account for the wave arrival azimuths (the lower right map).

Triangles in the lower maps mark the positions of observatories. The horizontal bars in the left lower edges of the maps show the color scales. The plots above and to the right of maps demonstrate the meridian and zonal distribution of thunderstorm intensities relevant to the initial (green lines) and reconstructed (blue lines) distributions. The lower maps also contain the cross-correlation coefficients of the initial and reconstructed distributions. These are $R = 0.46$ for the reconstruction applying the distance profiles of cumulative intensity and $R = 0.81$ for the partial profiles that account for azimuthal information. Also the cross-correlation coefficients are shown corresponding to the one-dimensional latitudinal and longitudinal profiles of intensities. These are equal correspondingly to $R = 0.42$ and $R = 0.86$ in the first case and $R = 0.94$ and $R = 0.98$ in the second case. Thus, the model data indicates a substantial improvement of reconstruction accuracy when the azimuthal information is exploited in the tomography.

The monthly averaged power spectra were computed for every UT hour covering the interval from August 1999 to may 2000, and the global maps of thunderstorm distribution were obtained (Shvets et al. 2009, 2010). The seasonal drift of global thunderstorms was deduced along the latitude that we reproduce in

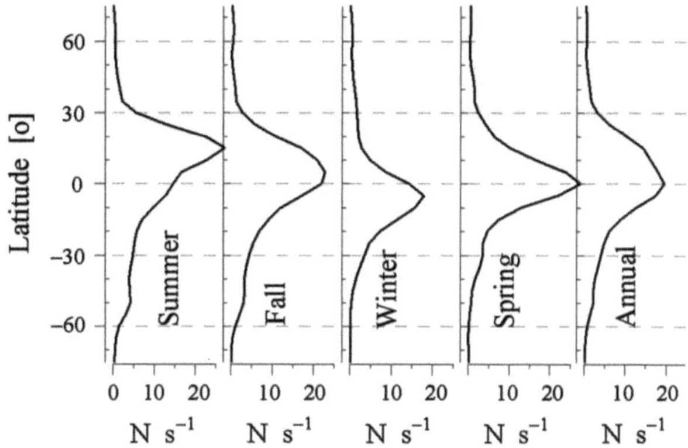

Fig. 10.14 Latitudinal distributions of the global lightning activity in different seasons recovered from the SR records

Fig. 10.14. The graphs show the average distribution of the global lightning activity against the latitude for different seasons: summer (August 1999), autumn (September, October, November 1999), winter (December 1999–February 2000) and spring (March, April and May 2000). The right plot depicts the resulting annual distribution of thunderstorms versus latitude. Obviously, the solution of the inverse problem based on the SR monitoring allows for inferring the seasonal drift of global thunderstorms. The summer peak of activity is shifted northward from the equator by 15°, and it is shifted to the southern hemisphere by 5° in the winter. This result is consistent with the climatologic data. It also confirms the findings of other studies in the SR band that we mentioned in Chap. 5.

In particular, the seasonal drift of global thunderstorms was studied by Sátori (2003) and Sátori et al. (2003, 2008). Diurnal frequency variations were used of SR observed at the Nagycenk observatory (Hungary). It has been shown that three types of diurnal patterns were recoded in the peak frequencies during the year. The first type "January" is relevant to the boreal winter, which is observed from November to April for about 165 days per year. Variations of the "July" type were observed from the late May to the late August during ~ 98 days. The "October" type variations were observed in the September–October period and last for ~ 60 days. Variations that cannot be attributed to any of the above types correspond to remaining 42 days of the year in the April—May period. During this time interval, the area occupied by thunderstorms substantially spreads reducing the frequency range of diurnal frequency variations (see Chap. 5). Changes in the shape of frequency variations are attributed to the meridianal shift of the global lightning activity in different seasons. Durations of the seasons are uneven, at least, in the context of the global lightning activity. Thunderstorms linger in the Southern hemisphere during the local summer, as compared with the Northern hemisphere: 165 days versus 98. In transitional spring and fall periods, the thunderstorms quickly move to the north or to the south and simultaneously spread over the planet. Despite the thunderstorm concentration over the continents, their northward or southward drift depends on the temperature of the tropical Pacific

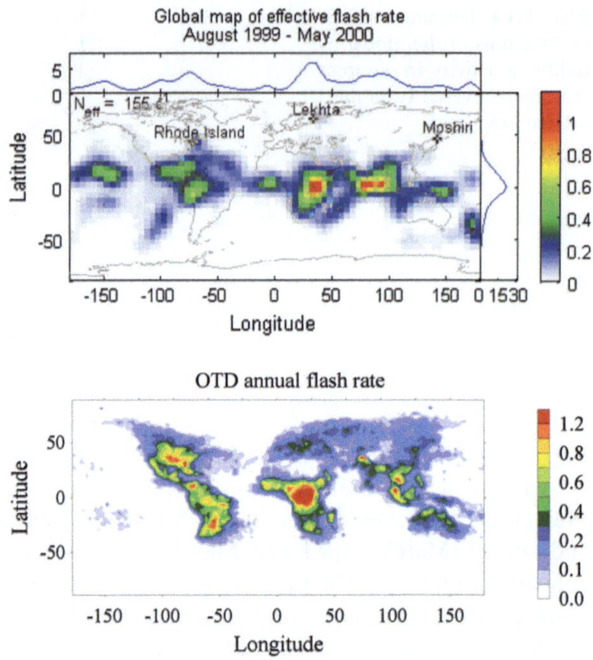

Fig. 10.15 The yearly averaged (August 1999–May 2000) global lightning distributions obtained from the SR records and observed by OTD

Ocean. Concurrently, the SR records showed some relation to the development of El Nino—La Nina oscillations (ENSO) (Sátori et al. 2008), described above.

Apparently, the average annual distribution of lightning is the most interesting as shown in Fig. 10.15 (Shvets et al. 2009, 2010). It was constructed by averaging the results of inversions of SR monitoring at three stations shown in figure by crosses. The data used was collected from August 1999 to May 2000. The above method of signal processing was used: the distance profiles of lightning intensity were found from SR data. Afterwards, the spatial distributions of the lightning rate were obtained by using the tomographic reconstruction with an account for the azimuthal data of the wave arrival. The plots of longitudinal and latitudinal distributions of activity are shown above and to the right of the map that were obtained by summing of the intensity in the 5° cells. The map of the global distribution of the lightning flashes is shown for comparison in the lower part of Fig. 10.15 that was observed by OTD spacecraft (Christian et al. 2003). By comparing the results of optical observations from the orbit and the "ground-based" tomographic reconstruction of global lightning activity, we must admit their unmistakable similarity.

Simultaneously, the maps are slightly different. The reconstructed longitude distribution of thunderstorms contains three major peaks corresponding to Central America, Africa and Asia. In the map based on the SR data a shift is visible of the "Asian" maximum toward the Indian subcontinent, which departs from the optical OTD observation. Besides, some areas are visible in the oceans where the thunderstorm activity is also concentrated. One of these is in the Gulf of Guinea. This result corresponds to the data on the location of sources of Q—bursts (Huang et al. 1999; Heckman et al. 1998; Nickolaenko and Hayakawa 2002). These contained a noticeable "tail" in the spatial distribution of sources extended from Africa to the South America. There are also a few active regions: one over the Sargasso Sea, and over some areas in the Pacific Ocean.

SR records in the particular observation period show no significant activity in the middle latitudes. That is, thunderstorms rarely move above the 30° latitude in both hemispheres. This seems to be in conflict with the results obtained by OTD, according to which there is a noticeable lightning activity in the middle and high latitudes of the Eurasian continent (see the lover map in Fig. 10.15). At the same time, the global location of the sources of ELF transients during the year (Füllekrug and Constable 2000; Heckman et al. 1998; Huang et al. 1999; Yamashita et al. 2009) provides the latitudinal distribution of thunderstorms having similar characteristic width as was obtained from the reconstruction based on the SR background.

The formally posed inverse problem of the SR was resolved in two stages. First, the power spectra of magnetic and electric fields were inversed giving the distance distribution of thunderstorms relative each observatory. Then, the distance profiles were applied as tomographic projections in the recovery of the spatial distribution of sources. The accuracy of the solution of inverse problem increases when the wave arrival information is exploited. Reconstruction based on the records of Lehta (Russia), Moshiri (Japan), and West Greenwich (USA) showed the diurnal/seasonal dynamics of the global distribution of lightning strokes, which fairly agree with the data of meteorology and optical observations from space. When designing the global network of ELF radiolocation of thunderstorms working in the real time, the main problem will be in establishing the reliable system of automated selection and the data transmission. Subsequent application of the tomographic reconstruction is quite obvious.

References

Ando Y, Hayakawa M, Shvets AV, Nickolaenko AP (2005a) Finite difference analyses of Schumann resonance and reconstruction of lightning distribution. Radio Sci 40(2). doi:10.1029/2004RS003153

Ando Y, Maltsev P, Sukhynyuk A, Goto T, Yamauchi T, Hobara Y, Sekiguchi M, Ikegami Y, Sera M, Korepanov V, Hayakawa M (2005b) New ELF observation system at Moshiri, Japan and assessment of acquired data. J Atmos Electr 25(1):29–39

Ando Y, Hayakawa M (2007) Use of generalized cross validation for identification of global lightning distribution by using Schumann resonance. Radio Sci 42(2). doi:10.1029/2006RS003481

Belyaev GG, Schekotov AYu, Shvets AV, Nickolaenko AP (1999) Schumann resonances observed using Poynting vector spectra. J Atmos Solar-Terr Phys 61:751–763

Betz HD, Schmidt K, Ottinger WP (2009) LINET-An international VLF/LF lightning detection network. In: Betz HD, Schumann U, Laroche P (eds) Lightning: principles, instruments and applications. Springer, Dordrecht, pp 115–140

Christian HJ, Blakeslee RJ, Boccippio DJ, Boeck WL, Buechler DE, Driscoll KT, Goodman SJ, Hall JM, Koshak WJ, Mach DM, Stewart MF (2003) Global frequency and distribution of lightning as observed from space by the optical transient detector. J Geophys Res 108(D1):4005. doi:10.1029/2002JD002347

Dowden RL, Brundell JB, Rodger CJ (2002) VLF lightning location by time of group arrival (TOGA) at multiple sites. J Atmos Solar-Terr Phys 64:817–830

Dowden RL, Holzworth RH, Rodger CJ, Lichtenberger J, Thomson NR, Jacobson AR, Lay E, Brundell JB, Lyons TJ, O'Keefe S, Kawasaki Z, Price C, Prior V, Ortega P, Weinman J, Mikhailov Y, Woodman R, Qie X, Burns G, Collier A, Pinto Junior O, Diaz R, Adamo C, Williams ER, Kumar S, Raga GB, Rosado JM, Avila EE, Clilverd MA, Ulich T, Gorham P, Shanahan TJG, Osipowicz T, Cook G, Zhao Y (2008) World-wide lightning location using VLF propagation in the Earth-ionosphere waveguide. Radio Sci Bull 327:39–53

Fraser-Smith AC, McGill PR, Bernardi A, Helliwell RA, Ladd ME (1991) Global measurements of the low frequency radio noise. In: Kikuchi H (ed) Environmental and space electromagnetics. Springer, Tokyo, pp 191–200

Füllekrug M, Fraser-Smith AC (1997) Global lightning and climate variability inferred from ELF magnetic field observations. Geophys Res Lett 24:2411–2414

Füllekrug M, Constable S (2000) Global triangulation of lightning discharges. Geophys Res Lett 27:333–336

Hayakawa M (2009) A review on direction finding of VLF/LF sferics. J Atmos Electr 29(1):35–52

Heckman SJ, Williams ER, Boldi R (1998) Total global lightning inferred from Schumann resonance measurements. J Geophys Res 103(31):31775–31779

Hobara Y, Iwasaki N, Hayashida T, Tsuchiya N, Williams ER, Sera M, Ikegami Y, Hayakawa M (2000) New ELF observation site in Moshiri, Hokkaido Japan and the results of preliminary data analysis. J Atmos Electr 20(2):99–109

Hobara Y, Hayakawa M, Williams E, Boldi R, Downes E (2006) Location and electrical properties of sprite producing lightning from a single ELF site. In: Füllekrug M, Mareev EA, Rycroft MJ (eds) Sprites, elves and intense lightning discharges. Springer, Dordrecht, pp 211–235

Huang E, Williams E, Boldi R, Heckman S, Lyons W, Taylor M, Nelson T, Wong C(1999) Criteria for sprites and elves based on Schumann resonance observations. J Geophys Res 104:16943–16964

Ishaq M, Jones DLl (1977) Method of obtaining radiowave propagation parameters for the Earth-ionosphere duct at ELF. Electronic Lett 13:254–255

Iwai A, Ohtsu J, Nishino M, Kashiwagi M (1969) A new direction finding network for locating the sources of atmospherics. Proc Res Inst Atmospherics Nagoya University 16:17–20

Iwai A, Kashiwagi M, Nishino M, Satoh M (1979) Triangulation direction finding network for fixing sources of atmospherics. Proc Res Inst Atmospherics Nagoya University 26:1–16

Jones D Ll (1969) The apparent resonance frequencies of the Earth-ionosphere cavity when excited by a single dipole source. J Geomagn Geoelectr 21:679–684

Jones D Ll (1970a) Numerical computations of terrestrial ELF electromagnetic wave fields in the frequency domain. Radio Sci. 5:803–809

Jones DL (1970b) Propagation of ELF pulses in the Earth-ionosphere cavity and application to slow tail sferics. Radio Sci 5:1153–1163

Korovkin NV, Chechurin VL, Hayakawa M (2007) Inverse problems in electric circuits and electromagnetics. Springer, Berlin, p 331

Krider EP, Noggle RC, Uman MA (1976) A gated, wideband magnetic direction finder for locating return strokes. J. Appl. Meteor. 15:301–306

Lawson C, Hanson LRJ (1974) Solving least squares problems, Chap. 23. Prentice Hall, New Jersey

Lee ACL (1986) An experimental study of the remote location of lightning flashes using a VLF arrival time difference technique. Q J Roy Meteor Soc 112:203–229

Nickolaenko AP (1997) Modern aspects of the Schumann resonance studies. J Atmos Solar-Terr Phys 59(7):805–816

Nickolaenko AP, Sátori G, Zieger B, Rabinowicz LM, Kudintseva IG (1998) Parameters of global thunderstorm activity deduced from long term Schumann resonance records. J Atmos Solar-Terr Phys 60:387–399

Nickolaenko AP, Hayakawa M (2002) Resonances in the Earth-ionosphere cavity. Kluwer Academic Publishers, Dordrecht-Boston-London, p 380

Nickolaenko AP, Hayakawa M (2007) Diurnal variations in Schumann resonance intensity in the local and universal times. J Atmos Electr 27(2):83–93

Nickolaenko AP, Hayakawa M (2010) Model disturbance of Schumann resonance by the SGR 1806-20 γ-ray flare on December 27, 2004. J Atmos Electr 30(1):1–11

Nickolaenko AP, Yatsevich EI, Pechony OB (2008) Diurnal and seasonal variations in the intensities and peak frequencies of the first three Schumann-resonance modes. Radiophys Quantum Electron 51(7):528

Nickolaenko AP, Yatsevich EI, Shvets AV, Hayakawa M, Hobara Y (2011a) Schumann-resonance records at three observatories and ELF universal- and local-time variations. Radiophys Quantum Electron 53(12):706–716. doi:10.1007/s11141-011-9263-y

Nickolaenko AP, Yatsevich EI, Shvets AV, Hayakawa M, Hobara Y (2011) Universal and local time variations deduced from simultaneous Schumann resonance records at three widely separated observatories. Radio Sci 46:RS5003. doi:10.1029/2011RS004663

Pechony O, Price C (2007) Schumann resonances: interpretation of local diurnal intensity modulations. Radio Sci 42(2). doi:10.1029/2006RS003455

Pechony O (2007) Modeling and Simulations of Schumann resonance parameters observed at the Mitzpe Ramon field station (Study of the day-night asymmetry influence on Schumann resonance amplitude records), Ph.D. Thesis, Tel-Aviv University, Israel, March 2007, p 92

Pechony O, Price C, Nickolaenko AP (2007) Relative importance of the day-night asymmetry in Schumann resonance amplitude records. Radio Sci 42(2):RS02–RS06. doi:10.1029/2006RS003456

Pinto O Jr, Pinto IRCA, Saba MMM, Naccarato KP (2009) Cloud-to-ground lightning observations in Brazil. In: Betz HD, Schumann U, Laroche P (eds) Lightning: principles, instruments and applications. Springer, Dordrecht, pp 209–229

Polk C (1969) Relation of ELF noise and Schumann resonances to thunderstorm activity. In: Coronoti SC, Hughes J (eds) Planetary electrodynamics, vol 2. Gordon and Breach, New York, pp 55–83

Sátori G (2003) On the dynamics of the north—south seasonal migration of global lightning. Proceeding of the 12th international conference on atmospheric electricity. Versailles, France, pp 1–4

Sátori G, Williams ER, Boccippio DJ (2003) On the dynamics of the north—south seasonal migration of global lightning. AGU fall meeting, San Francisco, 8–12 Dec 2003. P. AE32A-0166

Sátori G, Williams E, Lemperger I (2008) Variability of global lightning activity on the ENSO time scale. Atmos Res. doi:10.1016/j.atmosres.2008.06.014

Sentman DD, Fraser BJ (1991) Simultaneous observation of Schumann resonances in California and Australia: evidence for intensity modulation by local height of D region. J Geophys Res 96(9):15973–15984

Shvets AV (2001) A technique for reconstruction of global lightning distance profile from background Schumann resonance signal. J Atmos Solar-Terr Phys 63(10):1061–1074

Shvets AV, Hayakawa M, Sekiguchi M, Ando Y (2009) Reconstruction of the global lightning distribution from ELF electromagnetic background signals. J Atmos Solar-Terr Phys 71:1405–1412

Shvets AV, Hobara Y, Hayakawa M (2010) Variations of the global lightning distribution revealed from three station Schumann resonance measurements. J Geophys Res 115:A12316. doi:10.1029/2010JA015851
Shvets A, Hayakawa M (2011) Global lightning activity on the basis of inversions of natural ELF electromagnetic data observed at multiple stations around the world. Surv Geophys 32:705–732. doi:10.1007/s10712-011-9135-1
Tikhonov AN (1963) Solution of incorrectly formulated problems and the regularization method. Soviet Math Dokl 4:1035–1038. English translation of Dokl. Akad Nauk SSSR 151:501–504
Tikhonov A, Arsenin V (1977) Solutions of ill-posed problems. Wiley, New York
Troyan V, Hayakawa M (2002) Inverse geophysical problems. Terrapub, Tokyo, p 289
Yamashita K, Otsuyama T, Hobara Y, Sekiguchi M, Matsudo Y, Hayakawa M, Korepanov V (2009) Global distribution and characteristics of intense lightning discharges as deduced from ELF transients observed at Moshiri(Japan). J Atmos Electr 29(2):71–80
Yatsevich EI, Nickolaenko AP, Shvets AV, Rabinowicz LM (2006) Two component source model of Schumann resonance signal. J Atmos Electr 26(1):1–10

Chapter 11
SR and Global Temperature

SR oscillations are excited by electromagnetic radiation from the global lightning, whose intensity is proportional to the level worldwide thunderstorm activity. Since thunderstorms are driven by the air convection, one may expect that SR intensity is connected with the global soil temperature. When the surface temperature grows, the convection increases, the water vapor transport becomes stronger, and the clouds become thicker. Vertical motion of particles in a cloud leads to its electrification. Therefore, the thunderstorm activity should increase with the convection and the underlying surface temperature, and this effect must be observed as relevant increase in the SR intensity. After a short historical review, we formally compare the data of SR monitoring with the data on the global soil temperature from November 1998 to May 2002. By using the singular spectral analysis (SSA), we single out the annual, semi-annual, and interannual variations from the SR record and the temperature data within different latitude intervals and compare these. Analyses demonstrated that annual SR variations follow the temperature alterations within the mid-latitude interval. Simultaneously, the inter-annual variations in the resonance intensity are similar to the temperature anomaly of the tropical belt. The non-linear relations were derived that allowed us to compute SR cumulative intensity from the temperature data or to reconstruct the global temperature from the resonance monitoring.

11.1 SR Intensity and Soil Temperature

A typical SR spectrum is shown in Fig. 11.1 observed at Moshiri. The horizontal axis depicts frequency in Hz, and the vertical axis corresponds to the spectral amplitude of the south–north magnetic field component in pT/(Hz)^{½}. The geographic coordinate system is used. The resonant peaks of different modes (up to six) are seen in the spectrum. We will demonstrate how this and similar records might be used for evaluating a link between the soil temperature and the thunderstorm activity (Sekiguchi et al. 2004, 2006; Hayakawa et al. 2006, 2011;

A. Nickolaenko and M. Hayakawa, *Schumann Resonance for Tyros*, Springer Geophysics, DOI: 10.1007/978-4-431-54358-9_11,

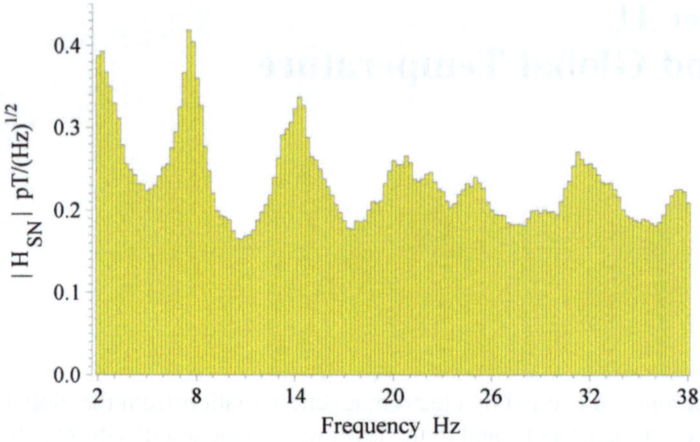

Fig. 11.1 Typical amplitude spectrum of the SR recorded at the Moshiri observatory (Japan)

Nickolaenko et al. 2008; Hobara et al. 2011). Though, there is a recent publication based on the Universal time (UT) variations extracted from the resonance records (see Chap. 10) rather than on the cumulative intensity of SR (Yatsevich et al. 2012).

Williams (1992) demonstrated similarity of the temporal variations in the long-term record of the SR amplitude and the tropical temperature anomaly, as seen in Fig. 11.2. Here, the months are plotted on the horizontal axis and the amplitude is plotted on the left vertical axis of the single horizontal magnetic field component. The anomaly of the tropical air temperature near the ground surface is plotted along the right vertical axis. It is obvious that that the 6-dB variation in the SR

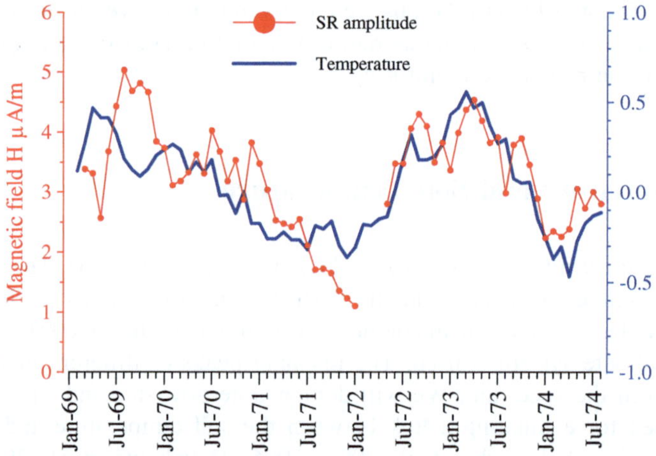

Fig. 11.2 Tropical temperature anomaly °K (*thick blue curve*) and the SR amplitude μA/m (*thin red line with dots*), adapted from *Williams* (1992)

amplitude corresponds to the temperature variation by 0.5° K. The H horizontal magnetic field component was measured by the Polk group, which is in fact the south–north component. A possible difference in the field amplitude is related to a small magnetic declination, which is about $-10°$ at the observatory. The SR observations were performed in the 1960–1970s at the West Greenwich observatory, Rhode Island (USA) during 5.5 years (Clayton and Polk 1977).

Below we use the ground surface temperature (GST) rather than the anomaly of the near-surface air temperature (Sekiguchi et al. 2004, 2006). The temperature is compared with the global thunderstorm activity estimated by the cumulative intensity of the global resonance (the field intensity integrated over three resonance modes). The observations are used that were performed in Japan from 1999 to 2002 and cover 43 months.

We obtain the quantitative characteristics of the relation between the ground temperature and the intensity of the SR. Various techniques of experimental data processing are used. For example, the degree of linear connection between the time series is estimated by the cross-correlation coefficient. Separating periodic annual and semi-annual variations from the inter-annual trends is performed by the principal component analysis (Danilov 1996; Danilov and Zhiglyavsky 1997; Golyandina et al. 2001; Troyan and Hayakawa 2002). The links between the principal components is established by using the regression analysis. This will allow us to "retrieve" one series from the record of the other one and compare the original and retrieved records. Their reciprocity justifies the approach and allows for evaluating the accuracy and reliability of the results obtained.

Similarly to Clayton and Polk (1977) and Williams (1992), we use a single horizontal magnetic field component H_{SN}. The magnetic field was oriented in the geographic rather than geomagnetic coordinates. Experimental data on the temperature were taken from global temperature distribution published on the site of National Climate Data Center (USA) http://www.ncdc.noaa.gov/oa/climate/research/ghcn/ghcn.html. This worldwide temperature distribution is given in a form of map with the cells of 15° along the longitude and 5° on the latitude. After superimposing the continental outline on this map, one may compute the soil temperature integrated in the symmetric latitude intervals (Sekiguchi et al. 2004, 2006; Hayakawa et al. 2006; Hobara et al. 2011; Yatsevich et al. 2012). Owing to symmetric intervals, the tropics were always included in the consideration as the area where the global thunderstorms are concentrated.

SR monitoring is continued at the Moshiri observatory, Japan (geographic coordinates 44.365° N and 142.24° E) since 1998. The map in Fig. 11.3 depicts the observatory with respect to the world thunderstorm centers in a special projection. The observer is located at the center of the cylindrical coordinate system, which clearly shows the directions to the African, American, and Asian thunderstorm centers. The angular pattern of the horizontal magnetic H_{NS} antenna is superimposed on the map. Obviously, the antenna is sensitive to the signals arriving from the east and west, i. e., from the African and American thunderstorms. The signals from Asian sources are also detectable due to their closeness to the observer, and this partially compensates the antenna pattern. Thus, the SR

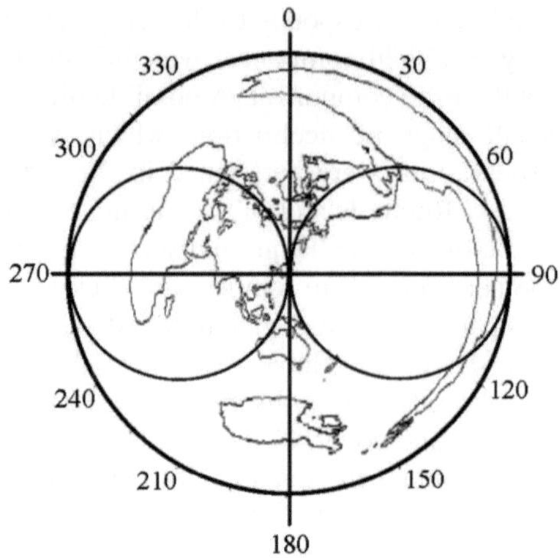

Fig. 11.3 The world map centered at the Moshiri observatory (Japan). The H_{SN} antenna pattern is shown

intensity is determined by the contributions from all three global thunderstorm centers. We use the continuous record of the H_{NS} field component from November 1998 to May 2002.

We performed the following signal processing. The SR power spectra were computed by using the FFT of the 10-s record segments. This provided the frequency resolution of 0.1 Hz and simultaneously allowed for efficient elimination of the spectra distorted by the interference. The selected elementary power spectra were averaged over the 10-min intervals and accumulated thus forming the "initial" data set. During the subsequent processing, the one-hour average of the power spectra was obtained (the corresponding amplitude spectrum is shown in Fig. 11.1). These spectra relevant to 24 h of the day were monthly averaged then to provide the "intermediate" data set. If we average the latter over the 24 h, we obtain the mean power spectrum of SR for each month of observations (the 43 months as a whole). The cumulative intensity of electromagnetic oscillations as a function of month was found by integrating these mean spectra.

The amplitude of resonance oscillations of separate modes depends on the intensity of the source, and also on the source–observer distance. To reduce this latter, we used the cumulative resonance power:

$$P = A_1^2 + A_2^2 + A_3^2 \tag{11.1}$$

where A_n is the amplitude of the mode $n = 1, 2, 3$. The total intensity was obtained as the integral of the power spectrum over the frequency range ±0.5 Hz from the peak frequencies of the first three resonance maxima.

Figure 11.4 shows the diurnal–seasonal variations of the cumulative resonance intensity over the entire period of observations. One may observe that contributions of all global thunderstorm centers are present in the record. The maximum of activity is related to the Asian thunderstorms in the vicinity of 5–8 UT, while contributions of the African and American thunderstorms dominate in the 14–16

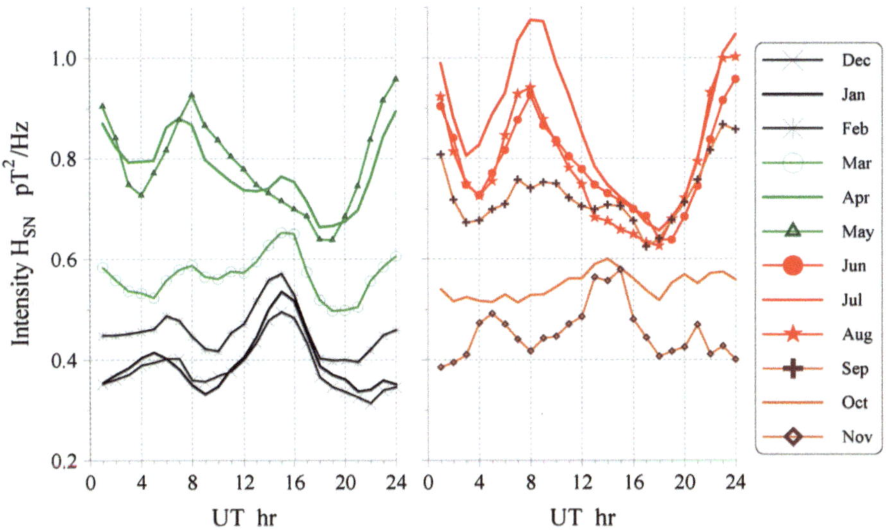

Fig. 11.4 Thunderstorm activity: diurnal variations in the total SR intensity in the H_{NS} field component recorded in Japan

UT and 20–24 UT intervals respectively. Contributions from the centers vary during a year, and their seasonal motion additionally alters the shape of diurnal pattern.

Seasonal variations in the SR intensity and the ground temperature are shown in Fig. 11.5. The figure contains two frames. The thick blue curve in the upper frame of Fig. 11.5 demonstrates the experimentally observed cumulative SR intensity (in dB relative to 1 pT^2). The red curve with diamonds shows the SR intensity deduced from the ground temperature in the latitude band ±60° (see below). The color curves in the lower frame of Fig. 11.5 present the ground surface temperature measured in the symmetric latitude intervals: 20° S–20° N; 40° S–40° N; 60° S–60° N; and 80° S–80° N. The violet line in the upper plot of the lower frame refers to the observational data in the interval of ±20° latitudes, and the magenta curve with diamonds depends the tropical temperature inferred from the electromagnetic data (see below). As one may note, there is a good correspondence between the ELF electromagnetic records and the data of space-borne monitoring of the global ground surface temperature.

Application of symmetric latitude intervals decreases the role of the seasonal thunderstorm drift from south to north and backward. Besides, the temperature is always present in the data of the equatorial belt where the global thunderstorms concentrate. As is seen in the lower frame of Fig. 11.5, the tropical zone is characterized by the significant semi-annual temperature variation. This component is also observed the optical observation of lightning activity from space (Christian et al. 2003). The semi-annual variations are conditioned by the distinction of the tropical climate with two rainy and two dry seasons following each

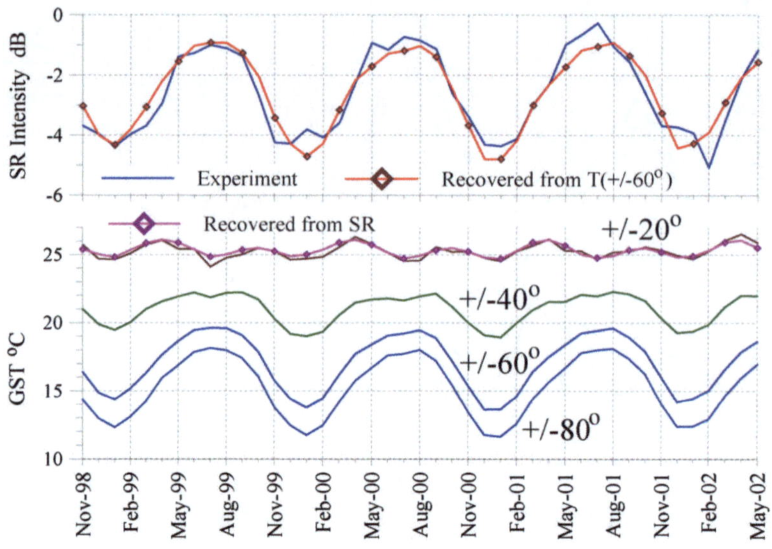

Fig. 11.5 Seasonal variations in the electromagnetic and temperature data

other during the year. Annual motion of the Sun in the sky is combined with the asymmetric global distribution of the continents, and significant annual temperature variations arise in the middle latitudes. This 12 months period exposes itself in the lower plots of Fig. 11.5, and it is also present in the statistics of optical lightning flashes: the semi-annual variation dominates at low latitudes, and it is hidden under the annual variation when the latitude interval increases. Simultaneously with these periodic variations, a clear decrease exists of the median temperature related to the widening of the latitude interval. This is not a surprise, but it is colder at the Poles than in the tropics.

Figure 11.6 shows the cross-correlation coefficient between the SR intensity and global surface temperature as a function of the latitude interval. The abscissa depicts the half-width of the latitude interval and the ordinate shows the relevant cross-correlation coefficient. The red curve with dots corresponds to the cross-correlation between the temperature and the SR intensity itself, while the blue line with crosses shows the cross-correlation between the temperature and the resonance intensity measured in decibels. One may observe that correlation is systematically higher in the second case. Therefore, we may state that the intensity of SR and the ground temperature are connected by the non-linear exponential law.

The cross-correlation coefficient exceeds the level of 0.5 when the latitude interval is greater than $\pm 30°$. The slope of the curves sharply decreases when the latitude exceeds the $\pm 45°$ interval, which is conditioned by a substantial decrease in the number of thunderstorms in the high latitudes.

The relation between the resonance intensity and ground temperature including the phase and frequency relationships is demonstrated in the clearest way by the

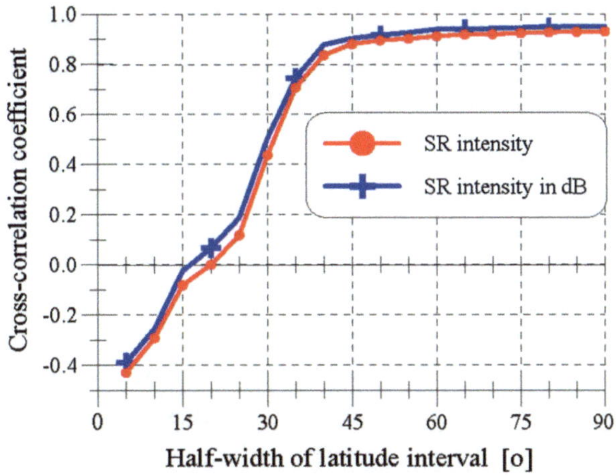

Fig. 11.6 Latitudinal dependence of the cross-correlation coefficient of the global surface temperature and the intensity of resonance oscillations

Lissajous figures (phasors) in Fig. 11.7. Here, the resonance intensity is plotted along the abscissa and the ground temperature is shown on the ordinate. The plot includes nine figures, and the unprocessed experimental data are shown in the upper line of frames.

Variations of the ground temperature in the interval from 20° S to 20° N and the SR intensity are compared in the upper plots of Fig. 11.7. The middle and right plots correspond to the intervals of 40° S–40° N and 60° S–60° N. The tendency is evident toward the coherent long-term variations of the ground temperature and the SR intensity. The middle and the bottom lines of plots in Fig. 11.7 illustrate the phasors of the principal components of the time series corresponding to the annual period and to the sum of annual and semi-annual variations. To single out the trends and periodic components from the data series, we applied the principal component analysis (the "Caterpillar" procedure already used before). Essentials of this method, which is called the singular spectral analysis (SSA), will be addressed in Chap. 14.

Patterns of seasonal variations in the ground temperature depend on the latitude interval and comprise two evident components. One of them is the semi-annual period related to the seasonal variations in the tropics. The other component is the annual variation dominating at the mid-latitudes. Two components are also present in the SR data. A significant semi-annual component was found in the resonant frequency of the basic mode in the vertical electric field oscillations, detected in Japan at the end of the 1960s—the beginning of the 1970s (see Chap. 5). It was shown that the range of diurnal variations of the first resonance frequency ΔF_1 is related to the effective size S of the area occupied by the thunderstorms and to the number of midnight lightning flashes N_F detected from the space by using optical transient detector (OTD). The analytical relation was obtained of the form $N_F \sim S^{1/3}$. Processing of the SR long-term records performed in Hungary also allowed for distinguishing the semi-annual and annual variations (see Chap. 5).

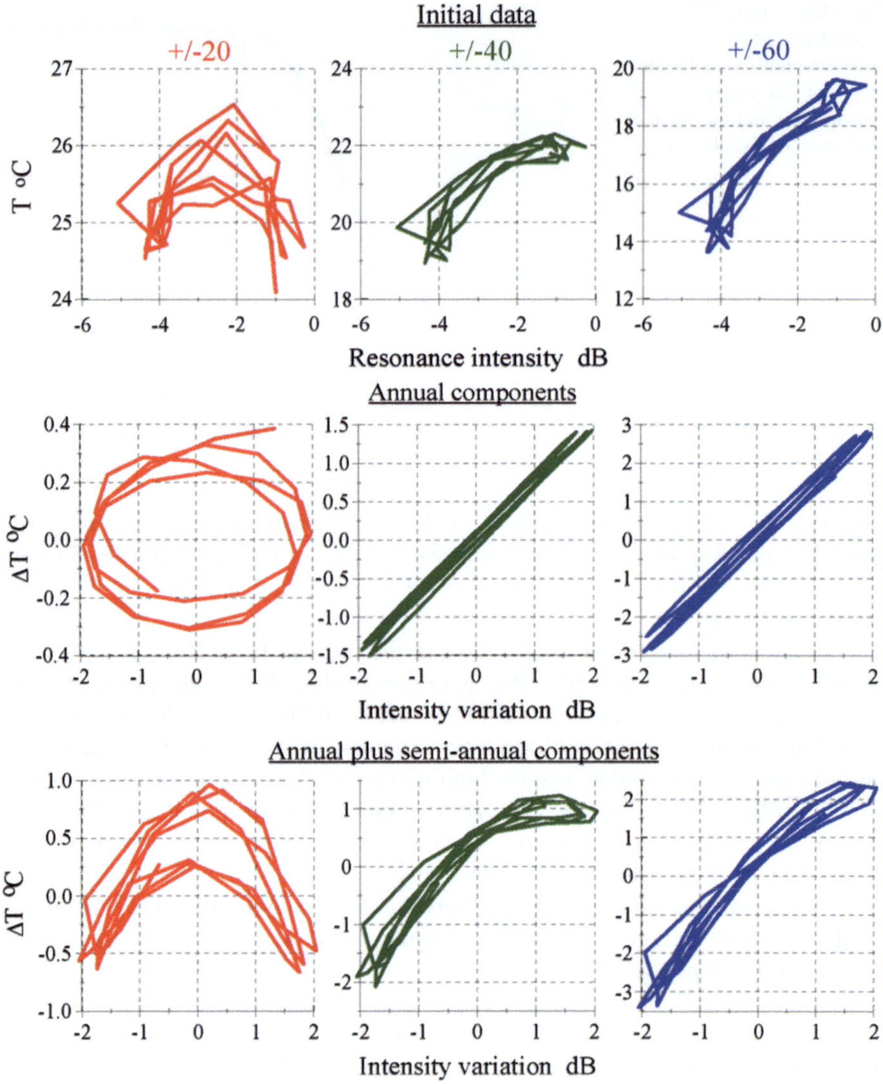

Fig. 11.7 Lissajous figures of variations in the ground temperature T versus the SR intensity in dB

In the data processing, we have chosen the caterpillar length $L_G = 12$ months, which facilitates separating the inter-annual trend from the annual and semi-annual variations. Trends were relatively small in the particular data sets therefore, the first and the most powerful pair of principal components PC_1 and PC_2 corresponds to the annual variation. Components PC_3 and PC_4 are usually relevant to the semi-annual period.

Table 11.1 Annual, semi-annual, and inter-annual components retrieved by the principal component analysis

Variable	Annual variation (%)		Semi-annual variation (%)		Inter-annual variation (%)
SR intensity	$PC_1 = 49.5$	$PC_2 = 45.6$	$PC_3 = 1.2$	$PC_4 = 1.1$	$PC_5 + PC_6 = 1.2$
Temperature in interval of ±20°	$PC_3 = 9.5$	$PC_4 = 8.2$	$PC_1 = 34.6$	$PC_2 = 33.7$	$PC_8 = 1.7$
Temperature in interval of ±40°	$PC_1 = 43.5$	$PC_2 = 40.6$	$PC_3 = 7.0$	$PC_4 = 6.7$	$PC_5 = 1.0$
Temperature in interval of ±60°	$PC_1 = 49.0$	$PC_2 = 46.1$	$PC_3 = 2.1$	$PC_4 = 2.0$	$PC_5 = 0.47$
Temperature in interval of ±80°	$PC_1 = 49.8$	$PC_2 = 46.8$	$PC_3 = 1.4$	$PC_4 = 1.4$	$PC_5 = 0.35$

Numbers show the relative intensity of separate components

The annual period is present in the SR records and in temperature variations. The semi-annual component is significant only in the temperature data. The results are collected in Table 11.1. The first column of this table lists the variables, columns 2 and 3 and columns 4 and 5 correspond to the annual and semi-annual variations, respectively. The last column shows the inter-annual variations (trend). Strictly speaking, the data from this particular column are the most interesting in the context of the global warming. However, they should be considered as preliminary estimate, since the duration of the time series was only 43 months, which is rather short for a reliable selection of the slow and minute trends.

The numbers in Table 11.1 characterize the relative intensity of the principal components in the initial signal. The semi-annual variation in the cumulative intensity of SR is small and it reaches 3 %, while the annual component provides 93 % of the total intensity. On the contrary, annual variations in the tropic temperature are insignificant: their contribution is 18 % against the 68 % of the semi-annual variation (the latter is represented by the components PC_1 and PC_2 in the tropics).

Let us return to Fig. 11.7 where the left column of plots shows the temperature variations from the ±20° interval against the SR intensity measured in dB. The horizontal axis shows that deviations of the resonant intensity from its mean value are ±2 dB. The ordinates show variations in the ground surface temperature. The middle and the right columns of plots correspond to the intervals ±40° and ±60°. The middle line of frames presents the Lissajous figures for the one-year principal components. The lower line of plots in Fig. 11.7 demonstrates variations of the sum of the annual and semi-annual components.

One may note that annual components of the global temperature and SR intensity vary in phase, provided that we consider the temperature in the middle and high latitude intervals. Simultaneously, the tropical land temperature leads against the annual variations of resonance intensity by approximately 90° or three months, thereby transforming the Lissajous figure into a circle at the middle left frame in Fig. 11.7. Owing to the small annual component in the tropical temperature and to the phase shift, the cross-correlation coefficient between the resonance intensity and tropical temperature is close to 0.1. Besides, as the lower left frame shows, the temperature and electromagnetic signals have the frequency ratio of 2:1, hence, the link of these processes is non-linear (quadratic). When the latitude interval is increased, the temperature begins to vary in phase with the intensity of SR. This is clearly seen from the middle and right lower frames where the phasor only slightly deviates from a straight line. Plots of the lower line deviate from the figures in the middle row due to the presence of semi-annual component, which distorts the in-phase variations of the base annual period.

The connection of annual changes between the SR intensity and the ground surface temperature might be readily established, with the help of linear regression. Let us choose for example the temperature variations in the latitude interval ±60° and derive the regression coefficients A_P and B_P in the relationship

$$P[dB] = A_P + B_P \cdot T_{60} \tag{11.2}$$

where P is the intensity of resonant oscillations and T_{60} is the monthly averaged ground surface temperature in the ±60° latitude interval. The linear-regression coefficients are calculated from the following relations:

$$B_P = \frac{N \cdot \sum_{k=1}^{N} (T_{60} \cdot P[dB]) - \sum_{k=1}^{N} P[dB] \cdot \sum_{k=1}^{N} T_{60}}{N \cdot \sum_{k=1}^{N} (T_{60})^2 - \left(\sum_{k=1}^{N} T_{60} \right)^2} \tag{11.3}$$

$$A_P = \frac{\sum_{k=1}^{N} P[dB] - B_P \cdot \sum_{k=1}^{N} T_{60}}{N} \tag{11.4.}$$

where N is the ensemble length of 43 months. The results of analysis are shown in Table 11.2.

Table 11.2 Results of the linear regression for the data on the SR intensity and the ground surface temperature in different latitude intervals

Type of data	Latitude interval	A_P	B_P	Cross-correlation coefficient R
Initial data	±20°	-7.01021	0.173658	0.0676055
	±40°	-25.7577	1.10382	0.876738
	±60°	-13.6444	0.648358	0.938440
	±80°	-11.7425	0.600135	0.946990
$PC_1 + PC_2$	±40°	-0.02433	1.32549	0.996715
	±60°	-0.007697	0.679027	0.995494
	±80°	-0.004666	0.617393	0.994625

The following features should be mentioned of the obtained parameters.

(1) The constant terms A_P are close to zero for the principal components as was expected, since the processing algorithm removes the trends and noise.
(2) The cross-correlation coefficient of the principal components is higher than that of the initial data. This is also related to the data filtering during the principal component analysis.
(3) The coefficients B_P are close to each other when computed for the initial data and the sums $PC_1 + PC_2$ in the fixed latitude intervals.

By using Table 11.2, we obtain the following relations for the ±60° interval:

$$P[dB] = -13.64 + 0.648 \cdot T_{60} \tag{11.5}$$

which is equivalent to

$$P[dB] = -0.077 + 0.679 \cdot \Delta T_{60} \tag{11.6}$$

The ground surface temperature was used in Eq. (11.5) from the latitude interval ±60°, while Eq. (11.6) employs the principal components ($PC_1 + PC_2$), i.e., the annual periodic deviations of this temperature from the average value $\langle \boldsymbol{T}_{60} \rangle = 16.96$ °C. Clearly, the linear relation (11.5) corresponds to the non-linear link between the ground temperature and the absolute intensity of the global electromagnetic resonance measured in pT^2:

$$P\left[pT^2\right] = P_0\left[pT^2\right] \cdot \exp\{\zeta_{60} T_{60}\} \tag{11.7.}$$

Here $P[pT^2]$ is the current intensity of SR in pT^2, the constant $P_0[pT^2] = 0.04325$, the coefficient $\zeta_{60} = 0.0648 \cdot \ln(10) = 0.1492$, and the median temperature for a particular month is taken in the ±60° latitude interval.

Equation (11.7.) relates the intensity of the horizontal magnetic field component H_{SN} at the Moshiri station to the global ground temperature in the interval from 60° S to 60° N. Naturally, the ζ coefficient depends on the selected latitude interval. However, it is sufficiently stable. For instance, we obtain correspondingly: $\zeta_{40} = 0.1104 \ln(10) = 0.2327$ and $\zeta_{80} = 0.06 \ln(10) = 0.1382$ in the ±40° and ±80° intervals. Relevant changes are rather small.

Table 11.1 and Eq. (11.6) allows us to calculate ("retrieval") the seasonal variations in the intensity of SR from the ground temperature variations in the ±60° interval. The upper plot in Fig. 11.5 demonstrates an outstanding agreement between the recovered curve with electromagnetic observations ((line with diamonds and blue smooth curve). These results are also in accord with the literature data (Sátori and Zieger 1996, 1999; Sátori 2003; Sátori et al. 2003), which show that variations of the peak frequencies and intensities of the first two SR modes are similar to the surface temperature pattern in the Northern and Southern hemispheres. Unfortunately, quantitative characteristics were not available in these works. By acknowledging the fact that thunderstorms occur above the continents (WMO 1956; Christian et al. 2003), we compared the SR observations with the ground surface temperature. Our 3-centigrade annual temperature

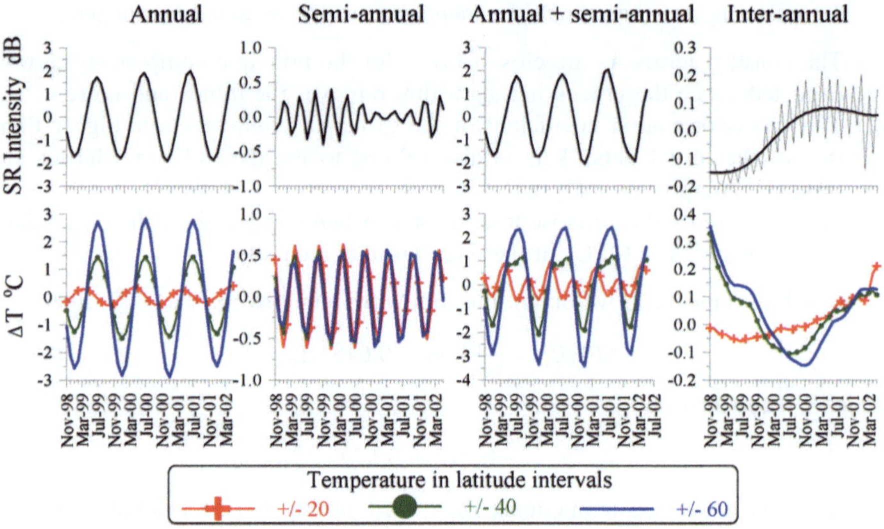

Fig. 11.8 Principal components of the time series of SR cumulative intensity (*upper plots*) and the global ground temperature (*lower plots*)

variation is close to the value of 2 °C obtained by Sátori (2003) and Sátori et al. (2003) relevant to the ocean surface.

Figure 11.8 illustrates temporal variations of dominating principal components in the resonance intensity and the global temperature. The period of time in months, is shown on the abscissas. The upper line of plots presents the electromagnetic data, namely, the annual and semi-annual components, their sum, and the inter-annual trend. The plots in the second line demonstrate the same principal components of the temperature pertinent to different latitude intervals. One may observe to what extent the data are similar, especially if we compare the SR annual variations and the temperature of the mid-latitudes. The "beat mode" is observed in the semi-annual component of SR intensity. The similar behavior was noted in Chap. 5 for other sets of data. Plots of semi-annual temperature variations demonstrate close coincidence in all latitude intervals, because this period is pertinent only to the tropical belt. The annual component arises and then dominates in the data of the ground surface temperature in the wider latitude intervals.

The right plots in Fig. 11.8 depict the inter-annual variations in the SR intensity (upper plot) and the ground surface temperature (lower plot). In fact, these are the anomalies of the temperature and the intensity of SR for the period from November 1998 to May 2002. These look similar to variations of Fig. 11.2 with an exception that our independent data were obtained 30 years after C. Polk's records. The expansion into principal components allowed for rigorous by separating the quantitative inter-annual terms. The inter-annual variation of SR is shown by two lines. The thin fluctuating line is the inter-annual principal component obtained directly from the observation, while the thick black line is the result of smoothing. Concerning the temperature anomalies in different bands, their outlines depend on the chosen latitude interval. We want to stress that only the tropical temperature anomaly behaves similarly to the SR intensity. Duration of the data series presented here is somewhat shorter than those used by Williams (1992), but still the final results are similar. It would be interesting to process likewise the data of other SR observatories.

11.2 Non-Linear Connections

The lower left frame in Fig. 11.7 depicts the Lissajous figure for a sum of annual and semi-annual components pertinent to the land temperature in the ±20° latitude interval. The characteristics of this curve in this frame indicate that the frequency ratio of two processes is equal to 2:1. This non-linear link is responsible for the low cross-correlation between the resonance intensity and tropic land temperature. We will compare the principal components of the annual and semi-annual variations in Fig. 11.9. The components are presented by obvious approximating functions:

$$\Delta T_{\mathrm{ANNO}} = -0.3\cos(M\pi/6)\ [^{\circ}\mathrm{C}] \tag{11.8}$$

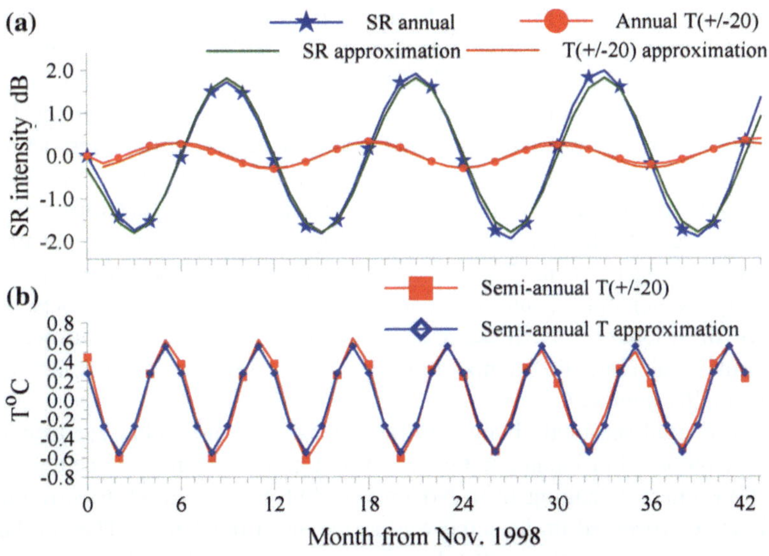

Fig. 11.9 Approximation of the principal components in annual variation of SR intensity and semi-annual changes of the ground surface temperature in the tropical interval of ± 20°

and

$$\Delta P_{\text{ANNO}} = -2\cos(M\pi/6)[\text{dB}]. \tag{11.9}$$

Here M is the month number counted from the beginning of the record (November of 1998).

The blue curve with stars in the upper frame of Fig. 11.9 shows the annual principal component derived from measurements of the SR intensity. The red line with dots depicts annual variation of the ground surface temperature observed in the ±20° latitude interval. Smooth lines in the upper frame of Fig. 11.9 represent the approximating functions (11.8) and (11.9). A close similarity is apparent of measurements and approximating functions. The lower frame of Fig. 11.9 shows the semi-annual variations of the ground surface temperature in the tropical belt. The principal component of observed temperature is shown by the red curve with squares, and the blue line with diamonds corresponds to the approximating dependence $\Delta T_{1/2} = 0.55\cos(M\pi/3)$. After performing a few simple transformations we obtain the following relation between the temperature anomaly and annual variation of SR intensity:

$$\begin{aligned}\Delta T &= 0.55\cos\left(\frac{\pi M}{3}\right) - 0.3\cos\left(\frac{\pi M}{6}\right)\\ &= 0.55\left[1 - 2\cdot\left(\frac{\Delta P_{\text{ANNO}}}{2}\right)^2\right] - 0.3\cdot\left(\pm\sqrt{1-\left(\frac{\Delta P_{\text{ANNO}}}{2}\right)^2}\right)\end{aligned} \tag{11.10}$$

The ± sign in front of the square root in Eq. (11.10) accounts for the change of the sign of the $\cos(\pi M/6)$ function. In particular, the square root is positive when $1 \le M < 3$, and it is negative when $3 \le M < 9$. Then, it is positive again ($9 \le M < 15$), etc. We used Eq. (11.10) for calculating the seasonal variations of the ground tropical temperature from the SR records performed at the Moshiri observatory. The "recovered" seasonal changes of the ground temperature in the ±20° interval are shown in the lower frame of Fig. 11.5. Two lines are compared there: the smooth violet line demonstrates the observations results and the pink curve with diamonds depicts the temperature variations "reconstructed" from the SR data by using Eq. (11.10). It is obvious that both these curves are practically coincident. Thus, the parameters of non-linear link between the intensity of resonance oscillations and the tropical temperature allowed us to successfully calculate the temperature variations relevant to regular seasonal changes of SR intensity.

We compared seasonal changes in the intensity of the global electromagnetic resonance recorded in Japan and the concurrent global ground surface temperature in the time interval ranging from November 1998 to May 2002. Electromagnetic records were processed in the same way as we did it in Chap. 5. The similarity of results obtained demonstrates the long-term stability of the SR parameters and their connection to periodic changes of the global temperature.

Similarly to the previously mentioned data, cumulative intensity of the horizontal magnetic field component H_{SN} varied by 5 dB during the year (practically a two-fold variation). The annual principal component varies with respect to the average value by ±2 dB. We must note that "modern" data on the SR intensity have a smaller semi-annual variation in comparison with the records of 1960–1970s. In both these cases, the semi-annual component looks like "beating mode". It is possible that recent observations coincided with some minimum of the semi-annual variations, and these latter are controlled by some undisclosed mechanism. Processing of longer time series of records might help in establishing the nature of these modulations.

The cross-correlation coefficient between the SR intensity and the ground surface temperature demonstrates a strong link between these data. Its values are equal to 0.05, 0.85, 0.92, and 0.95 for the latitude intervals ±20°, ±40°, ±60°, and ±80° correspondingly. Analogous results were obtained recently by Hobara et al. (2011) who applied the cumulative intensity of SR records at the Nakatsugawa (Japan) observatory. A comparison was also made between the global land temperature and the Universal time factors (see Chap. 10) derived from the concurrent records of three widely separated SR observatories (Yatsevich et al. 2012).

Regression analysis of the paired observational data allows for relating the resonance intensity with the ground surface temperature. By applying Eq. (11.5), we can reconstruct the experimentally observed seasonal variation of the SR intensity from the mid-latitude ground temperature. By comparing annual and semi-annual components we derived the quadratic approximating formula (11.10), which successfully restores the tropical temperature variation from the electromagnetic resonance records. All experimental data confirm that SR is indeed a special global thermometer, which allows for electromagnetic monitoring of the planetary surface temperature in different latitude intervals.

We must specially note that the inter-annual trend in the ELF electromagnetic intensity showed a similarity to the trend in the ground temperature of the tropical belt. In the context of global warming, this particular issue should be an objective of further investigations based on the longer time series of data.

References

Christian HJ, Blakeslee RJ, Boccippio DJ, Boeck WL, Buechler DE, Driscoll KT, Goodman SJ, Hall JM, Koshak WJ, Mach DM, Stewart MF (2003) Global frequency and distribution of lightning as observed from space by the optical transient detector. J Geophys Res 108(D1):4005. doi:10.1029/2002JD002347

Clayton M, Polk C (1977) Diurnal variation and absolute intensity of world-wide lightning activity September 1970 to, May 1971. In: Dolezalek H, Reiter R (eds) Electrical processes in atmospheres. West Germany, Darmstadt, pp 440–449

Danilov DL (1996) Proceeding statistical computing section of the American statistical association, p 156

Danilov DL, Zhiglyavsky AA (1997) Principal components of the time series: the "caterpillar" method. St. Petersburg State University, St. Petersburg (in Russian)

Golyandina N, Nekrutkin V, Zhiglyavsky AA (2001) Analysis of time series structure. CRC, Florida

Hayakawa M, Sekiguchi M, Hobara Y, Nickolaenko AP (2006) Intensity of Schumann resonance oscillations and the ground surface temperature. J Atmos Electr 26(2):79–93

Hayakawa M, Nickolaenko AP, Shvets AV, Hobara Y (2011) Recent studies of Schumann resonance and ELF transients. In: Wood MD (ed) Lightning: properties, formation and types. Nova Science Publishers, NY, pp 39–71

Hobara Y, Harada T, Ohta K, Sekiguchi M, Hayakawa M (2011) A study of global temperature and thunderstorm activity by using the data of Schumann resonance observed at Nakatsugawa, Japan. J Atmos Electr 32(2):11–19

Nickolaenko AP, Hayakawa M, Sekiguchi M, Hobara Y (2008) Comparison of the variations in the intensity of global electromagnetic resonance and ground surface temperature. Radiophys Quantum Electron 51(12):931–945

Price C, Rind D (1990) The effect of global warming on lightning frequencies. In: Proceeding 16th conference on severe storms and atmospheric electricity, AMS, Kananski Park, Alberta, Canada, pp 748–751

Price C (2000) Evidence for a link between global lightning activity and upper tropospheric water vapor. Lett Nat 406:290–293

Price C, Asfur M (2006) Can lightning observations be used as an indicator of upper-tropospheric water-vapor variability? Bull Amer Meteorol Soc 87(3):291–298. doi:10.1175/BAMS-87-3-291

Price C, Pechony O, Greenberg E (2007) Schumann resonance in lightning research. J Light Res 1:1–15

Sátori G, Zieger B (1996) Spectral characteristics of Schumann resonances observed in central Europe. J Geophys Res 101(D29):29663–29669

Sátori G, Zieger B (1999) El Niño related meridianal oscillation of global lightning activity. Geophys Res Lett 26:1365–1368

Sátori G (2003) On the dynamics of the North–South seasonal migration of global lightning. In: Proceeding 12th international conference on atmospheric electricity. Versailles, France, Global Lightning and Climate, pp 1–4

Sátori G, Williams ER, Boccippio DJ (2003) On the dynamics of the North–South seasonal migration of global lightning. AGU Fall Meeting, San Francisco, 8–12 Dec 2003, p AE32A-0166

Sekiguchi M, Hayakawa M, Hobara Y, Nickolaenko A, Williams E (2004) Links of Schumann resonance intensity with average global land temperature. Radiophys Electron 9(2):383–391 (in Russian)

Sekiguchi M, Hayakawa M, Nickolaenko AP, Hobara Y (2006) Evidence of a link between the intensity of Schumann resonance and global surface temperature. Ann Geophysicae 24:809–817

Troyan VN, Hayakawa M (2002) Inverse geophysical problems. TERRAPUB, Tokyo

Williams ER (1992) The Schumann resonance: a global tropical thermometer. Science 256:1184–1188

Williams ER (1994) Global circuit response to seasonal variations in global surface air temperature. Mon. Weather Rev 122:1917–1929

WMO (1956) World distribution of thunderstorm days, part 2: tangles of marine data and world maps, OMM-No. 21, TP. 21

Yatsevich EI, Nickolaenko AP, Shvets AV (2012) Analysis of annual and semiannual components of the Schumann resonance intensity based on data of three observatories, Reports on conference electromagnetic methods of environmental studies (EMES'2012), Kharkov, 25–27 Sep 2011, pp 112–115

Chapter 12
Signals in Adjoining Frequency Bands

The Earth–ionosphere spherical cavity occupies three dimensions. SR corresponds to radio waves circling the globe, to the 'longitudinal' propagation. There is the transverse (vertical) direction of radio wave propagation: between the ground surface and the lower edge of the ionosphere. In the present chapter we briefly describe the natural pulsed signals associated with the transverse resonance of the Earth–ionosphere cavity. This resonance occupies the ELF/VLF frequency band with the basic frequency about 1.6–1.7 kHz. When speaking in terms of subionospheric radio propagation, the transverse resonance frequencies correspond to the cut-off frequencies of the Earth–ionosphere duct. The pulsed transverse resonance signals arriving from great distances are also called 'tweek–atmospherics'. We use a record of typical tweeks to demonstrate the 'Kharkov technique' yielding the simultaneous source location and finding the waveguide effective height. In addition, we discuss properties of ionospheric Alfvén resonance (IAR) that is observed below the basic SR frequency. We present some experimental data and their interpretation in terms of the resonance of the plasma slab bounded by the lower ionosphere and the plasmapause.

12.1 Transverse Resonance

The three-dimensional Earth–ionosphere spherical cavity has the eigen-frequencies characterized by three quantum numbers: p, n and m. The quantity n is called the zonal number; which shows how many wavelengths fit around the circumference of the Earth. The figure m is called the azimuthal number; it varies from $-n$ to $+n$, and characterizes the waves traveling from west to east and from east to west. The azimuthal number m indicates the number of wavelengths laid along the equator. The numbers n and m are used when considering the resonance line splitting under the influence of the ionosphere anisotropy (see Chap. 8). The transverse quantum number p indicates how many half-wavelengths fit the ionosphere height. Eigen-values (resonance frequencies) of the spherical cavity formed by the perfectly conducting walls, satisfy the equation:

A. Nickolaenko and M. Hayakawa, *Schumann Resonance for Tyros*,
Springer Geophysics, DOI: 10.1007/978-4-431-54358-9_12,

$$k = \sqrt{\frac{n(n+1)}{a^2} + \frac{p^2\pi^2}{h^2}} \tag{12.1}$$

where $k = \omega / c$ is the wave number in free space, a is the Earth's radius, and h is the height of the ionosphere above the ground.

In the frequency range of SR, the transverse wave number $p = 0$, and one obtains the well-known Schumann formula: $(ka)^2 = n\ (n + 1)$. Transverse resonance corresponds to $p = 1, 2, 3$, etc. The relevant waves bounce from the Earth and the ionosphere like waves in a Fabry–Perot interferometer. These signals are regarded as tweek atmospherics, which were popular in 1960s. When treating the transverse resonance, one can consider the Earth–ionosphere cavity as being flat. Obviously, the basic resonance frequency of the global electromagnetic (Schumann) resonance ($p = 0$) are determined by the terrestrial circumference, and it should be about 10 Hz. The eigen-value of transverse resonance depends on the height of the ionosphere and becomes the multiples of 1,500 Hz when $h = 100$ km.

A typical tweek atmospheric is shown in Fig. 12.1. It was recorded on board the "Academician Vernadsky" scientific vessel in the Guinea Gulf at the point with geographic coordinates 19° S and 7° E on January 21, 1991, at 20:31:46 UT.

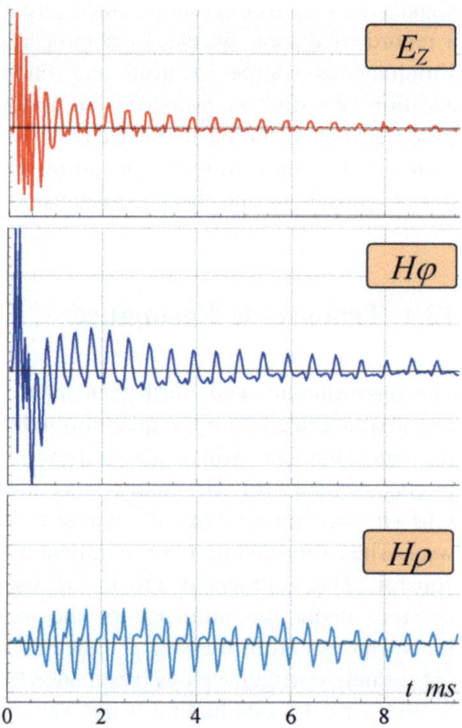

Fig. 12.1 Three field components of the nighttime electromagnetic pulse recorded in the Gulf of Guinea on January 21, 1991 at 20 h 31 m 46 s UT

Transverse resonances were measured at a great number of places, including high altitude balloons (see e.g. Simoes et al. (2009) and references therein).

Three plots are shown in Fig. 12.1. They have the common abscissa where the time is shown in milliseconds, and the ordinates show the signal amplitude in arbitrary units. The upper plot depicts the record of vertical electric field component, and the middle and the lower plots show relevant pulses in two orthogonal horizontal magnetic fields measured in the same arbitrary units. The coordinate system was rotated in such a way that the recorded magnetic field was reduced to the radial H_ρ and the transverse H_φ components. For this purpose, the wave arrival angle was found from the Poynting vector in the time domain. Similar processing was mentioned when we discussed the ELF transient.

The field components were recorded with the help of the vertical electric ball antenna and two orthogonal air core loop antennas. The linear receiver had three identical channels with the bandwidth from 300 to 13 kHz. The tunable gain of receiver could reach 50 dB. Electromagnetic pulses that exceeded a fixed threshold were recorded by the computer through the 12-bit analog–digital converter (ADC) that worked with the 100 kHz sampling frequency. Three time realizations containing 2,048 samples each were stored in the computer memory (approximately 20 ms).

As Fig. 12.1 demonstrates, the vertical electric $E(t)$ and the transverse magnetic field $H_\varphi(t)$ contain a high amplitude rapidly varying onset, which is practically absent in the 'longitudinal' plot $H_\rho(t)$. The "head" of the tweek atmospheric is almost linearly polarized since it is formed by a superposition of the direct radio wave and of the waves bounced from the boundaries of the Earth–ionosphere cavity a few times. The radial field component $H_\rho(t)$ will be absent in the cavity with isotropic boundaries. It appears due to depolarization of the wave multiply reflected from the anisotropic ionosphere. Therefore, the $H_\rho(t)$ amplitude gradually increases and then decreases together with the other field components.

As the time goes, the waveforms turn into a sequence of distinct widening pulses. The repetition period corresponds to the basic frequency (about 1.7 kHz) of the first transverse resonance, which is also regarded as the first cut-off frequency of the Earth–ionosphere wavegiude. Formation of such a regular tail is clarified by the scheme of Fig. 12.5. The sequential part of atmospheric reminds us of classical multiply reflected atmospherics (see, e.g., Iwai and Nakai 1953; Iwai et al. 1953, 1969; Kimpara 1953, 1955, 1956; Iwata et al. 1970; Yano et al. 1991; Yedemsky et al. 1992 and references therein). More than 20 reflections are seen in the radial horizontal magnetic field component indicating on a high reflection coefficient of the lower ionosphere in ambient night condition.

By performing the Fourier transform, we obtain the amplitude spectra of the field components depicted in Fig. 12.2, which contains three panels. The abscissa depicts the frequency in kHz and the ordinates show the spectral amplitude in arbitrary units. A characteristic feature of the amplitude spectra is a strong interference pattern. It arises due to 'beating' between the particular wavegiude modes with modes of higher order. Each pattern starts just above the cut-off frequency corresponding to $p = 2$ and 3, and the latter are shown by vertical lines.

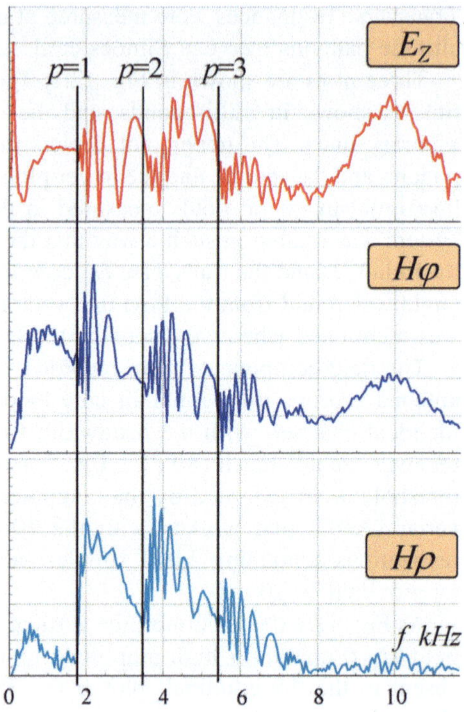

Fig. 12.2 Amplitude spectra of three field components for the atmospheric in Fig. 12.1

Variations are very fast initially, and the spectral modulations slow down with increasing frequency. When the frequency exceeds the next cut-off, the fast oscillations re-appear again.

The concept of inter-modal interference was the background for the so-called "Kharkov technique" of solving the inverse problem (Rafalsky et al. 1995a, b; Shvets and Hayakawa 1998; Brundell et al. 2002). Modal interference explains all the features of amplitude spectra, and it simultaneously allows for finding the distance to the parent stroke of lightning. It is obvious from Fig. 12.2 that a pronounced interference starts in the amplitude spectrum of the radial H_ρ field component only from the second mode, i.e., when $f > f_p|_{p=2}$. Simultaneously, the amplitude is small at frequencies below the first cut-off. This is no surprise, since the H_ρ component may appear due to anisotropy of the ionosphere boundary or from a considerable horizontal current in the lightning discharge. At any rate, the small amplitude of the first mode in the H_ρ component explains why the interference is absent at the first mode frequencies: the "wave beating" structure appears only when two propagating modes exist of approximately equal amplitude. This condition is satisfied at the frequency above the second cut-off.

The modal interference provides the initial rapid oscillations. A new-borne Brillouin wave (Wait 1962) appears at the cut-off frequency, which is characterized by the complex sine function $S_p = [1 - (p\pi/kh)^2]^{1/2}$ starting from zero. Here

k is the free space wave number, h is the height of the wavegiude, and the parameter kS_p is the 'longitudinal' wave number involved in the $\exp(-ikS_pD)$ dependence. The new mode is formed by the couple of plane waves incident on the wavegiude boundaries at almost normal angle (Budden 1961; Wait 1962), so that S_p is close to zero. Since the wave front is parallel to the boundaries, the phase does not vary all along the waveguide. The $S_p(f)$ sine function initially grows fast with frequency. The relevant phase delay due to the propagation $ikS_p(f)D$ increases rapidly for the new mode, and it varies not so fast for the old modes having almost constant $S_p(f)$. As the frequency grows, the new $S_p(f)$ function also transforms into a constant, and the wave catches up in phase with other propagating modes.

When the frequency of the signal is high enough, the low order modes acquire the vertical phase fronts; they travel along the boundaries and have the equal distance phase delays. Hence, the initial phase difference of two modes depends solely on the source distance. When the frequency increases, we expect a gradual decrease of the phase difference approaching to zero at an infinitely great frequency. As a result, the number of oscillations in the interference pattern provides an estimate for the source–observer distance, measured in the wavelengths at the basic cut-off frequency (the doubled wavegiude height). We suppose of course that the slowly varying pattern at higher frequencies is not spoiled by the interference caused by the novel modes of higher orders.

Since the spectra in Fig. 12.2 exhibit 6 or 7 complete oscillations starting from the cut-off frequency of about 1.7 kHz, the distance lies between 1,000 and 1,300 km (Rafalsky et al. 1995a, b; Shvets and Hayakawa 1998). Workability of the modal concept is illustrated by decomposition of the atmospheric spectrum. Let us consider the zero and the first order modes and use the distance estimate (1,000–1,300 km) obtained with the above 'thumb rule' as the starting point. Strictly speaking, the spectral decomposition belongs to the ill-posed problem, so that we have to imply some assumptions concerning the two sought complex functions forming the oscillation pattern. We postulate that the zero order mode had the following properties:

(a) The phase velocity of the mode is a linear function of frequency,
(b) Its amplitude decreases exponentially above the first cut-off frequency.

The phase velocity of the zero order mode varies very slowly with frequency, so the first condition is a reasonable first-order approximation. The second condition is also rational because it is in accord with the propagation model by Greifinger and Greifinger (1978) and with calculations by Harth (1982) and Porrat et al. (2001). The algorithm of decomposition (being the Kharkov technique itself) is constructed in the following way (Rafalsky et al. 1995a, b):

1. The phase spectrum of a signal below the first cut-off is used to find the best-fit linear approximation for the zero order mode phase velocity as the function of frequency. The result is extrapolated to the frequency above the cut-off.
2. The amplitude spectrum below the first cut-off is smoothed with a spline.

3. The smoothed spectrum is extrapolated to the higher frequencies with an exponentially decreasing function.
4. The complex function obtained in steps 1–3 is treated as the spectrum of the zero order mode, which is subtracted from the original experimental data.
5. Parameters of the exponent are varied to obtain the smoothest remainder function above the first cut-off frequency.
6. The remainder function is treated in this band as the first order mode similarly to the zero order mode described above.

Figure 12.3 presents the result of decomposition procedure applied to the $H\varphi$ spectrum shown in Fig. 12.2. The decomposed amplitude spectra of two smoothed modes are plotted in the top graph. Their complex sum provides the interference pattern matching the experimental spectrum fairly well, as might be seen from the bottom graph. The source distance was found to be 1,100 km.

Let us turn to the experimental detection of transverse resonance. Until now we mentioned only the mode theory and the cut-off frequencies. In addition, the reader can hardly agrees that the amplitude spectra shown in Fig. 12.2 illustrates a resonance phenomenon, specially, as there is an amplitude minimum at the first cut-off frequency instead of conventional resonance maximum. To reveal the transverse resonance, one has to apply the dynamic spectra (sonograms) shown in Fig. 12.4. The transverse resonance is seen in the "tail" of the record where the regular pulse succession emerges. This regular sequence is formed by reflections from the wavegiude interfaces and corresponds to the radio waves "trapped" in the Earth–ionosphere gap. Here, the multiply reflected pulses form a pronounced resonance structure (e.g. Hayakawa and Shimakura 1992; Ohta et al. 1994; Cummer and Inan 2000; Porrat et al. 2001).

One may observe that the pulsed signal occupies initially the wide frequency band. Here, the electromagnetic radiation from the causative discharge (direct

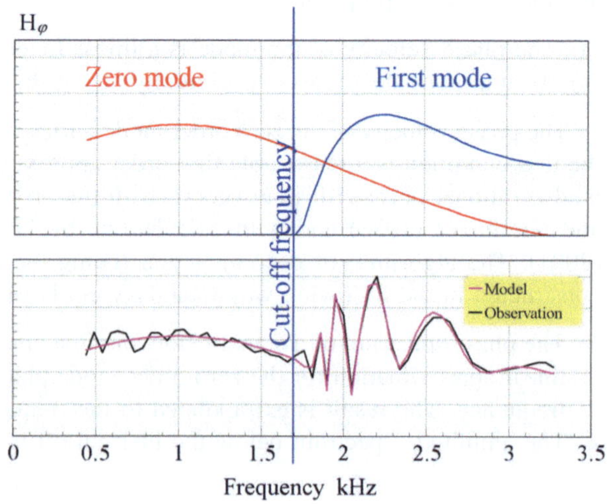

Fig. 12.3 Decomposition of an atmospheric spectrum. Model amplitude spectra of the zero and first modes (*top graph*). Comparison of the model spectrum with the experimental one (*bottom graph*)

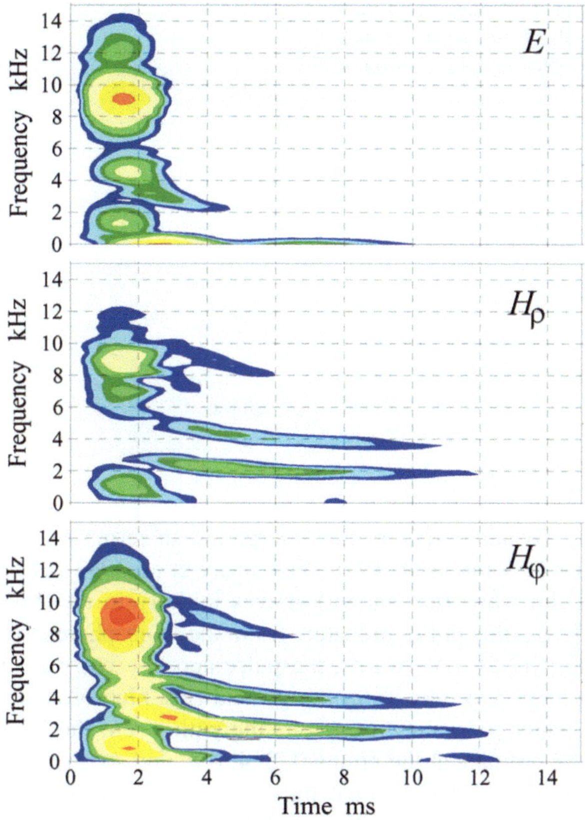

Fig. 12.4 Dynamic spectra of the atmospheric shown in Fig. 12.1. The tail of the pulsed signal contains multiple reflections forming a series of spectral ridges each approaching the transverse resonance frequency and its harmonics

pulse) merges with the initial reflections from the boundaries. This is the part of the signal that forms a vertical "stem" in the sonogram, which reflects the modal interference. Later, a system of the narrow-band harmonics appears (the 'branches') that gradually advance toward the frequency of the transverse resonance ~ 1.7 kHz or toward its harmonics. The number of branches corresponds to the number of interference patterns seen in Fig. 12.2. So we can conclude that each branch in the sonogram is related to a particular waveguide mode. This is not surprising since the mode number p indicates by definition the number of half-waves fitting the height of the Earth–ionosphere wavegiude.

By returning to the waveforms of Fig. 12.1, we find that time delays where the branches appear correspond to the multiply reflected pulses. These pulses have the repetition period determined by the time delay of successive reflections from the nighttime ionosphere and the ground. The ionosphere height is readily estimated as 88 km, which corresponds to the 1.7 kHz basic transverse resonance frequency. The number of harmonics depends of the sub-pulse waveform: "structured" pulses provide a greater number of "branches" in the dynamic spectrum.

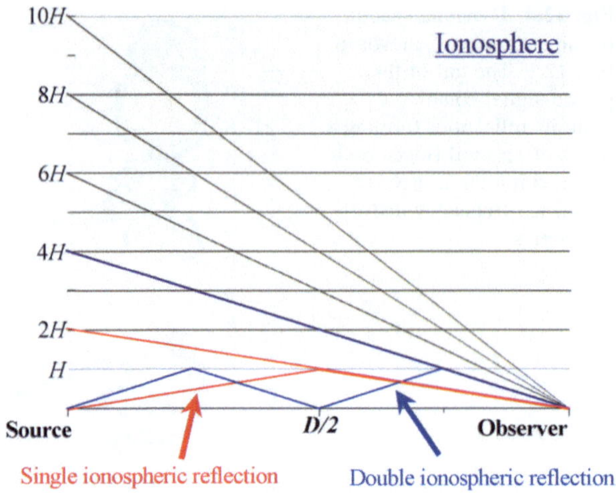

Fig. 12.5 Scheme of the pulse bouncing in the flat Earth–ionosphere waveguide

Dynamic spectra are interpreted in terms of multiply reflected waves (Wait 1962; Yano et al. 1991; Yedemsky et al. 1992) as shown in Fig. 12.5. The direct radio wave (not shown) arrives initially to an observer through the waveguide. Then, the pulse reflected from the ionosphere arrives (the red broken line in Fig. 12.5). Afterwards, the pulse appears that was reflected twice from the ionosphere and once from the ground (the blue broken line), etc. An equivalent and clearer presentation exploits the paths from the imaginary sources positioned on the ordinate at the points 0, 2, 4, 6 h, and so on. Such a scheme correctly explains the delay of individual pulses, as well as a more rapid decrease in the vertical electric amplitude. Elevated imaginary sources radiate the equally decreasing vertical electric and horizontal magnetic fields. However, the former are additionally tilted with respect to the observatory vertical antenna thus providing a faster decrease of reflected pulses. This is why the sonogram 'branches' are noticeably shorter in the spectra of E–field. The equal length of branches in horizontal magnetic components also suggests that multiply reflected pulses tend to have a circular polarization due to the anisotropy of the night ionosphere.

We demonstrated that the dynamic spectra reveal the transverse resonance phenomenon. The resonance structure does not appear in the common amplitude spectra of a complete pulsed waveform: these are masked by the modal interference. Side branches in the sonogram similar to those we had shown are often observed in the dynamic spectra of whistlers. This occurs when an observer is positioned at some distance from the end point of magnetospheric duct carrying the whistler (e.g., Hayakawa 1995). In this case, the wave "falling" into the subionospheric waveguide has to withstand a series of reflections from the ground and ionosphere before it reaches the observer. These hops are observed in the dynamic spectra as characteristic side branches asymptotically approaching the frequencies of transverse resonance (Shimakura et al. 1991; Sukhorkov et al. 1992; Ohta et al. 1994).

The physics of transverse resonance is obvious: this is the radio wave arrested between two parallel mirrors. The resonance takes place when the distance between the mirrors is equal to an integer number of λ / 2. From the viewpoint of the waveguide propagation theory, the cut-off of traveling wave occurs when the frequency is equal to that of transverse resonance. The relevant power flow along the waveguide is terminated, and the radio wave becomes trapped: it bounces along the normal to the reflecting walls. There is no difference between the TE and TM polarized waves for the captured waves since these simply have different orientations. The resonant oscillations might be generated exclusively by the electric currents parallel to the walls (horizontal currents).

An observer, remote from the lightning stroke, cannot detect a pure transverse resonance because the radio wave has to reach the field site somehow. This might be achieved only by a series of reflections. The ionosphere that forms the upper boundary of the waveguide becomes a medium with substantial losses in the ambient day condition. Therefore, the multiple reflections and pertinent transverse resonance pattern disappear during the day, as well as pronounced interference in the amplitude spectra of individual events. Averaging the amplitude spectra of the night ELF-VLF pulses that arrive from different distances will smooth the interference frequency oscillation, while the transverse resonance is always present as an abrupt increase in the signal amplitude just above the basic 1.7 kHz frequency (e.g. Lazebny et al. 1988; Porrat et al. 2001). The amplitude minimum is observed at the resonance frequency itself. We must also note here that the basic frequency of transverse resonance is clearly seen in the space-borne records (Toledo-Redondo et al. 2012). It is seen on the night side dynamic spectra as a stable minimum in the field spectrum at 1.7 kHz frequency.

The asymptotes of peak frequencies of different transverse resonances in the dynamic spectra of the night atmospherics are not directly proportional to the mode number p. This allows us to estimate the height conductivity profile. However, this task demands much more sophisticated techniques in the signal processing. An interested reader might find the relevant information in the recent papers (see e.g. Simoes et al. 2009; Shvets and Gorishnya 2010, 2011 and references therein).

12.2 Anisotropy: Ionospheric Alfvén Resonance

Ionospheric Alfvén resonance (IAR) was predicted in the works by Polyakov (1976) and Polyakov and Rapoport (1981), which was discovered in the measurements by Belyaev et al. (1987, 1989, 1990). The IAR is observed in the Ultra Low Frequency (ULF) band and it was explained by the Alfvén waves trapped in the magnetized plasma of the ionosphere. The resonator volume is bounded below by the lower edge of the ionosphere and at the top by the transition to the less dense plasma of magnetopause. The upper boundary of the region occupied by the resonant oscillations corresponds to the violation of the geometrical optics condition for the Alfvén waves (Polyakov 1976). The IAR frequencies are determined by the "electric thickness" of the ionosphere, and the typical distance between the spectral peaks is 0.5–2.8 Hz (Belyaev et al. 1987). The phenomenon is local, and it is observed at night. After the resonance was detected by various research groups, a considerable number of publications appeared. Extensive information on the IAR is collected in a special issue of Journal of Atmospheric and Solar-Terrestrial Physics (2000 Vol. 62, No. 2). The review of recent data and the bibliography might be found in the papers by Schekotov et al. (2011), Demekhov (2011), and Pilipenko (2011).

The horizontal magnetic induction coil sensors are used for detection of IAR. In the experiments by Belyaev et al. (1987, 1989), they used two orthogonal coil antennas of 100,000 turns composed of the permalloy core 2 m long and with diameter of 2 cm. The IAR is observed at night in the quiet geomagnetic conditions as a sequence of distinct peaks in the power spectrum of the ULF–ELF radio noise. The lowest peak frequency is equal to a few parts of a Hz, and usually a series of peaks is observed. The resonant frequencies increase during the night owing to the decrease of the ionospheric plasma density. Sometimes, the highest resonance peaks reach the first SR frequency (8 Hz) and even higher.

The spectral maxima of IAR are observed in the power spectra of both the horizontal magnetic field components, provided that the level of local noise is low. The peaks are most visible in the polarization spectra of the field (see Fig. 12.6 and

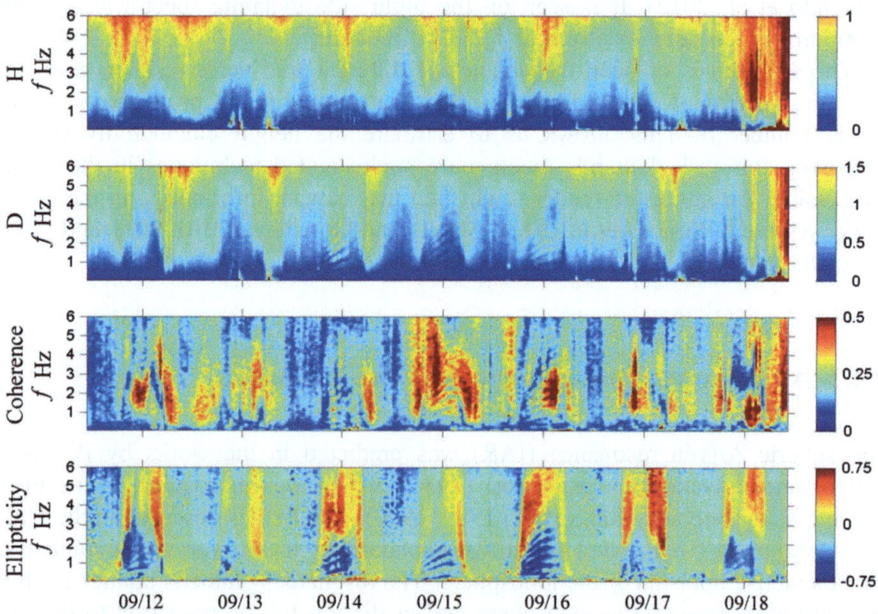

Fig. 12.6 Ionospheric Alfven resonance (IAR) observed during a week in September 2000. Resonance is visible at night as the bands of increasing frequency

Chap. 7). The property is conditioned in part by using the cross-spectra in the polarization measurement, and these suppress the incoherent interference. Maxima in the IAR amplitude correspond to the almost circular field polarization.

The ULF–ELF monitoring was carried out at the Karymshino observatory, Kamchatka (53° N, 157° E) (Molchanov et al. 2004; Hayakawa et al. 2004). Figure 12.6 shows the dynamic spectra in the frequency range from 0.1 to 6 Hz recorded during a week in September 2000. The abscissa shows the UT day, and the local midnight (when the IAR is observed) corresponds to the middle of the daily interval in the Universal time. The ordinate depicts the signal frequency in Hz. The spectra and signal polarization (ellipticity) are shown by color.

Two upper graphs in Fig. 12.6 demonstrate the "usual" dynamic spectra of orthogonal magnetic field components H and D. Here the Alfven resonance is visible during some nights, but not very clearly. The H-field component lies in the magnetic meridian plane, and D-component is perpendicular to this plane. Since the magnetic declination at Karymshino is only about 6.5°W, one can accept that H- and D- fields practically coincide with the geographic projections H_{SN} and H_{WE}. The third graph shows the dynamic spectra of the wave coherence, and the lower plot shows the wave polarization (ellipticity). Here the IAR is observed in the clearest way.

Figure 12.7 illustrates in more detail the data on the wave polarization similar to those shown in Fig. 12.6. The frequency range spans the IAR area and the first SR mode. The plot is divided into two frames for convenience. Each of them is a part of the complete dynamic spectrum that was "cut" into two parts along the 4 Hz frequency. The resulting parts were moved apart vertically. One may see that the field ellipticity is close to -1 in the vicinity of the first SR frequency during the whole day. Since the Karymshino observatory utilizes the geomagnetic coordinate system, the signal ellipticity is negative. At the frequency around 9 Hz and higher, the polarization might change the sense, and the ellipticity might vary with time of a day due to the diurnal motion of global thunderstorm activity. At frequencies below 4 Hz, a characteristic sequence of peaks appears of the IAR observed practically every night. The peak frequencies of IAR increase during the night. The lowest graph depicts a separate spectrum of polarization, and the vertical arrow shows its correspondence to the record.

Such dynamic spectra are typical, although the IAR phenomenon is not observed every night even in the wave polarization. The resonance is observed in all plots of Figs. 12.6 and 12.7 in the form of oblique bright-dark strips observed in the night hours. The frequency increase in time is explained by the change in the ionosphere plasma density. Since the polarization of the resonance field is nearly circular, the field components H and D have the phase shift about 90°. This is why the resonance pattern is more evident in the imaginary part of the cross-spectrum than in its real part (not shown).

A commonly accepted explanation for the IAR phenomenon was not elaborated in spite of the fact that the phenomenon was discovered long ago and it is observed at all latitudes (Belyaev et al. 1999; Bösinger et al. 2002; Molchanov et al. 2004; Semenova and Yahnin 2008; Semenova et al. 2008). It is clear that the signal itself hardly might be attributed to the Alfven plasma waves in the ionosphere because of the high energy observed: This contradicts the transformation of plasma waves into the electromagnetic field at the plasma–air interface. The nature of the field source is also unclear. There are three possible candidates. The first one is the electromagnetic waves coming from the space through the ionosphere (e.g. Nickolaenko et al. 2004). The second mechanism involves various plasma instabilities arising in the ionosphere (Trakhtengertz and Feldstein 1991; Trakhtengerts et al. 2000; Surkov et al. 2004). And finally, the resonance energy is supplied by the radio emission from the lightning strokes (Sukhorukov and Stubbe 1997; Demekhov et al. 2000; Trakhtengerts et al. 2000; Molchanov et al. 2004; Fedorov et al. 2006; Surkov et al. 2006; Shalimov and Bösinger 2008; Demekhov 2011; Schekotov et al. 2011). In most of the works observations are interpreted as electromagnetic resonance in the plasma of the ionosphere.

All theoretical works successfully explain the time dependence of resonance frequencies during the night that arise from the change in the plasma density. For example, the calculated and experimental dynamic spectra are compared in Fig. 12.7 of September 29, 2000 (Nickolaenko et al. 2004). This is the particular day of observations marked by the vertical arrow in Fig. 12.8. The model dynamic

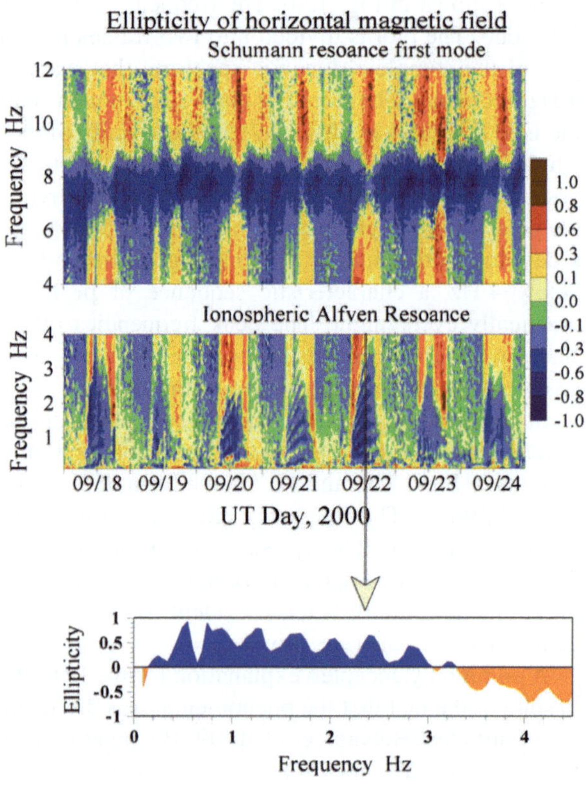

Fig. 12.7 Dynamic spectrum of the IAR in the field polarization

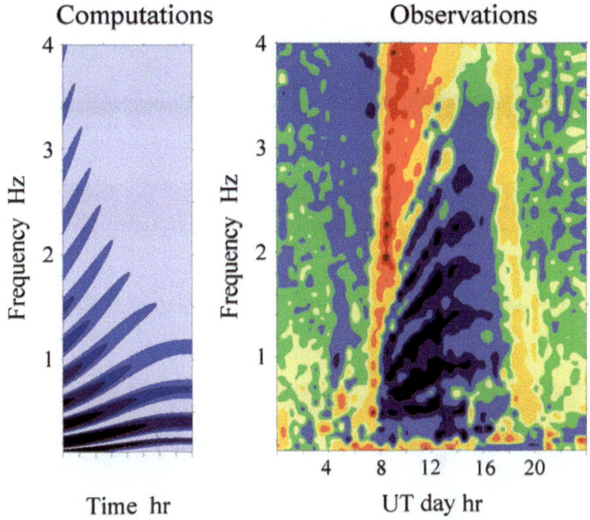

Fig. 12.8 Model and experimental temporal changes in the IAR ellipticity

spectrum includes only the night hours, because the IAR is not observed during the day. Obviously temporal changes of ellipticity are very similar.

Concerning the source of oscillations, we should mention the work by Schekotov et al. (2011) based on the Karymshino observations as seen in Fig. 12.9. The figure depicts the magnetic field components H and D as a function of time. The time is plotted on the abscissa in seconds. The upper panel shows the time dependence observed in the frequency band 6–20 Hz together with the histograms of amplitude distribution shown to the right of the plots. This is the usual time realization of the low-frequency signal in the SR band after elimination of industrial interference.

The middle line of plots in Fig. 12.9 shows the same signal after passing through the low-pass filter of 0.25–4 Hz band. The temporal realizations contain a pronounced ULF activity starting from 1700 s. Again, the histogram of observed amplitudes is shown to the right of plots.

The bottom line of graphs indicates the wavelet spectra of the signal. The graphs to the right of dynamic spectra show the amplitude spectra of the whole realization. The observational data demonstrate that the characteristic "striped" structure of the Alfvén resonance is associated with the emergence of a series of pulses in the time realization of the field. Schekotov et al. (2011) concluded that the horizontal electric discharges in the atmosphere served as a source of these oscillations. Unfortunately, such an explanation is not very convincing for two reasons. The first is that owing to the local nature of the IAR, the nocturnal electrical activity must be observed at all latitudes including the Arctic. The such a source. The horizontal electric current might be located in the troposphere.

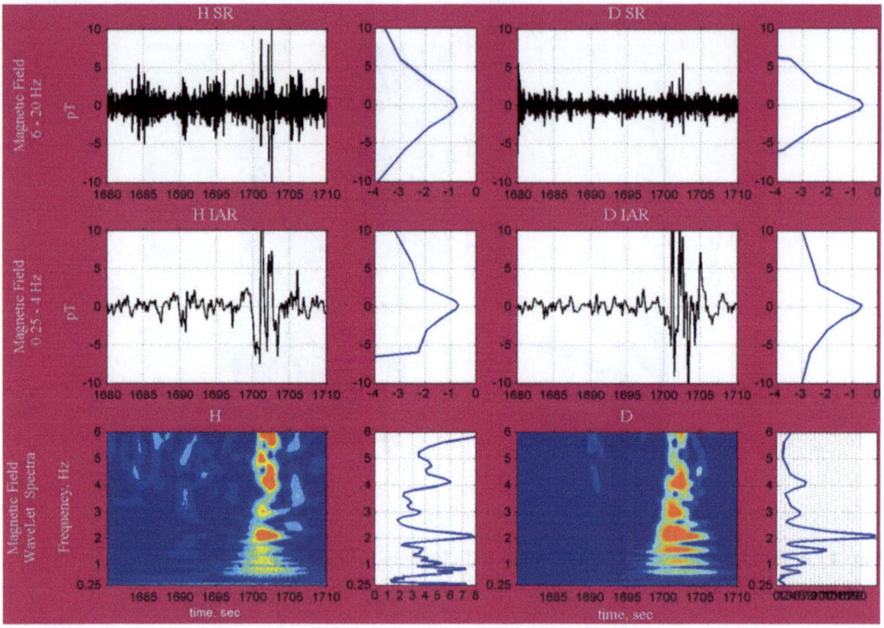

Fig. 12.9 IAR and its relationship to the repetitive ULF pulses

However, such a source becomes inefficient at the frequency of a fraction of one Hz because there is the current reflection in the well-conducting ground. The field radiated by the real current is compensated by the similar radiation from its reflection in the ground. To obtain the observed amplitudes, one has to imply extraordinary large source currents. Simultaneously, the radio emission from such a powerful source was not registered in the SR band where the relevant Q–burst should appear. Thus, the nature of the IAR source remains unclear.

At any rate, the IAR is one of the most popular areas in the low-frequency research (Hobara and Hayakawa 2012). We may expect novel and interesting results in this exciting field of radio science.

References

Belyaev PP, Polyakov SV, Rapoport VO, Trakhtengertz VY (1987) Discovery of the resonance spectrum structure of atmospheric electromagnetic noise background in the range of short-period geomagnetic pulsations. Dokl Akad Nauk SSSR 297:840–846

Belyaev PP, Polyakov SV, Rapoport VO, Trakhtengertz VY (1989) Experimental studies of resonance structure of the spectrum of atmospheric electromagnetic noise in the range of short-period geomagnetic pulsations. Izv. VUZov, RADIOPHYSIKA 32(6):663–672 (in Russian)

Belyaev PP, Polyakov SV, Rapoport VO, Trakhtengertz VYu (1990) The ionospheric Alfven resonator. J Atmos Terr Phys 52:781–788
Belyaev PP, Bösinger T, Isaev SV, Kangas J. (1999) First evidence at high latitudes for the ionospheric Alfven resonator. J Geophys Res 104:4305–4317
Bösinger T, Haldoupis C, Belyaev PP, Yakunin MN, Semenova NV, Demekhov AG, Angelopoulos V (2002) Spectral properties of the ionospheric Alfven resonator observed at a low-latitude station (L = 1.3). J Geophys Res 107:1281. doi:10.1029/2001JA005076
Brundell JB, Rodger CJ, Dowden RL (2002) Validation of single-station lightning location technique. Radio Sci 37(4):12-1–12-9. doi:10.1029/2001RS002477
Budden KG (1961) Radio waves in the ionosphere. Cambridge University Press, Cambridge
Cummer SA, Inan US (2000) Ionospheric E-region remote sensing with ELF radio atmospherics. Radio Sci 35:1437–1444
Demekhov AG, Trakhtengertz VY, Bösinger T (2000) Pc 1 waves and ionospheric Alfven resonator: generation or filtration? Geophys Res Lett 27(23):3805–3808
Demekhov AG (2011) Coupling at the atmosphere–ionosphere–magnetosphere interface and resonant phenomena in the ULF range. Space Sci Rev. doi:10.1007/s11214-011-9832-6
Fedorov E, Schekotov AYu, Molchanov OA, Hayakawa M, Surkov VV, Gladichev VA (2006) An energy source for the mid-latitude IAR: world thunderstorm centers, nearby discharges or neutral wind fluctuations? Phys Chem Earth 31:462–468
Greifinger C, Greifinger P (1978) Approximate method for determine ELF eigen-values in the Earth-ionosphere waveguide. Radio Sci 13:831–837
Harth W (1982) Theory of low frequency wave propagation. In: Volland H (ed) Handbook of atmospherics, vol II. CRC Press, Boca Raton, pp 133–302
Hayakawa M, Shimakura S (1992) High accuracy method of locating distant lightning by means of tweek-atmospherics. 11th International Wroclav Symposium on Electromagnetic Compatibility, Wroclav, Poland, pp 566–569
Hayakawa M, Ohta K, Baba K (1994), Wave characteristics of tweek atmospherics deduced from the direction finding measurement and theoretical interpretation. J Geophys Res 99:10733–10743
Hayakawa M (1995) Whistlers. In: Volland H (ed) Handbook of Atmospheric Electrodynamics, vol II, Chap. 7. CRC Press, Boca Raton, pp 155–193
Hayakawa M, Molchanov OA, Schekotov AY, Fedorov E (2004) Observation of ionospheric Alfven resonance at a middle latitude station. Adv Polar Upp Atmos Res 18:65–76
Hobara,Y, Hayakawa M (2012) The effects of lightning on the ionosphere/magnetosphere. In: Cooray V (ed) Lightning electromagnetics, Chap. 17. Institute Engineering and Technology, London, pp 647–685
Iwai A, Nakai T (1953), Unidirectional direction finder of atmospherics. In: Proceeding Research Institute Atmospherics, Nagoya University, vol 1, pp 50–53
Iwai A, Ito K, Tanaka T, Ebuchi T (1953) Direction finder of atmospherics. In: Proceeding Research Institute Atmospherics, Nagoya University, vol 1, pp 54–62
Iwai A, Ohtsu J, Nishino M, Kashiwagi M. (1969) A new direction finding network for locating the sources of atmospherics. In: Proceeding Research Institute Atmospherics, Nagoya University, vol 16, pp 17–20
Iwata A, Ishikawa H, Takagi M (1970) Slow tail atmospherics and their origin. In: Proceeding Research Institute Atmospherics, Nagoya University, vol 17, pp 13–21
J Atmos Solar-Terr Phys (2000), 62(2), Special issue on the IAR.
Kimpara A (1953) The waveform of atmospherics in the daytime. In: Proceeding Research Institute Atmospherics, Nagoya University, vol 1, pp 1–11
Kimpara A (1955) Atmospherics in the far east, Proceeding Research Institute Atmospherics, Nagoya University, vol 3, pp 1–28
Kimpara A (1956) The waveform of atmospherics in the daytime and at night. In: Proceeding Research Institute Atmospherics, Nagoya University, vol 4, pp 1–19

Lazebny BV, Nickolaenko AP, Rafalsky VA, Shvets AV (1988) Detection of transverse resonances of the Earth-ionosphere cavity using average spectra of VLF atmospherics. Geomag Aeron 28:329–330 (in Russian)

Molchanov OA, Schekotov AYu, Fedorov EN, Hayakawa M (2004) Ionospheric Alfven resonance at middle latitudes: Results of observations at Kamchatka. Phys Chem Earth Parts A/B/C 29:649–655

Nickolaenko AP, Rafalsky VA, Shvets AV, Hayakawa M (1994) A time domain direction finding technique for ELF-VLF atmospherics. J Atmos Electr 12:97–107

Nickolaenko AP, Rabinowicz LM, Shvets AV, Schekotov AYu (2004) Polarization characteristics of low frequency resonances. Izv. VUZOV, Radiofizika 47(4):267–291 (in Russian)

Ohta K, Shimizu A, Hayakawa M (1994) The effect of subionospheric propagation on whistlers as deduced from direction finding measurements. Geophys Res Lett 21:89–92

Pilipenko VA (2011) Impulsive coupling between the atmosphere and ionosphere/magnetosphere. Space Sci Rev. doi:10.1007/s11214-011-9859-8

Polyakov SV (1976) On the properties of the ionospheric Alfvén resonator, KAPG Symposium on Solar-Terrestrial Physics, 3rd edn. Nauka, Moscow, pp 72–73

Polyakov SV, Rapoport VO (1981) Ionospheric Alfvén resonator. Geomag Aeron 21:816–822 (In Russian)

Porrat D, Bannister PR, Fraser-Smith AC (2001) Modal phenomena in the natural electromagnetic spectrum below 5 kHz. Radio Sci 36:499–506

Rafalsky VA (1991) Resonance phenomena in cross-section of the Earth-ionosphere duct and their influence on radio wave excitation and propagation, Ph.D. Thesis, Kharkov (in Russian)

Rafalsky VA, Nickolaenko AP, Shvets AV, Hayakawa M. (1995a) Location of lightning discharges from a single station. J Geophys Res 100(D10):20829–20838

Rafalsky VA, Shvets AV, Hayakawa M (1995b) One-site distance-finding technique for locating lightning discharges. J Atmos Terr Phys 57(11):1255–1261

Schekotov AYu, Pilipenko V, Shiokawa K, Fedorov E (2011) ULF impulsive magnetic response at mid-latitudes to lightning activity. Earth Planets Space 63:1–10

Shalimov S, Bösinger T (2008) On distant excitation of the ionospheric Alfven resonator by positive cloud-to-ground lightning discharges. J Geophys Res 113:A02303

Shimakura S, Moriizumi M, Hayakawa M (1991) Propagation mechanism of very unusual low-latitude whistlers with additional traces of the earth-ionosphere waveguide propagation effect. Planet Space Sci 39:611–616

Shvets AV (1994) Experimental study of propagation of ELF-VLF radio waves and dynamics of the global thunderstorm activity, Ph.D. Thesis, Kharkov (in Russian)

Shvets AV, Hayakawa M (1998) Polarization effects for tweek propagation. J Atmos Solar-Terr Phys 60:461–469

Shvets AV, Hobara Y, Hayakawa M (2010) Variations of the global lightning distribution revealed from three station Schumann resonance measurements. J Geophys Res 115:A12316. doi:10.1029/2010JA015851

Shvets AV, Gorishnya YuV (2010) Location of lightning strokes and estimation of the ionosphere height by using the dispersion properties of tweek-atmospherics. Radiophys Electron 16(4):53–59 (in Russian)

Shvets AV, Gorishnya YV (2011) A technique for lightning location and estimation of the lower ionosphere parameters using tweek-atmospherics. Telecommun Radio Eng 70(11):1013–1026

Semenova NV, Yahnin AG (2008) Diurnal behavior of the ionospheric Alfven resonator signatures as observed at high latitude observatory Barentsburg ($L = 15$). Ann Geophys 26:2245–2251

Semenova NV, Yahnin AG, Vasil'ev AN, Amm O (2008) Specific features of resonance structures in spectra of ULF electromagnetic noise at high latitudes (Barentsburg Observatory). Geomagn Aeron 48:36–44

Simoes F, Berthelier J–J, Godefroy M, Yahi S (2009) Observation and modeling of the Earth-ionosphere cavity electromagnetic transverse resonance and variation of the D-region electron density near sunset. Geophys Res Lett 36:L14816. doi:10.1029/2009GL039286

Sukhorukov AI, Shimakura S, Hayakawa M (1992) On the additional dispersion of a whistler in the earth-ionosphere waveguide. Planet Space Sci 40:1185–1191

Sukhorukov AI, Stubbe P (1997) Excitation of the ionospheric Alfven resonator by strong lightning discharges. Geophys Res Lett 24:829–832

Surkov VV, Pokhotelov OA, Parrot M, Fedorov EN, Hayakawa M (2004) Excitation of the ionospheric resonance cavity by neutral winds at middle latitudes. Ann Geophysicae 22:2877–2889

Surkov VV, Hayakawa M, Schekotov AY, Fedorov EN, Molchanov OA (2006) Ionospheric Alfvén resonator excitation due to nearby thunderstorms. J Geophys Res 111:A01303. doi:10.1029/2005JA011320

Toledo-Redondo S, Parrot M, Salinas A (2012) Variation of the first cut-off frequency of the Earth-ionosphere waveguide observed by DEMETER. J Geophys Res 117:A04321. doi:10.1029/2011JA017400

Trakhtengerts VY, Feldstein AY (1991) Turbulent Alfvén boundary layer in the polar ionosphere, 1, Excitation conditions and energetics. J. Geophys Res 96:19363–19374

Trakhtengerts VY, Demekhov AC, Belyaev PP, Polyakov SV, Ermakova EN, Isaev SV (2000) A mechanism of anti-correlation in the occurrence of ULF electromagnetic noise resonance structure and Pc1 magnetic pulsations through the solar activity cycle. J Atmos Solar-Terr Phys 62:253–256

Wait JR (1962) Electromagnetic Waves in Stratified Media. Pergamon Press, Oxford

Yano S, Ogawa T, Hagino H (1991) Dispersion characteristics and waveform analysis of tweek atmospherics. In: Kikuchi H (ed) Environmental and space electromagnetics. Springer, Tokyo, pp 227–236

Yedemsky DY, Ryabov BS, Schekotov AY, Yarotsky VS (1992) Experimental investigation of the tweek field structure. Adv Space Res 12(6):251–254

Chapter 13
Extraordinary ELF Signals

In the present chapter we consider unusual manifestations of SR or the ELF signals that might originate from unusual sources. We describe the most curious cases here. First of all, we present the signatures of SR in the observed spectra of the wave arrival angle. This effect is interpreted as an impact of a complicated spatial distribution of the worldwide lightning activity, which interacts with the modal structure of the field. Such spectra may present additional information on positions and dynamics of the global lightning activity. We also show the data on the SR signatures in the spectra of short radio wave signals. The observed modulation might appear due to non-linear interaction of two waves in the ionosphere plasma. Model computations are presented demonstrating that a rocket start is able to generate the ELF noise. The signal appears due to fluctuations of the length of the conducting rocket flare in the outer fair weather field. Properties are discussed of resonance signals that were excited by a magnetospheric ELF source. Power spectra of these signals have a fine structure around SR maxima, a kind of 'splitting' seen directly in the amplitude spectra. Recent orbital observations are discussed of the SR and transverse resonance patterns by two different middle altitude satellites. These unexpected data provide deep impression, although they were not interpreted yet.

13.1 SR Spectra in the Wave Arrival Azimuth

Spectra of the source bearings obtained by the Poynting vector contain sometimes a replica of SR pattern (Belyaev et al. 1999). We collected a few samples of such arrival azimuth patterns in Fig. 13.1. Each plot in this figure contains two lines. The fluctuating curves present the spectra of the wave arrival angle, and the horizontal straight lines show the source bearings found from the Poynting vector in the time domain. Particular fields were recorded at the Lehta observatory (64° N and 34° E) on July 18 (upper plots) and July 25, 1998 (lower plots). Individual spectra were accumulated in the intervals 08–09, 14–15, and 20–21 h UT. It was noticed that the resonance structure tends to appear in these particular time

A. Nickolaenko and M. Hayakawa, *Schumann Resonance for Tyros*,
Springer Geophysics, DOI: 10.1007/978-4-431-54358-9_13,

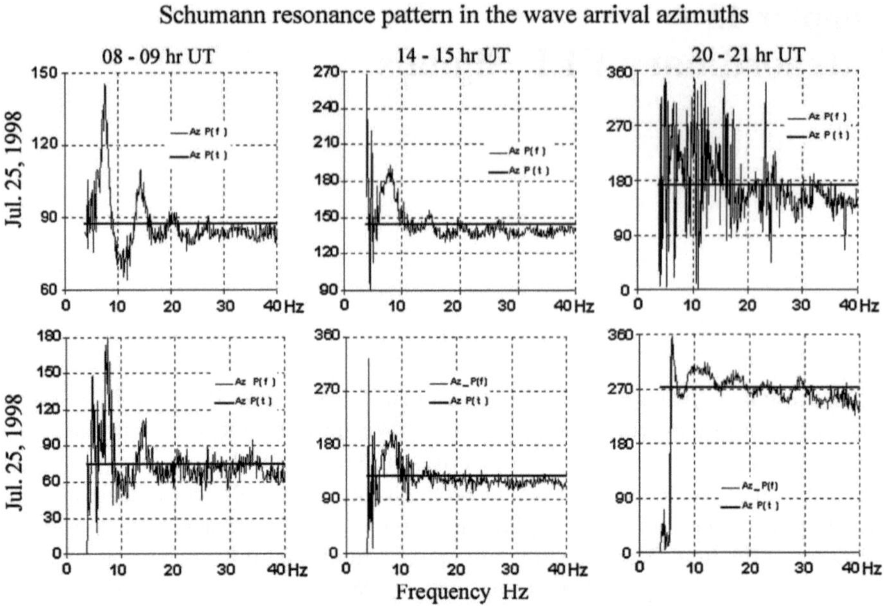

Fig. 13.1 SR observed in the spectra of the apparent source azimuths

intervals, which readily suggests an explanation by a simultaneous comparable activity at two global thunderstorm centers.

To check whether the spatial distribution of sources is able to cause a resonance pattern in the source beating, we considered the simplest disposition of two point sources having different distances from the field site and separated in the azimuth. We choose Africa as the place for the first compact source, and its center is placed at the point 10° N and 34° E, which is the central region of Sudan. The corresponding spherical coordinates are $\theta_1 = 4\pi/9$ and $\varphi_1 = 1.7\pi/9$. The propagation path is characterized by the 6 Mm distance and the source azimuth of 180°. The second point source is placed in the equatorial America, with its geographic coordinates being 0° N and 56° W or $\theta_2 = \pi/2$ and $\varphi_1 = 2.3\pi/9$ (Amazon basin). The distance from the observer is 10 Mm and the source bearing equals to 270°. Such a geometry of sources is possible in the UT evening hours.

The thunderstorm activity at both the centers has the same level, so that the resulting wave arrival angle is a subject of trade-off between the contributions from these sources. In this particular case, the power flux (Poynting vector amplitude) will depend on the frequency because of the nodal structure of the SR fields. The azimuth estimates will indicate a particular center at the frequency where its power flux dominates. The other source may govern the source bearing at other frequencies where its contribution prevails. Particular frequencies and azimuth patterns depend on the angular and distance separation with respect to the

Poynting vector nodal lines. Our goal is check that such a model provides the resonant pattern in the source bearing.

The following expressions are valid for the components of the Poynting vector in the uniform isotropic Earth–ionosphere cavity (see also Chap. 4):

$$P_X = -A_1 \operatorname{Re}\{\langle e_1(\omega)\, h_1^*(\omega)\cos(B_1)\rangle\} \tag{13.1}$$

and

$$P_Y = A_2 \operatorname{Re}\{\langle e_2(\omega)\, h_2^*(\omega)\sin(B_2)\rangle\} \tag{13.2}$$

Here the amplitudes $A_1 = A_2$ characterize the level of thunderstorm activity at two point sources having the bearings separated by 90°, the asterisk denotes the complex conjugate, the elementary electric and magnetic fields are found from the relations $e_{1,2}(\omega) = \frac{i\nu\,(\nu+1)}{\omega\varepsilon_0\,\sin(\pi\nu)} P_\nu\,[\cos(\pi - \theta_{1,2})]$ and $h_{1,2}(\omega) = -\frac{P_\nu^1\,[\cos(\pi - \theta_{1,2})]}{\sin(\pi\nu)}$, and the source bearings are correspondingly $B_1 = 180°$ and $B_2 = 270°$.

By choosing the right angle separation in the source azimuths, we simplified significantly the formulas. The particular Poynting vector components are "tied" to one of two sources, because every projection depends on its own source only.

Figure 13.2 depicts the results of model computations. The left plot demonstrates the Lissajous figure relevant to the frequency variations of the complete Poynting vector. The P_X component of the power flux is depicted on the abscissa and the relevant P_Y is shown on the ordinate. Frequency is the parameter, and the color dots denote a few characteristic positions of the Poynting vector for $f = 7.5$, 7.75, 8.0, and 8.25 Hz. Corresponding frequencies are printed in color. The deceptive 'elliptical polarization' of the Poynting vector has nothing to do with the anisotropy of the ionosphere: the loops in the excursions of Poynting vector are the exclusive product of the spatial separation of field sources.

The right plot in Fig. 13.2 shows the frequency dependence of the wave arrival angle that was conditioned by oscillations of the Poynting vector components. The actual source azimuth is found to deviate from the orientation of the power flux 180°, so that the Poynting vector corresponds to ~210° direction to the source

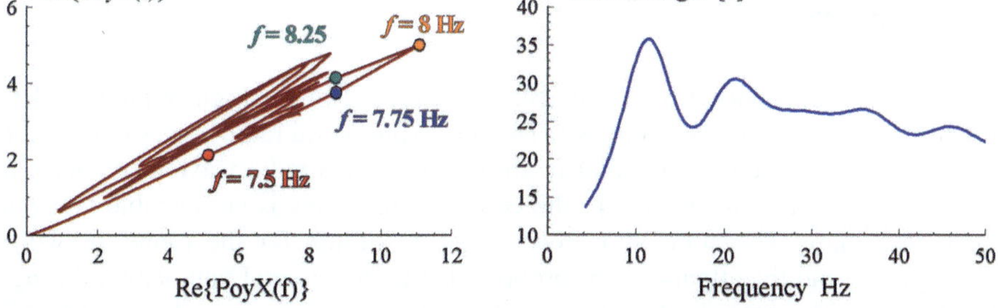

Fig. 13.2 Model spectra of the apparent source azimuth in the model of African and American point sources

(or $A_z \approx -150°$). By comparing Figs. 13.1 and 13.2, we find that the model behaves similarly to the observation.

The spatial distribution of sources is more complicated in reality: individual thunderstorms "work" concurrently and cover significant areas within particular zones. The model of two distinct point sources is very simple, however, it predicts the frequency modulation in the wave propagation direction comparable with the observations. Discrepancies arise from the nodal structure of resonance fields.

The nodal lines form the systems of concentric zones around the source, and this structure depends on the frequency. If we have a couple of separated thunderstorm centers working simultaneously, one of them will dominate, at some frequency, due to a particular source–observer distance. The major portion of power will arrive from the direction of this particular center. The second center might dominate at other frequency owing to the relevant source–observer angular separation. The power will arrive from the second source at these frequencies. In a case when the sources are azimuthally separated, oscillations might appear in the wave arrival azimuth.

We know that the nodal lines of a given mode are most apparent in the band around the peak frequency. They become fuzzy and irresolute at the intermediate frequencies. Consequently, all global thunderstorms tend to equally contribute to the signal received, at these "intermediate" frequencies, and the wave arrival angle corresponds to the 'weighted' source position. Similar reasoning was mentioned in Chap. 4 when describing the intermediate $\sim$11 Hz frequency band where the spatial structure of SR provides minor impact on the global thunderstorm monitoring. We see now that the modal spatial distribution of resonance field may explain the frequency variations of the source azimuth.

The described impact of SR on the median wave arrival angle might be caused by a complicated spatial distribution of the worldwide lightning activity, which interacts with the modal structure of the field. This means that such spectra may supply some additional information concerning positions and dynamics of the lightning activity.

13.2 Non-linear Interaction of SR and Short Radio Wave Signals

Modulation was reported by Yampolski et al. (1997) of the high frequency (HF) round-the-world radio signals by SR. The round-the-world radio waves in the short wave (SW) band were recorded. The absorption of this radio wave takes place in the D region of ionosphere where the collision frequency is comparable with the radio frequency. The attenuation reaches its maximum for the round-the-world propagation, and the attenuation is proportional to the integral path length through the D-layer. The probability of successful detection of the round-the-world HF radio signal increases when the propagation path is close to the solar terminator.

The RWM radio station was used as the source of continuous radio wave of the 14.996 MHz frequency. It is a part of the time-and-frequency service, and radiates a stable signal from the antenna situated near Moscow. The receiver was located near Kharkov (Ukraine) at the radio telescope UTR-2 of the Institute of Radio Astronomy. The direct radio path from Moscow to Kharkov has a length about 700 km, and it is close to the meridian. This distance is within the "dead zone" for the direct SW radio signal. The measurements were carried out around the period of autumnal equinox from 25 to 30 September 1994.

The large antenna of the UTR-2 radio telescope was connected to a multi-channel coherent Doppler receiver. The telescope selects the radio waves in both the azimuth and elevation angles. The receiving antenna beam was oriented southward along the meridian, which is the most probable arrival direction of the round-the world signal from the RWM transmitter.

The spectral analysis of the round-the-world HF signal provided the amplitude spectra in the frequency band of ±25 Hz around the carrier frequency. The similar spectral processing was applied to the ELF background noise at the site when the transmitter was not working. A subtraction of the noise spectra from the spectra of the signal reduced the impact of the amplitude-frequency characteristics of the filters.

We reproduce in Fig. 13.3 the spectra showing distinct peaks at the frequencies around 8–14 Hz that were detected in the SW radio signal (the red line). The blue line presents the conventional SR observed in the vertical electric field component. Exceptionally good correspondence is evident between the spectra, which demonstrate a feasible modulation of the HF radio wave by SR. The modulation index was estimated as the ratio of the spectral density at 8 Hz frequency to the spectral density of the HF carrier, being the value between −46 and −60 dB.

A theoretical model was suggested by P.V. Bliokh explaining the origin of the cross-modulation due to the plasma electron heating in the D-region of the ionosphere by the SR radio waves. Plasma heating by ELF radio wave leads to the time dependent modulation of the HF attenuation rate via modification of the electron collision frequency. It is interesting to note that this effect exists only in the presence of the fair weather field which comes into the lower region together with the SR signal. The fair weather field plays the role similar to that of the

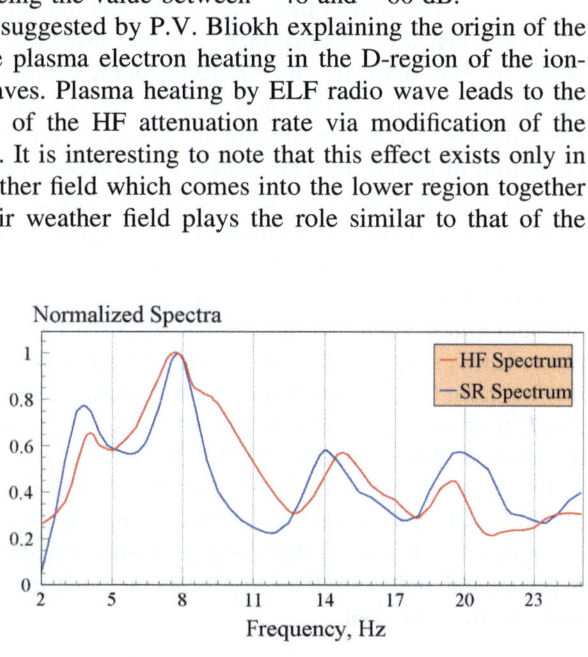

Fig. 13.3 An example of simultaneously measured HF and ELF spectra [adopted from Yampolski et al. (1997)]

heterodyne signal in super-heterodyne receivers. The presence of considerable static fair weather field and the smaller quasi-static resonance wave results in the linear dependence of the instant electron temperature on the amplitude of SR.

The interaction of electromagnetic waves with the ionosphere becomes non-linear when the amplitude of the wave exceeds the plasma threshold field (Ginzburg 1970):

$$E_p = \sqrt{3\kappa\, T \frac{m_e}{e^2} \delta \left(\omega^{\,2} + \nu_{eff}^2\right)} \tag{13.3}$$

where $\kappa = 1.4 \times 10^{-23}$ J/K is the Boltzmann's constant, m_e and e are respectively the electron mass and charge, $\delta = m_e/M$ is the fraction of energy lost by electrons when colliding with heavy particles, ω is the angular frequency of incident field, and ν_{eff} is the electrons effective collision frequency.

Direct comparison of the SR signal ($\sim$1 mV·m^{-1}·(Hz)$^{-1/2}$) with the estimates for the threshold field ranging from 30 to 300 mV·m^{-1}·(Hz)$^{-1/2}$ leads to the pessimistic conclusion that no effect is expected. Impact of large quasi-static and alternative electric fields on the mesosphere (lower ionosphere) was investigated in detail in connection with the "red sprites", blue jets, ELVES, haloes, and other transient luminous events (TLEs). The corresponding bibliography is enormous; to start with it, we recommend the book edited by Füllekrug et al. (2006), and then the papers (Gurevich et al. 1992; Pasko et al. 1999; Cho and Rycroft 2001; Price et al. 2004; Yashunin et al. 2007; Asano et al. 2009; Kudintseva et al. 2009; Hiraki 2010; Rycroft 2012; Williams et al. 2012; and references therein). The particular phenomenon that we discuss is different, because it is associated with a combined effect of the stationary fair weather field and the superimposed SR wave.

An original way was invented to avoid this obstacle and to explain the results observed. The following hint was found in the experimental data. The ELF heating modifies the electron collision frequency, which determines the attenuation rate of SW radio signals and thus modulates the round-the-world HF radio signals. Since alterations of the electron temperature are proportional to the intensity of an outer field, the doubled SR frequencies must be expected in the spectra of Fig. 13.3. This contradicts with the observation. The discrepancy might be removed when the plasma heating is driven by the combination of SR field with the static fair weather field. In this case, the electron temperature is modulated by the first harmonic of the sum $[E_0 + E(t)]^2 \approx (E_0)^2 + 2E_0{\cdot}E(t)$, so that no frequency doubling takes place. It was supposed that amplitude of the fair weather field is high enough, and the first term $(E_0)^2$ provides a constant heating of electrons, while the second term $2E_0{\cdot}E(t)$ exceeds the plasma threshold field and therefore initiates the linear modulation of the round-the-world HF radio wave.

Estimates demonstrated that both the fields are able to penetrate up to the 50–80 km altitude depending on time of the day. Then, the modulation index was computed for the round-the-world propagation path. A comparison between the experimental and theoretical results showed that the mechanism suggested agrees with the measurement.

13.3 ELF Fields Produced by a Rocket Flare

In the present section we consider conversion of the fair weather field into the extremely low frequency radio emission by the flare of a starting rocket. Generation of such a field is interesting from the physical point of view. Additionally, it permits us to obtain the vertical profile of the fair weather field and to scrutinize simultaneously the rocket propelling system during the launch.

The Earth–ionosphere spherical capacitor carries the voltage of about 250 kV. This field, its dynamics and dispersion:distribution are an objective of permanent attention (see, e.g. Rycroft et al. 2008, 2012; Rycroft and Harrison 2011; and references therein). An important feature of the fair weather electricity is the altitude profile of the field that reflects physical properties and structure of the atmosphere. We consider below a specific situation when the fair weather field transforms into the ELF radio signal when a rocket starts. The idea exploits electromagnetic radiation by a conducting body of fluctuating length, which is immersed into the fair weather field. The rocket corpus together with its flare serves as such a converter. The geometry of the problem is shown in Fig. 13.4.

A perfectly conducting spheroid acquires the electric dipole moment when interposed into the outer electrostatic field (Landau and Lifshits 1957) that we divide into constant and fluctuating parts:

$$M(t) = M_0 + M_1(t) = \frac{4\pi\varepsilon_0 E L^2}{[\ln(2L/b) - 1]}\left\{\frac{L}{3} + l(t)\right\} \tag{13.4}$$

Here M_0 and $M_1(t)$ are the constant and varying parts of the spheroid's vertical electric dipole moment, ε_0 is the permeability of free space, E is the amplitude of the fair weather field, L and b are the constant vertical and the horizontal half-axes of the spheroid $L \gg b$, $l(t)$ is the fluctuating portion small in comparison with L: $L(t) = L + l(t)$ and $L \gg l(t)$.

The fluctuating length of the rocket flare modifies the moment of the vertical dipole which excites a radio noise. The plane geometry is used in the solution of the problem (see Fig. 13.4), and we account for the static, induction, and radiation fields produced by a fluctuating conductor:

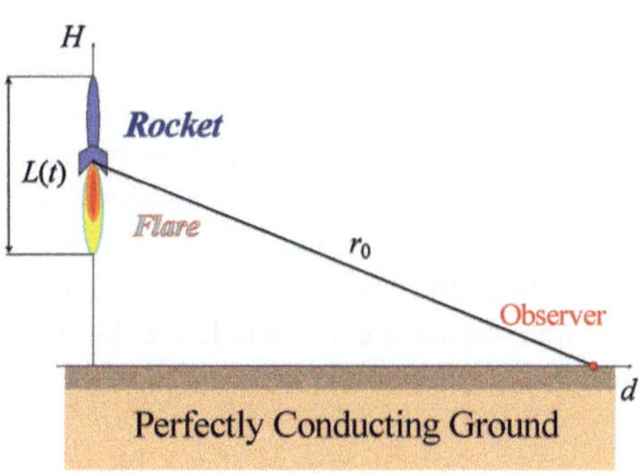

Fig. 13.4 Geometry of the fair weather field conversion problem

$$4\pi\varepsilon_0 E(\omega) = M_1(\omega)[Y_1 + Y_2 + Y_3] \tag{13.5}$$

where

$$Y_1 = \left[r_0^2(kd)^2 + ikr_d^2 r_0 + r_d^2\right] \frac{\exp(-ikr_0)}{r_0^5} \tag{13.6}$$

$$Y_2 = \sum_{n-1}^{\infty} R_i^n \left[(r^+)^2(kd)^2 + ik(D^+)^2 r^+ + (D^+)^2\right] \frac{\exp(-ikr^+)}{(r^+)^5} \tag{13.7}$$

$$Y_3 = \sum_{n-1}^{\infty} R_i^n \left[(r^-)^2(kd)^2 + ik(D^-)^2 r^- + (D^-)^2\right] \frac{\exp(-ikr^+)}{(r^-)^5} \tag{13.8}$$

$k = \omega/c$ is the free space wave number, R_i and h are the reflection coefficient and the effective height of ionosphere, d is the horizontal distance from the observer to the start position, H is the rocket altitude above the perfectly conducting ground, and the slanted distances are found from the following relations: $r_0 = [H^2 + d^2]^{1/2}$, $r^{\pm} = [(2nh \pm H)^2 + d^2]^{1/2}$, $r_d = [2H^2 - d^2]^{1/2}$, and $D^{\pm} = [2(2nh \pm H)^2 - d^2]^{1/2}$. Equation (13.6) is the radiation field, while Eqs. (13.7) and (13.8) are the induction and static fields respectively. The presence of the ionosphere is accounted by introducing the image sources with the amplitudes proportional to the ionosphere reflection coefficient $|R_i| < 1|$, and n denotes the number of ionospheric reflections.

The spectrum of fluctuating moment of the spheroid $M_1(\omega)$ is equal to:

$$M_1(\omega) = 4\pi\varepsilon_0\, E \frac{L^2\, l(\omega)}{\ln(2L/b) - 1} \tag{13.9}$$

The condition $kd \ll 1$ is satisfied for the nearby observer position, and the modulated field contains only the static component (13.8). We have $kd \gg 1$ in the far zone of the source where the radiation field (13.6) prevails. In this zone the sphericity of the Earth–ionosphere cavity should be accounted for, and one obtains the familiar formula for the vertical electric field component:

$$E(\omega) = \frac{i\nu\,(\nu + 1)}{4\varepsilon a^2 h \sin\pi\,\nu} M_1(\omega)\, P_\nu\left[\cos(\pi - \theta)\right] \tag{13.10}$$

where $\theta = \pi d/a$ is the angular source-observer distance, a is the radius of the Earth, $\nu(\omega)$ is the propagation constant, and $P_\nu[\cos(\pi - \theta)]$ is the Legendre function.

It is interesting to estimate the field amplitude arising from the flare of a starting rocket and to evaluate its dynamics during the vehicle elevation. We consider the close vicinity of a launching site where $kd \approx kr \ll 1$, so that the static term dominates and the exponential distance factors might be replaced by unity. Two possible cases are modeled: the launch of a 'big' and of a 'small' rocket. The following parameters are used for these vehicles:

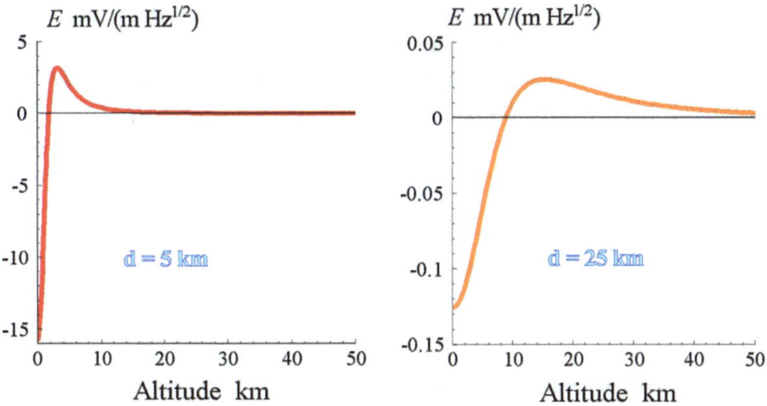

Fig. 13.5 Amplitude of the vertical electric field at 25 Hz frequency induced by the big rocket start at the horizontal distances of 5–25 km from the launch facility. For the "small rocket" (see the text) the scale of the ordinate should be reduced by the factor of 1,000

- The big rocket body has the length $L = 270$ m, and the fluctuating length is $l(\omega) = 135$ m/Hz$^{1/2}$ with spectral density uniformly distributed over the ELF band
- The small one has $L = 27$ m and $l(\omega) = 13.5$ m/Hz$^{1/2}$.

A conducting prolate spheroid produced by the rocket and its flare has the geometry characterized by $\ln(2L/b) = 3$ at an arbitrary frequency. It is easy to see that the source moment of the small rocket is three orders of magnitude smaller than that of the big one, so that relevant fields deviate by the same 10^{-3} factor.

The "day" and "night" models were used of the ionosphere. We utilize the effective height of 60 km and the reflection coefficient 0.6 for the ambient day conditions and have the height $h = 90$ km and $R = 0.9$ for the nighttime. We also assume that the fair weather field is equal to 100 V/m and is independent of altitude for simplicity.

The computed results are shown in Fig. 13.5. The abscissa shows the rocket altitude in km and the ordinate depicts the amplitude of spectral component at the 25 Hz frequency in mV/m/Hz$^{1/2}$. The effect is independent of frequency in our model. Two distances are shown from the launching site being equal to 5 and 25 km (the left and right plots correspondingly). Decrease of the field with distance indicates that the static field plays a dominant role in the model. Reflections from the ionosphere are insignificant at such short distances, and the computed results coincide for both the day and night conditions.

The amplitude at 25 Hz frequency in Fig. 13.5 is initially "negative". This means that the signal detected is in anti-phase with the variations of the source moment $M_1(\omega)$ since the parameter r_d^2 is negative at small altitudes. Variations of the flare and that of the field occur in phase when the $r_d^2 = 2H^2 - d^2 \geq 0$.

Computed field amplitudes show that the flare of a big rocket is able to produce detectable ELF signals exceeding the level of natural noise.

The curve similar to that we show in Fig. 13.5 may be used for deducing the altitude profile of the fair weather field above the launching site from the experimental records of the ELF noise made during the start, provided that the trajectory of the rocket is known as well as the engine regime. On the other hand, such signals carry information on the frequency spectrum of the flare fluctuations. These might be helpful at the ground facilities either for investigations or for the control of the rocket propelling system.

13.4 Magnetospheric Sources

The global thunderstorm activity is the main source of SR. However, one knows that the power density of ELF radio noise in the magnetosphere exceeds that under the ionosphere by an order of magnitude at least. Electromagnetic signals of the ULF band can penetrate down to the ground. For instance, periodical pulsations Pc1 ranging from 0.1 Hz to a few Hz frequencies arrive from the magnetosphere (e.g., Alperovich and Fedorov 2007). This is why an excitation of SR by the cosmic sources was addressed at least theoretically (see Bliokh et al. 1977, 1980; Nickolaenko and Hayakawa 2002 and references therein). We discuss below some distinctive features of ELF signals arriving from the magnetosphere.

A characteristic of waves from the magnetospheric source is the dual path of the signal propagating to the ground. ELF waves in the magnetized plasma propagate along the magnetic field lines, and most probably these will be the Alfven waves transformed into the EM wave at the lower edge of the ionosphere. We avoid this sophisticated problem and simply postulate that there are two propagation paths from the magnetospheric source, which terminate at the magnetically conjugate points of the lower boundary of ionosphere. Our goal is to find out whether such a simple "symmetric" excitations of cavity results in the distinctive properties of ELF radio signal (see Fig. 13.6).

The path in the Northern and the Southern hemispheres may have different lengths, so that a time delay (a phase shift) appears between two currents exciting the cavity oscillation. It is clear that the meridional component of the current density is important at SR frequencies, which is described by the following formula:

$$\vec{j} = \vec{e}_\theta \cdot j_\theta(r, \theta, \varphi) \tag{13.11}$$

where

$$j_\theta\,(r,\ \theta\ \varphi) = \frac{\delta\,(r-b)}{b^2 \sin\theta}\left[j_1(t)\Phi_1(\varphi\,)\delta\,(\theta-\theta_1) + j_2(t)\Phi_2(\varphi)\delta\,(\theta-\pi\ +\theta_1\,)\right] \tag{13.12}$$

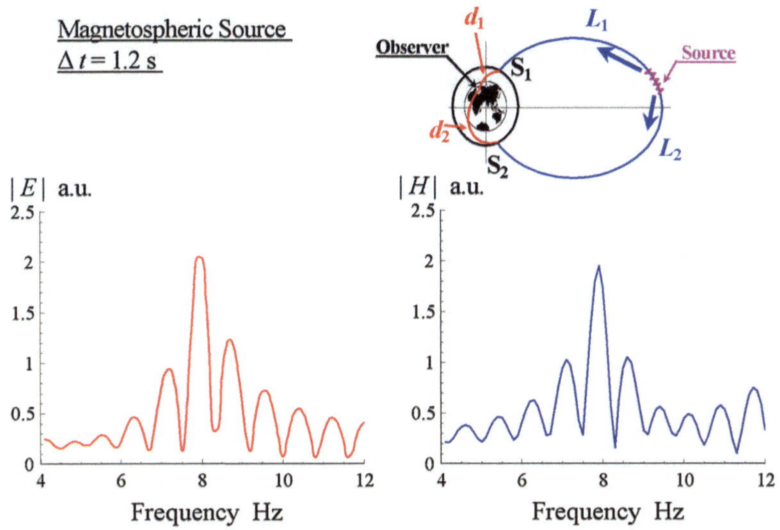

Fig. 13.6 Amplitude spectra of SR excited by a cosmic source

Here b is the height of lower ionosphere boundary, $\delta(x)$ is the Dirac's delta-function, θ is the co-latitude, the functions $j_{1,2}(t)$ describe the current waveforms at the conjugate source areas at the lower ionosphere, and the functions $\Phi_{1,2}(\varphi)$ accounts for the longitudinal distribution of the source currents.

After doing some algebra (Nickolaenko and Hayakawa 2002) one obtains the following source function for the n-th SR mode:

$$I_{n,m}(\omega) = -A_{nm}\frac{P_n^1(\cos\theta_1)\, w_n'(k_n b)}{b}\left[\int_0^{2\pi} \Phi_1(\varphi)\exp(im\varphi)d\varphi \int_{-\infty}^{\infty} j_1(t)\exp(-i\omega t)\,dt \right.$$
$$\left. +(-1)^{n+1}\int_0^{2\pi} \Phi_2(\varphi)\exp(im\varphi)d\varphi \int_{-\infty}^{\infty} j_2(t)\exp(-i\omega t)\,dt\right] \quad (13.13)$$

The equalities were used: $\cos(\pi - \theta_1) = -\cos(\theta_1)$ and $-d/d\theta\,\{P_n(\cos\theta)\} = P_n^1(\cos\theta) = (-1)^{n+1} P_n^1(-\cos\theta)$ (Gradstein and Ryzhik 1963).

Efficiency of the horizontal source current is reduced by a factor of about $\frac{w_n'(k_n b)}{w_n(k_n b)} \approx k_n(b-a) \approx 10^2$ in comparison with the vertical electric current. Here $w_\nu(kr) = u_\nu(kr)\, v_{\nu'}(ka) - v_\nu(kr)\, u_{\nu'}(ka)$ is the function describing the radial dependence of SR field, $u_\nu(x)$ and $v_\nu(x)$ are expressed through spherical Hankel functions: $u_\nu(x) = x\, h_\nu^{(1)}(x)$ and $v_\nu(x) = x\, h_\nu^{(2)}(x)$ (e.g., Nickolaenko and Hayakawa 2002). Therefore only powerful magnetospheric events may noticeably influence the SR observations, and we know this from many experiments. It is natural to suggest that currents in both the hemispheres are equal in amplitude and deviate by a temporal delay Δt due to different path lengths L_1 and L_2 in Fig. 13.6. Thus, we obtain two coherent sources of ELF radio waves placed at the geomagnetic conjugate points and having the current moments:

$$M_C^{1,2} = kh\,F(\theta_1,\ \varphi\,)S^{(1,2)}(\omega\,) \tag{13.14}$$

where the function $F(\theta_1, \varphi)$ accounts for the spatial distribution of the current at the ionosphere boundary and $S^{(1,2)}(\omega)$ describes the complex spectrum of the source: $S^{(2)}(\omega) = S^{(1)}(\omega)\cdot\exp\,(-\omega\Delta t)$.

A couple of coherent sources will excite the following fields in the uniform and isotropic Earth–ionosphere cavity:

$$E_r(\omega) = E_0\left\{P_\nu\left[-\cos\left(\frac{\pi\,d_1}{a}\right)\right] + P_\nu\left[-\cos\left(\frac{\pi\,d_2}{a}\right)\right]\cdot\exp(-i\omega\,\Delta\,t)\right\} \tag{13.15}$$

$$H_\varphi\,(\omega) = H_0\left\{P_\nu^1\left[-\cos\left(\frac{\pi\,d_1}{a}\right)\right]\cos\alpha_1 + P_\nu^1\left[-\cos\left(\frac{\pi\,d_2}{a}\right)\right]\cdot\exp(-i\omega\,\Delta\,t)\cos\alpha_2\right\} \tag{13.16}$$

The complex field amplitudes are $E_0 = \frac{i\nu\,(\nu+1)}{\omega\,\sin\pi\,\nu}\frac{M_C^1(\omega)}{4\,h\,a^2\varepsilon}$ and $H_0 = \frac{-M_C^1(\omega)}{4ha\sin\pi\nu}$, $d_{1,2}$ are the distances from the observer to the sources No. 1 and No. 2 at the lower ionosphere boundary, and $\alpha_{1,2}$ are the angles between the direction of the wave propagation and the geographic parallel at the observatory.

It is clear that the magnetospheric source provides a couple of coherent currents, which generate an interference pattern superimposed on the ordinary SR curve. This interference pattern is the major distinction of the SR signal when excited from the magnetosphere. Owing to irregularities of the magnetosphere, two currents maintain their mutual coherence over the finite time interval T_C, which results in the following substitution in Eqs. (13.15 and 13.16): $\exp(-i\omega\,\Delta\,t) \to \exp(-i\omega\,\Delta\,t)^{\sin\omega\,T_C}/_{\omega\,T_C}$ [compare with Bliokh et al. (1980)].

We demonstrate in Fig. 13.6 the amplitude spectra of the SR excited from the magnetosphere when the global thunderstorm activity is absent. The observer is placed in the middle latitudes of the Northern hemisphere at the point $\theta_0 = 45°$ and $\varphi_0 = 0°$. The point sources at the ionosphere occupy the position $\theta_1 = 20°$ and $\varphi_1 = 90°$ in the Northern and $\theta_2 = 160°$ and $\varphi_2 = 90°$ in the Southern hemispheres. Corresponding source-observer distances are: $d_1 = 5.4$ and $d_2 = 14.6$ Mm. The signals of the southern source are delayed against those of the northern currents by the constant value of 1.2 s.

Computations show the interference of two coherent signals in the SR spectra. The period of frequency modulations depends on the mutual delay of the sources, while the change in the source position at the ionospheric boundary will produce alterations in the geometry parameters d_1, d_2, α_1, and α_2. This may cause more complicated and slow frequency variations of amplitude spectra.

Low frequency waves are generated in the magnetosphere plasma due to the interaction between the plasma waves and the trapped particles (Sazhin and

Hayakawa 1992; Trakhtengerts and Rycroft 2008). As a rule, this occurs in close proximity to the equatorial plane, so that the delay in the ELF source current is conditioned by a slight displacement of this zone from the equator. The delay depends on the velocity of plasma waves in the magnetosphere, which is about 1,000 km/s. Therefore, the 1 s retardation corresponds to the displacement of about 1 Mm. When the signal originates say at the fifth L-shell, the zone of generation is nearby the equatorial plane at the distance $L = 5a \approx 26$ Mm from the ground surface. It is clear that 1 Mm displacement is relatively small.

The zone of wave-particle interaction might periodically oscillate from North to South. In this case, a periodic modulation of the time delay will arise. Simultaneously, the wave arriving along one of the traces might become 'compressed' by the Doppler effect, while the conjugate wave will be 'expanded' in time. Relevant periodic changes in the delay turns into the infinite sum of Bessel functions involved in Eqs. (13.5) and (13.6). Formally, one has to substitute $\Delta t \to T_M \cos \Omega t$ in these formulas where T_M denotes the amplitude and Ω is the "slow" frequency shifts of magnetospheric source (Nickolaenko 1972; Bliokh et al. 1980). After this replacement, one has to employ the following identity (Gradstein and Ryzhik 1963):

$$\exp(-i\omega\,\Delta\,t) = \exp(-i\omega\,T_M\cos\Omega\,t) = \sum_{k=-\infty}^{\infty} (-i)^k J_k(\omega\,T_M)\exp(ik\Omega\,t) \quad (13.17)$$

where $J_k(x)$ is the Bessel function of the first kind. Obtaining the final expressions for the signals is rather clear, but they are too cumbersome to show them here.

Physically, the swing of magnetospheric source may completely change the spectrum outline in comparison with that from a fixed source. Individual spectral lines might appear with amplitudes depending on the delay (through the Bessel functions) and with the frequency separation determined by the period of swing. The modification of SR spectrum will become especially apparent when the plasma source generates the narrow-band signals; say, the proton cyclotron resonance. In this case, the spectral 'splitting' may become noticeable in the amplitude spectrum.

We demonstrated that magnetospheric ELF emissions may cause peculiar fine structures in the SR curves recorded at high latitudes. A similarity between the experimental and model spectra may serve as an indication of a possible magnetospheric origin of the stable sub-peaks found in the SR (Egeland and Larsen 1968). Though the final and convincing confirmation of their connections to the magnetosphere is still desired.

13.5 Orbital Signatures of SR

SR signatures were detected in space (Simões et al. 2011). Measurements were performed at the Communications/Navigation Outage Forecasting System (C/NOFS) satellite orbiting the Earth in the altitude region of 400–850 km, which is far above the Earth–ionosphere cavity. The cosmic observations of SR might be explained by the energy leakage through the ionosphere, and this demands a revision of models of ELF fields in the ionosphere.

The C/NOFS satellite was launched in April 2008 aiming at the low latitude irregularities and scintillations in the ionosphere. Its elliptical orbit has the 401 km perigee and 852 km apogee, and the inclination is 13°. Among the other instruments, the vehicle carries three orthogonal pairs of 20 m tip-to-tip booms to perform the vector measurements of both DC and AC (or wave) electric fields. The ELF electric fields are digitized with the 1,024 samples/s rate and 16-bit resolution. Successive data are averaged by 2 in the receiver pass through the low pass, dual octave Butterworth filter with the 3 dB cut-off frequency ~192 Hz. The resulting waveforms are sent to the ground with the 512 sample/s rate. The sensitivity of the electric field channel is ~10 $\text{nV·m}^{-1}\text{·Hz}^{-1/2}$.

Figure 13.7 [adopted from Simões et al. 2011)] depicts the 0–80 Hz spectrogram of the total electric field recorded on the orbit 666 on 31 May 2008. The SR is seen as the horizontal bright lines below 50 Hz (shown by the white arrows) with peaks at 7.8, 14.0, 20.4, 26.7, and 33.0 Hz (Simões et al. 2011). There are also present other signals, e.g. ELF hiss (see Hayakawa and Sazhin 1992), which are highly variable due to changes in geomagnetic field and plasma conditions along the orbit. The narrow regions of broadband electrostatic irregularities are also seen starting from the DC range. However, the SR pattern is observed as faint regularly spaced and rather stable horizontal spectral lines observed on the night side of the globe. A horizontal black bar below the sonogram in Fig. 13.7 indicates when the satellite was in complete shadow. The lowest panel shows the satellite's path (black line) and altitude (blue line).

Such resonance patterns were usually observed over the nighttime hemisphere indicating that the resonant energy can escape from the cavity through the more transparent plasma on the night side of the globe. Simões et al. (2011) report that the resonance structure does not depend on the time when the satellite enters the shadow. Their amplitude increases at the lower altitudes of the satellite. Seasonal variations have also been noted. The investigation of these interesting effects should be continued.

The average spectra are shown in Fig. 13.8 corresponding to the entire data of the nighttime observations in Fig. 13.7. The lowest seven modes are readily recognized. Two kinds of spectra are shown separately. The upper panel demonstrates the "raw" spectrum together with smooth curves representing the estimated background. The lower panel depicts the same spectrum with the background subtracted. The Gaussian fits of experimental data (points) are shown by thin lines. Parameters of spectral peaks derived from the fits correspond to the measurements on the ground. The peak frequencies are 7.8, 14.0, 20.4, 26.7, and 33.0 Hz, and the Q–factors are 3.5, 4.5, 6.2, 7.7, and 8.2 for the first five modes. These were found from the full width at half power level of the resonance line fits.

It is interesting to note that the wave detected in the plasma was a transverse one: no electric field was observed collinear to the geomagnetic field. The electric field components were comparable in the "vertical" direction (i.e., the perpendicular to

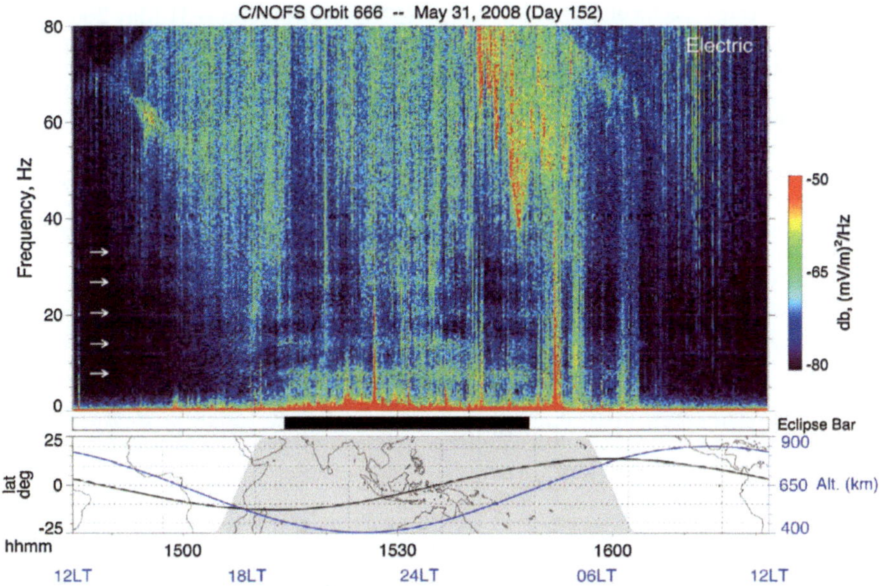

Fig. 13.7 Dynamic spectrum of the space-borne ELF electric field data observed at the C/NOFS satellite after Simões et al. (2011)

the magnetic field line lying in the meridianal plane) and in the zonal (i.e., East–West) direction. This suggests the circular polarization and either the whistler mode propagation or the extraordinary electromagnetic wave. The polarization of the waves observed is underway (Simões et al. 2011).

SR was a common phenomenon, observed throughout the 3-year satellite lifetime. SR signatures were detected on the night side throughout the altitude range between 400 and 850 km. Many peaks are usually observed. In some cases, up to 10 resonance peaks have been observed on a single orbit. Averaging the data from multiple orbits extends the number of observed peaks even further.

The electric spectral density at the first peak is $\sim 0.25\ \mu V \cdot m^{-1} \cdot Hz^{-1/2}$ or three orders of magnitude lower than in the ground based measurements (Sentman 1995; Nickolaenko and Hayakawa 2002). The difficulty of resonance observations on the dayside of the globe on the space-based platform may be explained by higher density of the daytime plasma. Similar nighttime leakage was noted by Madden and Thompson (1965). The altitude factor contributes to the damping of the resonant wave energy on its way to the satellite. For instance, SR records in the stratosphere onboard balloons showed a scale height of $\sim$50 km at the 25 km altitude (Ogawa et al. 1979), emphasizing that SR signatures indeed decrease with altitude. These authors showed that at 25 km during fair weather conditions, the electric field associated with the SR is about half that observed on the ground despite a quite small atmospheric conductivity in the troposphere.

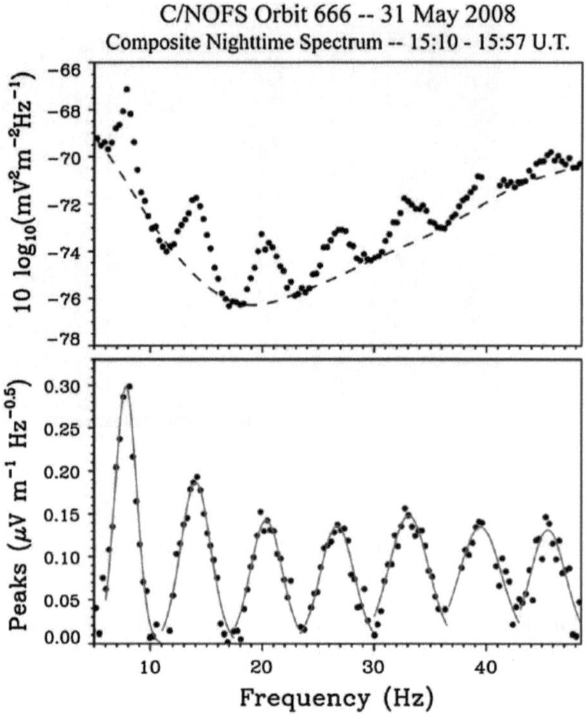

Fig. 13.8 Average spectra of the ELF electric fields for the nighttime portion of the data in Fig. 13.7: (*top*) the complete spectra including the background levels and (*bottom*) the same data with the background contributions removed and superimposed Gaussian fits of each mode

The presence of SRs in the orbital records requires an adequate interpretation. The anisotropy and vertical non-uniformity may play a key role in the energy leakage (Madden and Thompson 1965; Grimalsky et al. 2005, Simões et al. 2009). One neglects the anisotropy of the ionosphere in models used now and, consequently, assumes that waves are confined to the altitudes below ~100 km. Adequate models should be elaborated to explain the ELF radio wave propagation into the ionosphere, and the geomagnetic field must be taken into account.

The success of the space-borne observations of SR might be explained by adequate continuous measurements afforded by the C/NOFS satellite, by rather long antennas and a sensitive receiver working in the broadband. The orbit of low inclination might also play the positive role since it enables longer periods at night at low latitudes.

The detection of SR in space suggests a new turn in the studies of other planets with atmospheres starting from Venus to Neptune and Jupiter, and even the Saturn's moon, Titan (see e.g. Nickolaenko and Rabinowicz 1982, 1987; Sentman 1990; Molina-Cuberos et al. 2001, 2004, 2006; Tokano et al. 2001; Nickolaenko et al. 2003; Pechony and Price 2004; Yang et al. 2006; Béghin et al. 2007; Navarro et al. 2007; Simões et al. 2007, 2008a, b; Morente et al. 2008; Fischer and Gurnett 2011 and references therein). The leakage from the global electromagnetic cavity might offer means for orbital studies of planetary electric activity combined with the measurements by descending probes, buoyant vessels, or landers. SR at other planets and moons with ionospheres is a tool for sensing the dielectric properties and conductivity of the ground, atmosphere, ionospheres, etc., provided that there exists a lightning activity. To conclude the sub-section, we must remark that industrial narrow-band signals at 50–60 Hz frequencies were not visible in orbital records. This is an interesting feature that deserves an explanation.

13.6 Transverse Resonance in Space

The first frequency of transverse resonance of the Earth–ionosphere cavity (1.6–1.8 kHz) was detected in orbit by DEMETER satellite (Toledo-Redondo et al. 2012). DEMETER is a French scientific micro satellite, which was launched in June 2004 aiming at the studies of ionospheric precursors of EQs. It has a nearly circular sun-synchronous orbit and the up-going half-orbits correspond to the night time (22.30 LT) whereas the down-going half-orbits correspond to the day time (10.30 LT). In other words, measurements by DEMETER were performed either around 22.30 LT or 10.30 LT, the latitude or orbit number being insignificant. The polar orbit has an inclination of 98°, and the initial altitude was 710 km that was corrected to 660 km in December 2005. The vehicle performs 14 orbits per day and surveys the geomagnetic latitudes between ±65°. The rest of time is used for maintenance. There are two operational modes: survey and burst. The burst mode is activated over sensible EQ areas, and the whole time realizations are sent to the ground, including the onboard computed spectra. Only spectra are downloaded in the survey mode. The payload includes the electric field sensor (ICE), the magnetic search coil (IMSC), the Langmuir probe (ISL), the plasma analyzer, and the detector of energetic particles.

The electric field sensor contains four spherical electrodes with embedded preamplifiers. The sensors are placed at the booms 4 m long. The signal processing and A/D conversion is performed in four bands: DC/ULF (0–15 Hz), ELF (15 Hz–1 kHz), VLF (15 Hz–17.4 kHz), and HF (10 kHz–3.175 MHz). We reproduce the VLF data below (see Toledo-Redondo et al. 2012). These data were colleted with the sampling frequency of 40 kHz with the 16-bit ADC. The sensibility of the sensors in the range 15–17.4 kHz is $\sim$0.05 $\mu V \cdot m^{-1} \cdot Hz^{-1/2}$. The frequency resolution is 19.53 Hz in the onboard spectra. These are averaged in groups of 40, so that the time resolution is 2.048 s. The averaged spectra are normalized by the maximum, converted into 8 bits and sent to the ground. A single component of the electric field is measured in the survey mode. The data were collected in the years 2007–2010 and correspond to the E_{12}, which is approximately parallel to the East–West direction on the ground.

The VLF spectra were collected over the night hemisphere (due to the Sun-synchronous orbit, measurements were performed around 22:30 LT), and about 15,000 orbits were used. During the observational interval, the solar activity was low and the ionosphere was very different in the day and night time. An example

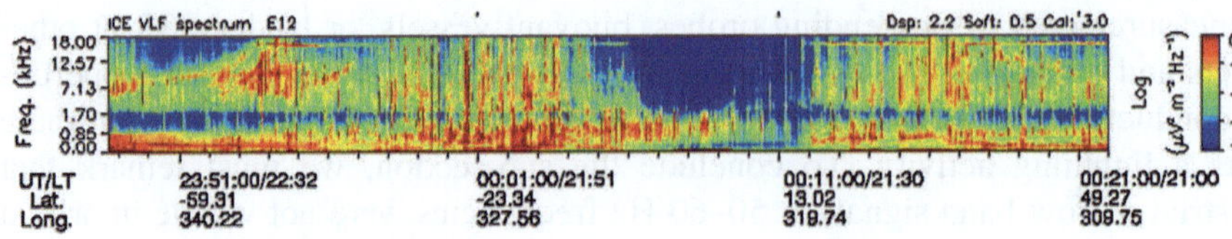

Fig. 13.9 The local nighttime dynamic spectrum recorded by DEMETER over the Atlantic Ocean on October 1, 2010. The basic frequency of the transverse resonance is apparent around 1.7 kHz seen as a spectral minimum

of the data can be seen in Fig. 13.9, where the received ICE-VLF spectrogram of DEMETER for one half-orbit is depicted. The cut-off frequency is clearly distinguishable at around 1.7 kHz.

We will not go into the details of signal processing on board the DEMETER satellite, but the relevant information might be found in the paper by Toledo-Redondo et al. (2012). We will demonstrate instead that the blue spectral line in Fig. 13.9 is really the trace of the transverse resonance.

For this purpose, we compare two plots in Fig. 13.10. The red axes and the red line here reproduce the spectrum of the transverse resonance in the H_ρ field component that we demonstrated in Fig. 12.2. The frequency band was reduced to 5 kHz in Fig. 13.10. The superimposed frame with the black lines presents the DEMETER data that we adopted from Fig. 3 of the paper by Toledo-Redondo

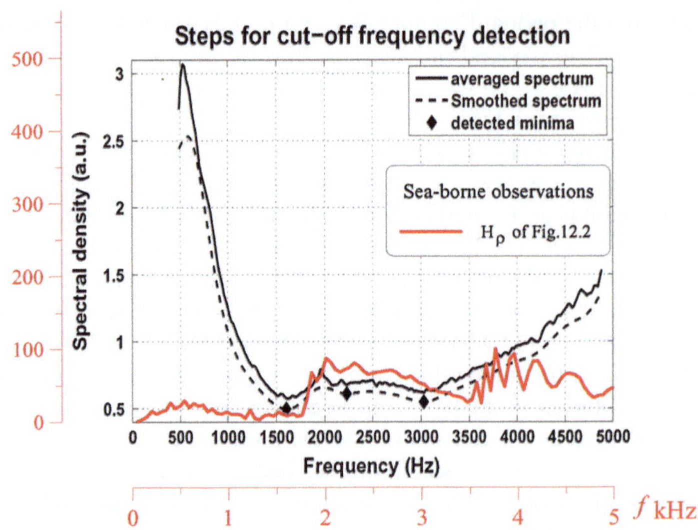

Fig. 13.10 Comparison of DEMETER spectra over the Atlantic Ocean on October 1, 2010. The basic frequency of the transverse resonance is apparent around 1.7 kHz seen as a spectral minimum

et al. (2012). One may observe that the average spectral pattern in the space is exceptionally close to the spectrum of individual atmospheric recorded in the sea (Shvets and Hayakawa 1998). Correspondence of these curves leaves no doubt that the spectra recorded by DEMETER reflect the basic frequency of the transverse resonance, and hence, the effective nocturnal ionosphere height. The new-borne wave with $p = 1$ appears as an increase in the field amplitude just above the minimum frequency of about 1.7 kHz.

A large data set of electric field measurements from DEMETER has been employed to draw average maps of the first transverse resonance frequency with dependence on the position. We described in the previous Chapter how to evaluate the effective height of the Earth–ionosphere cavity from the transverse resonance frequency. Such estimates were obtained recently from the ground observations (e.g. Outsu 1960; Hayakawa et al. 1994, 1995; Ohya et al. 2006; Kumar et al. 2009; Ostapenko et al. 2010; Shvets and Gorishnya 2011). The similar technique was applied to the data from the spacecraft. Thus, the global maps of the resonance frequency were obtained with the 2° resolution in the ±60° latitude range. The main result of this survey was the behavior of the nighttime ionosphere effective height, which decreased over oceans during the "boreal winter", i.e. from November to March (Toledo-Redondo et al. 2012). The first mode frequency varied around 1.7 kHz, which corresponds to the effective height of 88 km. The month after month monitoring over three years revealed the seasonal patterns repeating in all the years for the certain geographical location. In particular, the basic frequency of transverse resonance shifted upwards by 100–200 Hz (the ionosphere goes down to 83–79 km) over the South Pacific Ocean during the southern "warm season", i.e. between May and October. Another effect was observed over the Indian and Atlantic Oceans where the frequency increases in the area of roughly ±15° geomagnetic latitude twice per year. The amplitude of semi-annual variation was about one-half of the annual component.

Toledo-Redondo et al. (2012) interpreted these very interesting observational data (the shifts in ionosphere height appear mainly over the oceans) as an impact of the sea-salt aerosol or by changing the conductivity of the seawater. We think that the other, more realistic reason was overlooked here. The solar irradiation was correctly mentioned as a cause of the changes in the transverse resonance frequency. However, the annual and semi-annual variations in its impact on the lower ionosphere were unnoticed.

It is important that DEMETER data are relevant to the night ionosphere. The ionization of the nighttime D region is maintained by the galactic ionizing radiation (galactic background). The long-term observations of the VLF transmissions showed that the lower ionosphere boundary moves upward during the periods of high solar activity. Such "illogical" behavior is explained by the wiping out of the galactic background radiation by the solar wind, so that the nighttime ionization rate is reduced in the D region and it "moved upward". We think that the semi-annual variations of the sub-tropical ionosphere height are explained not by the "space weather", but by the "space climate". The seasonal drift of the Sun in the sky provides variable conditions for the interaction of the solar wind and the

Earth's magnetosphere. We speak about the stationary component of the solar wind. As a result, there are two yearly components in the velocity of the solar wind coming to the Earth. The annual component is probably conditioned by the changes in the distance from the Sun due to the eccentricity of the Earth's orbit.

The tilt of the Earth's rotation axis causes semi-annual variations. Thus, the solar wind has a higher speed in the boreal summer (the distance from the Sun increases), and it blows out the galactic background more actively, so that the D-region height increases (the resonance frequency decreases). The semi-annual variation arises from the different orientation. During the equinox periods, the solar wind arrives at the Earth in the equatorial plane, i.e. its velocity becomes perpendicular to the geomagnetic field. Therefore, the strength of solar wind reduces, and the minimum ionosphere heights (maximum resonance frequencies) are registered in the tropical zone through March–April and September–October equinox periods.

Pronounced modulations in the ionosphere height observed exclusively over the oceans might be explained by the relatively small thunderstorm activity in the seas. High lightning rate at the continents might modify the ionosphere height and thus mask the seasonal effect of solar radiation.

The studies by the DEMETER satellite imply that information about the D region of ionosphere is scarce, and it is practically absent over the seas (the exception are the works by Shvets and Hayakawa (1998) and Shvets and Gorishnya (2011). The lower ionosphere is mainly studied by incoherent scatter radars and by in situ measurements of rockets. Therefore, experimental data are relevant to a few land areas, and not available for the oceans. Thus, the observations from space proved to be helpful in obtaining the global coverage of the lower ionosphere boundary on the night side of the globe.

References

Alperovich LS, Fedorov EN (2007) Hydromagnetic waves in the magnetosphere and the ionosphere. Springer, p 426

Asano T, Suzuki T, Hiraki Y, Mareev E, Cho MG, Hayakawa M (2009) Computer simulations on sprite initiation for realistic lightning models with higher-frequency surges. J Geophys Res 114:A02310. doi:10.1029/2008JA013651

Béghin C, Simões F, Karsnoselskhikh V, Schwingenschuh K, Berthelier JJ, Besser B, Bettanini C, Grard R, Hamelin M, López-Moreno JJ, Molina-Cuberos GJ, GJ, Tokano T (2007) A Schumann-like resonance on Titan driven by Saturn's magnetosphere possibly revealed by the Huygens probe. Icarus 191:251–266

Belyaev GG, Schekotov AYu, Shvets AV, Nickolaenko AP (1999) Schumann resonances observed using Poynting vector spectra. J Atmos Solar-Terr Phys 61:751–763

Bliokh PV, Nickolaenko AP, Filippov YF (1977) Global electromagnetic resonances in the Earth–ionosphere cavity. Naukova Dumka, Kiev, p 199 (in Russian)

Bliokh PV, Nickolaenko AP, Filippov YF (1980) Schumann resonances in the Earth–ionosphere cavity. Paris, Peter Perigrinus, London, New York, p 168

Cho M, Rycroft MJ (2001) Non-uniform ionization of the upper atmosphere due to the electromagnetic pulse from a horizontal lightning discharge. J Atmos Solar-Terr Phys 63:559–580

Egeland A, Larsen TR (1968) Fine structure of the Earth–ionosphere cavity resonances, J Geophys Res, 73, pp 4986–4989

Fischer G, Gurnett DA (2011) The search for Titan lightning radio emissions. Geophys Res Lett 38:L08206. doi:10.1029/2011GL047316

Füllekrug M, Mareev EA, Rycroft MJ (eds) (2006) Sprites, elves and intense lightning discharges. NATO Science Series, Springer, Dordrecht

Ginzburg VL (1970) The propagation of electromagnetic waves in plasmas. Pergamon Press, Oxford, London, New York

Gradstein C, Ryzhik IM (1963) Tables of integrals, sums, series and products. Phys-Math Publ, Moscow

Grimalsky V, Koshevaya S, Kotsarenko A, Enriquez RP (2005) Penetration of the electric and magnetic field components of Schumann resonances into the ionosphere. Ann Geophys 23:2559–2564. doi:10.5194/angeo-23-2559-2005

Gurevich AV, Milikh GM, Roussel-Dupre R (1992) Runaway electron mechanism of air breakdown and preconditioning during a thunderstorm. Phys Lett A165:463

Hayakawa M, Sazhin SS (1992) Mid-latitude and plasmaspheric hiss: a review. Planet Space Sci 40:1325–1338

Hayakawa M, Ohta K, Baba K (1994) Wave characteristics of tweek atmospherics deduced from the direction finding measurement and theoretical interpretations. J Geophys Res 99:10733–10743

Hayakawa M, Ohta K, Shimakura S, Baba K (1995) Recent findings on VLF/ELF sferics. J Atmos Terr Phys 57:467–477

Hiraki Y (2010) The phase transition theory of the sprite halo. J Geophys Res 115, A00E20. doi:10.1029/2009JA014384

Kudintseva IG, Nickolaenko AP, Hayakawa M (2009) Spatial fine structure of model electric pulses in the mesosphere above a Γ-shaped stroke of lightning. J Atmos Solar-Terr Phys 71:603–608

Kumar S, Deo A, Ramachandran V (2009) Nighttime D-region equivalent electron density determined from tweeks sferics observed in the south pacific region. Earth Planets Space 61:905–911

Landau LD, Lifshits EM (1957) Electrodynamics of continuous media. Gostekhizdat, Moscow (in Russian)

Madden T, Thompson W (1965) Low frequency electromagnetic oscillations of the Earth–ionosphere cavity. Rev Geophys 3:211–254

Molina-Cuberos GJ, Lammer H, Stumptner W, Schwingenschuh K, Rucker HO, Lopez-Moreno JJL, Rodrigo R, Tokano T (2001) Ionospheric layer induced by meteoric ionization in Titan's atmosphere. Planet Space Sci 49:143–153

Molina-Cuberos GJ, Porti J, Besser BP, Morente JA, Margineda J, Lichtenegger HIM, Salinas A, Schwingenschuh K, Eichelberger HU (2004) Schumann resonances and electromagnetic transparence in the atmosphere of Titan. Adv Space Res 33:2309–2313

Molina-Cuberos GJ, Morente JA, Besser BP, Portí J, Lichtenegger H, Schwingenschuh K, Salinas A, Margineda J (2006) Schumann resonances as a tool to study the lower ionospheric structure of Mars. Radio Sci 41, RS1003

Morente JA, Portí JA, Salinas A, Navarro EA (2008) Evidence of electrical activity on Titan from the Schumann resonances sent by Huygens probe. Icarus 195:802–811

Navarro EA, Soriano A, Morente JA, Portí JA (2007) A finite difference time domain model for the Titan ionosphere Schumann resonances. Radio Sci 42, RS2S04. doi:10.1029/2006RS003490

Nickolaenko AP (1972) On the ELF noise spectra peculiarities when excited by cosmic sources. Geomag Aeronomia 12:458–463 (in Russian)

Nickolaenko AP, Hayakawa M (2002) Resonance in the Earth–ionosphere cavity. Boston, London, Kluwer Academic Publishers, Dordrecht, p 380
Nickolaenko AP, Rabinowicz LM (1982) Possible electromagnetic resonances on the planets of solar system. Cosm Res 20:67–71 Plenum Publishing Corporation
Nickolaenko AP, Rabinowicz LM (1987) Applicability of ultralow-frequency global resonances for investigating lightning activity on Venus. Cosm Res 25(1):239–243 Plenum Publishing Corporation
Nickolaenko AP, Besser BP, Schwingenschuh K (2003) Model computations of Schumann resonance on Titan. Planet Space Sci 51:853–862
Ogawa T, Kozai K, Kawamoto H, Yasuhara M, Huzita A (1979) Schumann resonances observed with a balloon in the stratosphere. J Atmos Terr Phys 41:135–142. doi:10.1016/0021-9169(79)90005-9
Ohya H, Nishino M, Murayama Y, Igarashi K, Saito A (2006) Using tweek atmospherics to measure the response of the low-middle latitude D-region ionosphere to a magnetic storm. J. Atmos. Solar-Terr. Phys 68:697–709
Ostapenko AA, Titova EE, Nickolaenko AP, Turunen JMT, Raita T (2010) Characteristics of VLF atmospherics near the resonance frequency of the Earth–ionosphere waveguide 1.6–2.3 kHz by observations in the auroral region. Ann Geophys 28:193–202
Outsu J (1960) Numerical study of tweeks based on waveguide mode theory. Proc Res Inst Atmospherics, Nagoya Univ 7:58–71
Pasko VP, Inan US, Bell TF (1999) Mesospheric electric field transients due to tropospheric lightning discharges. Geophys Res Lett 26(9):1247–1250
Pechony O, Price C (2004) Schumann resonance parameters calculated with a partially uniform knee model on Earth, Venus, Mars, and Titan. Radio Sci 39, RS5007
Price C, Greenberg E, Yair Y, Sátori G, Bor J, Fukunishi H, Sato M, Israelevich P, Moalem M, Devir A, Levin Z, Joseph JH, Mayo I, Ziv B, Sternlieb A (2004) Ground-based detection of TLE-producing intense lightning during the MEIDEX mission on board the space shuttle Columbia. Geophys Res Lett 31(20):L20107. doi:10.1029/2004GL020711
Rycroft, M.J., K.A. Nicoll, K.L. Aplin, and R.G. Harrison (2012) Recent advances in global electric circuit coupling between the space environment and the troposphere. J Atmos Solar-Terr Phys. http://dx.doi.org/10.1016/j.jastp.2012.03.015
Rycroft MJ, Harrison RG (2011) Electromagnetic atmosphere-plasma coupling: The global atmospheric electric circuit. Space Sci Rev 140:363–384. doi:10.1007/s11214-011-9830-8
Rycroft MJ, Harrison RG, Nicoll KA, Mareev EA (2008) An overview of Earth's global electric circuit and atmospheric conductivity. Space Sci Rev 137:83–105. doi:10.1007/s11214-008-9368-6
Sazhin SS, Hayakawa M (1992) Magnetospheric chorus emissions: a review. Planet Space Sci 40:681–697
Sentman DD (1990) Electrical conductivity of Jupiter's shallow interior and the formation of a resonant planetary-ionospheric cavity. Icarus 88:73–86
Sentman DD (1995) Schumann Resonances. in Volland H (ed) Handbook of Atmospheric Electrodynamics. CRC Press, Boca Raton, London, Tokyo, vol 1, pp 267–298
Shvets AV, Gorishnya YV (2011) A technique for lightning location and estimation of the lower ionosphere parameters using tweek-atmospherics. Telecommun Radio Eng 70(11):1013–1026
Shvets AV, Hayakawa M (1998) Polarization effects for tweek propagation. J Atmos Solar-Terr Phys 60:461–469
Simões F, Grard R, Hamelin M, Lopez-Moreno JJ, Schwingenschuh K, Beghin C, Berthelier JJ, Besser B, Brown VJB, Chabassiere M, Falkner P, Ferri F, Fulchignoni M, Hofe R, Jernej I, Jeronimo JM, Molina-Cuberos GJ, Rodrigo R, Svedhem H, Tokano T, Trautner R (2007) A new numerical model for the simulation of ELF wave propagation and the computation of eigenmodes in the atmosphere of Titan: did Huygens observe any Schumann resonance? Planet. Space Sci 55:1978–1989

Simões F, Rycroft M, Renno N, Yair Y, Aplin KL, Takahashi Y (2008a) Schumann resonances as a means of investigating the electromagnetic environment in the solar system. Space Sci Rev 137:455–471. doi:10.1007/s11214-008-9398-0

Simões F, Grard R, Hamelin M, López-Moreno JJ, Schwingenschuh K, Béghin C, Berthelier JJ, Lebreton JP, Molina-Cuberos GJ, Tokano T (2008b) The Schumann resonance: a tool for exploring the atmospheric environment and the subsurface of the planets and their satellites. Icarus 194:30–41. doi:10.1016/j.icarus.2007.09.020

Simões F, Berthelier JJ, Godefroy M, Yahi S (2009) Observation and modeling of the Earth–ionosphere cavity electromagnetic transverse resonance and variation of the D-region electron density near sunset. Geophys Res Lett 36:L14816. doi:10.1029/2009GL039286

Simões F, Pfaff R, Freudenreich H (2011) Satellite observations of Schumann resonances in the Earth's ionosphere. Geophys Res Lett 38:L22101. doi:10.1029/2011GL049668

Tokano T, Molina-Cuberos GJ, Lammer H, Stumptner W (2001) Modeling of thunderclouds and lightning on Titan. Planet Space Sci 49:539–560

Toledo-Redondo S, Parrot M, Salinas A (2012) Variation of the first cut-off frequency of the Earth–ionosphere waveguide observed by DEMETER. J Geophys Res 117:A04321. doi:10.1029/2011JA017400

Trakhtengerts VY, Rycroft MJ (2008) Whistler and Alfvén Mode cyclotron masters in space. Cambridge Univ, Press 354p

Williams E, Kuo CL, Bór J, Sátori G, Newsome R, Adachi T, Boldi R, Chen A, Downes E, Hsu RR, Lyons W, Saba MMF, Taylor M, Su HT (2012) Resolution of the sprite polarity paradox: the role of halos. Radio Sci 47, RS2002 doi:10.1029/2011RS004794

Yampolski YM, Bliokh PV, Beley VS, Galushko VG, Kascheev SB (1997) Non-linear interaction between Schumann resonances and HF signals. J Atmos Solar-Terr Phys 59:335–342

Yang H, Pasko VP, Yair Y (2006) Three-dimensional finite difference time domain modeling of the Schumann resonance parameters on Titan, Venus, and Mars. Radio Sci 41(2), RS2S03. doi:10.1029/2005RS003431

Yashunin SA, Mareev EA, Rakov VA (2007) Are lightning M components capable of initiating sprites and sprite halos? J Geophys Res 112:D10109. doi:10.1029/2006JD007631

Chapter 14
Supplementary Material

This chapter contains material used in the previous chapters; explanations and the program listing. We describe the formalism connecting the vertical profiles of atmosphere conductivity with ELF propagation constant. Formulas are given for the frequency domain and time domain solutions of pulsed radio propagation in the Earth–ionosphere cavity. These equations were used when obtaining the model results used throughout the book, and these are exploited in the relevant FORTRAM listings in this chapter. Spectra and waveforms are shown. We discuss the impact of frequency characteristics of typical SR receiver on the spectra and the waveforms of ELF transients. Compensating the phase distortions is addressed. Separate sections present formulas of spherical trigonometry and we demonstrate the essentials of singular spectral analysis (SSA) ('Caterpillar' algorithm). The listings are given for the field computations in the time and frequency domains.

14.1 ELF Propagation Constant $\nu(f)$ and Vertical Profiles of Atmosphere Conductivity

The height profile of ionosphere conductivity is unnecessary for the calculations of ELF electromagnetic fields in the Earth–ionosphere cavity. One needs the frequency dependence of the complex propagation constant of ELF radio waves. There was a belief at the dawn of ELF research that the conductivity profile of the ionosphere based on the numerous observations of the very low frequency (VLF: 10–20 kHz) radio propagation will allow prediction of the ELF field amplitude along different paths. However, the serious difficulties were met on this way, and the ionosphere profiles were not used in the practice of ELF radio communications. The necessary frequency dependence $\nu(f)$ was found directly from the measurements of the SR and then extrapolated to the desired frequency range. It is generally recognized that the most accurate (f) model based on the observed SR spectra is the model by Ishaq and Jones (1977). In this model, the complex sine S_ν

A. Nickolaenko and M. Hayakawa, *Schumann Resonance for Tyros*,
Springer Geophysics, DOI: 10.1007/978-4-431-54358-9_14,

of the Brillouin waves in the Earth–ionosphere cavity is the following function of frequency f:

$$S_\nu = \frac{c}{V_{ph}} - i(5.49)\frac{\alpha}{f}. \tag{14.1}$$

The ratio of the light velocity c to the phase velocity V_{ph} of the ELF radio wave is:

$$\frac{c}{V_{ph}} = 1.64 - 0.1759\ln(f) + 0.0179[\ln(f)]^2. \tag{14.2}$$

and the wave attenuation

$$\alpha = 0.063 f^{0.64}. \tag{14.3}$$

The propagation constant $\nu(f)$ is expressed through the complex sine function S_ν as:

$$\nu(\nu + 1) = (kaS_\nu)^2. \tag{14.4}$$

being equal to:

$$\nu = \sqrt{\frac{1}{4} + (kaS_\nu)^2} - \frac{1}{2}. \tag{14.5}$$

Here $k = \omega/c$ is the free space wave number, $\omega = 2\pi f$ is the circular frequency, and a is the Earth's radius.

A simpler but efficient model of (f) is the linear dependence: $\nu(f) = A_F{\cdot}f - B_F$. One of two following models is used having the parameters: $A_F = 1/6 - i/100$ and $B_F = -1/3$ or $A_F = 1/6 - i/70$ and $B_F = -1/3$. The first dependence $(f) = (f - 2)/6 - i{\cdot}f/100$ was derived from the cross-spectra of SR, and the second one $\nu(f) = (f - 2)/6 - i{\cdot}f/70$ corresponds to the power spectra (Nickolaenko and Hayakawa 2002). There are other much more sophisticated models based on the experimental resonance data (see e.g. Hynninen and Galyuk 1972; Bliokh et al. 1977).

The height profile of atmosphere conductivity is required when the observational data are absent of the global electromagnetic resonance. For example, when one estimates the expected resonance parameters at some other planet. The first attempts appeared to adjust the Greifinger and Greifinger (1978) model for the other planets (Nickolaenko and Rabinowicz 1982, 1987; Sentman 1990a, b). Experimental data on the terrestrial electromagnetic (Schumann) resonance were used for testing the elaborated approach and verify its applicability (Nickolaenko and Rabinowicz 1982, 1987). In the SR investigations, the vertical conductivity profile becomes necessary when studying the effect of the space weather on resonance. For example, when we estimate an impact of ionosphere modifications by a solar proton event (SPE) on the SR spectra.

The Greifinger model made a revolution in the ELF radio propagation. The exponential profile of atmospheric permittivity ε (h) has been widely applied long before their work (e.g. Galejs 1961, 1970, 1972; Wait 1962). An advantage of the exponential profile is its relevance to the rigorous solution in the terms of cylindrical functions. However, the radio waves reflected from the ionosphere are absent in this case. The correct model should incorporate a complex dielectric constant ε (h) containing the real part equal to 1 and some imaginary part varying with height (the conductivity depending on altitude above the ground).

Let us turn to the essentials of Greifinger and Greifinger (1978). The influence of ionosphere becomes noticeable from altitudes where the real and imaginary parts of air permittivity become equal to each other $\mathrm{Re}[\varepsilon(h)] = \mathrm{Im}[\varepsilon(h)] = 1$. The condition $\sigma(h_E) = \varepsilon_0\omega$ was used by Greifingers for finding the so-called "electric" height $h_0 = h_E$. Here σ is the air conductivity, ε_0 is the permittivity of free space, and ω is circular frequency of incident radio wave. This is the well-known height where the displacement current becomes equal to the conductivity current.

The merit by Greifinger and Greifinger (1978) was the introduction of the magnetic height $h_1 = h_M$, the suggestion of how to find this altitude, and formulation of the link between the electric and magnetic heights with the ELF propagation constant. The magnetic altitude is a greater height to which the magnetic field penetrates due to its "diffusion". It is found from the relation: $\sigma(h_M) = \frac{1}{4\mu_0\omega\zeta_M^2}$. Here μ_0 is the permeability of the free space and ζ_M is the scale height of the conductivity profile in the vicinity of this upper characteristic altitude h_M. The propagation constant is found as the following combination of four parameters of conductivity profile:

$$\nu(\nu + 1) = (ka)^2 \frac{h_M + i\zeta_M\pi/2}{h_E - i\zeta_E\pi/2}. \tag{14.6}$$

where h_E and h_M are electric and magnetic heights correspondingly and ζ_E and ζ_M are the relevant scale heights.

This technique was suggested by Greifinger and Greifinger (1978) for the monochromatic (man-made) waves emitted from an ELF transmitter. It implies that one manually finds the characteristic heights h_E and h_M from the σ (h) profile at the given frequency f. Concurrently, the scales ζ_E and ζ_M are derived, and the propagation constant $\nu(f)$ is computed by using Eq. (14.6). When the radio frequency changes, we expect the changes in the heights h_E and h_M together with ζ_E and ζ_M. The electric height increases with frequency and the magnetic height reduces (see e.g. Nickolaenko and Hayakawa 2002).

Such a procedure is tiresome when dealing with the spectra of natural ELF radio noise or the SR. Therefore, the Greifinger's model was extended to the wide band signals (Nickolaenko and Rabinowicz 1982, 1987; Sentman 1990a, b; Füllekrug 2000). The simplest model introduces the reference height G_h and the reference frequency F_g together with the single scale height ζ_E. The lower and the upper characteristic heights are found from the analytical expressions $h_E = G_h +$

$\zeta_E \cdot \ln(f/F_g)$ and $h_M = h_E - 2\zeta_E \cdot \ln(2\ k\zeta_E)$ at an arbitrary frequency f where $k = 2\pi\ f/c$.

The parameters G_h (reference height) and F_g (reference frequency) might be chosen in different ways. In the early studies, the "lower" reference point was used of $G_h \approx 45$ km and $F_g \approx 1$ Hz. Later, the upper reference point was used, which allows for obtaining the ambient day and the night conductivity profiles as the mere amendment of the scale height of the profile (Nickolaenko and Hayakawa 2002).

The drawback of the single scale model is that the parameter ζ_E is "imposed" to the entire profile. In particular, the magnetic height h_M is determined through the scale ζ_E, and the scale height at this altitude is also $\zeta_M = \zeta_E$. A further step is obvious: the model must have a postulated lower reference point and the electric scale ζ_E. The magnetic scale is calculated as above, by using the electric height h_E and the lower scale height ζ_E, while the upper scale height ζ_M is postulated separately.

The exponential model by Kirillov (1993, 1996, 2000), Kirillov et al. (1997), and Kirillov and Kopeykin (2002) was a further development of Greifinger's approach. These works formulated a formal criterion for selecting the magnetic height h_M. The upper characteristic height h_M came from the typical ionosphere profile. The following equation was used (Kirillov 1993, 1996, 2000; Kirillov et al. 1997; Kirillov and Kopeykin 2002; Mushtak and Williams 2002):

$$\omega_0\left(h_m^*\right) = \left\{\frac{\sqrt{\omega \nu_e}}{1.78 \cdot k}\left[\frac{1}{\zeta_e} + \frac{1}{\zeta_\nu}\right]\right\}. \tag{14.7}$$

Here ω_0 is the plasma frequency, ν_e is electron collision frequency and ζ_e and ζ_ν are the scale heights of vertical profiles of electron density and the electron collision frequency correspondingly. Since the penetration depth of the magnetic field into the ionosphere is a function of frequency, the magnetic scale ζ_M depends also on the frequency (Kirillov 1993):

$$\zeta_M = \zeta_{mp} + B_m^*\left(\frac{1}{f} - \frac{1}{F_{mp}}\right). \tag{14.8}$$

where ζ_{mp}, F_{mp}, and B_m^* are the constant parameters.

The "knee model" of the vertical conductivity profile was developed to match the observed peak frequencies and the Q-factors. The profile combines two exponential functions of altitude that intersect at a height regarded as the "knee altitude" (see Fig. 14.1). The reference frequency is chosen so that the reference height is positioned at the knee altitude. The major property of this model is that the basic SR mode has the electric altitude below the knee. When the frequency increases, the electric height h_E crosses the knee altitude h_{KNEE}, and the scale height of profile changes. The works by Mushtak and Williams (2002) and by Greifinger et al. (2007) insist that knee models accurately approximate the behavior of resonance Q-factors. Further development of the knee model resulted in the partially uniform knee (PUK) model by Pechony and Price (2004) and

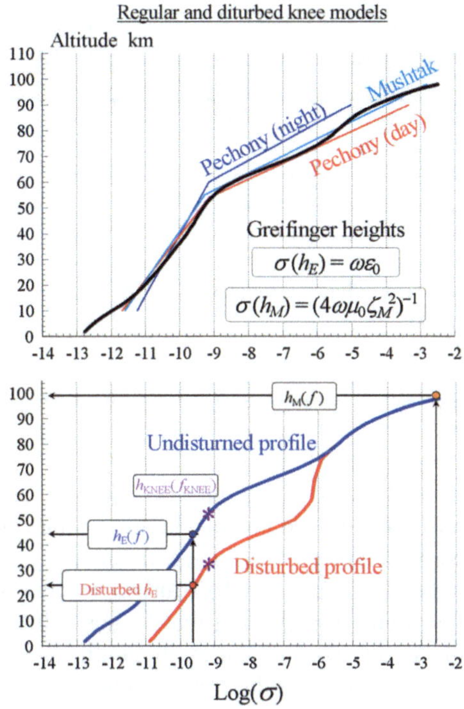

Fig. 14.1 The knee profile of air conductivity and its characteristic heights

Pechony (2007). The ionosphere is described by two slightly different knee profiles at the day and night sides of the globe in this model.

Figure 14.1 demonstrates variations of the air conductivity with altitude. Black curve in the upper frame of Fig. 14.1 shows the conductivity profile by Jones and Joyce (1989) and Jones and Knott (1999). Approximating knee models are also shown (Mushtak and Williams 2002; Pechony and Price 2004). Plots are given in the way commonly used in the atmospheric studies: the argument h (the height above the ground) is plotted along the ordinate in kilometers and the logarithm of air conductivity is shown along the abscissa. Parameters of the knee models are listed in Table 14.1. The green broken line in the upper frame depicts the Mushtak knee profile, and the red and blue lines show profiles of PUK model by Pechony (2007). We also mention below the uniform (average) knee model relevant to PUK, but it is not shown in the figure for simplicity. The upper frame also carries the relations introduced by Greifinger and Greifinger (1978) to estimate the characteristic heights for a particular circular frequency ω.

The lower frame in Fig. 14.1 demonstrates the regular and the disturbed profiles together with the scheme of finding the electric and magnetic heights. Two profiles are shown there. The blue curve depicts the regular profile by Jones and Joyce (1989) and Jones and Knott (1999) and the red line shows the disturbed profile with the lower part shifted downward by 20 km. This profile was used in the model

Table 14.1 Parameters of knee models introduced in the literature and used in computations

Parameter of knee profile	Knee model	Partial uniform knee model (day)	Partial uniform knee model (night)	Day–night average
Knee frequency f_{KNEE} Hz	10	13	13	13
Knee height h_{KNEE} km	55	54	60	57
Upper scale height ζ_a km	2.9	2.7	3.8	3.25
Lower scale height ζ_a km	8.3	7.5	9.1	8.3
Magnetic reference height h_m^* km	96.5	97.5	99	98.3
Magnetic reference frequency f_m^* Hz	8	6	6	6
ζ_{mp} km	4	3.7	3.54	3.6
F_{mp} Hz	8	6	6	6
B_m^* km	20	5		4.5

computations of the galactic gamma flare in Chap. 6. Violet asterisks mark the positions of knee heights on the regular (blue line) and disturbed (red line) profiles corresponding to the knee frequency $f_{KNEE} = 10$ Hz. The blue and red circles mark the regular and disturbed electric heights pertinent to the first SR mode $f = 8$ Hz. It is obvious that h_E lies below the knee altitude for the first resonance mode, and the relevant scale height ζ_E is larger than at the upper part of profile.

The arrows in the lower frame of Fig. 14.1 demonstrate the electric and magnetic heights at the first SR frequency 8 Hz. The left arrows correspond to the electric height h_E. Formal relations for the characteristic heights indicate that the function $h_E(f)$ grows and $h_E(f)$ decreases when the frequency f increases.

The propagation constant of ELF radio waves is found in the knee model by using the following expressions for the *complex* electric and magnetic heights:

$$h_E(f) = h_{KNEE} + \zeta_a \ln\left(\frac{f}{f_{KNEE}}\right) + \frac{1}{2}(\zeta_a - \zeta_b)\ln\left[1 + \left(\frac{f}{f_{KNEE}}\right)^2\right] + i\left[\frac{\pi}{2}\zeta_a - (\zeta_a - \zeta_b)\tan^{-1}\left(\frac{f_{KNEE}}{f}\right)\right]. \tag{14.9}$$

$$h_M(f) = h_M^* - \zeta_M \ln\left(\frac{f}{f^*}\right) - i\frac{\pi}{2}\zeta_M(f). \tag{14.10}$$

where f_{KNEE} is the knee frequency, h_{KNEE} is the knee height, ζ_a is the upper scale height, ζ_b is the lower scale height, h_m^* is the magnetic reference height, and f_m^* is the magnetic reference frequency. The magnetic scale height is found from Eq. (14.8).

The propagation constant of ELF radio wave $\nu(f)$ is obtained from the following expression [compare with Eq. (14.6)]:

Fig. 14.2 Typical models of ELF propagation constant

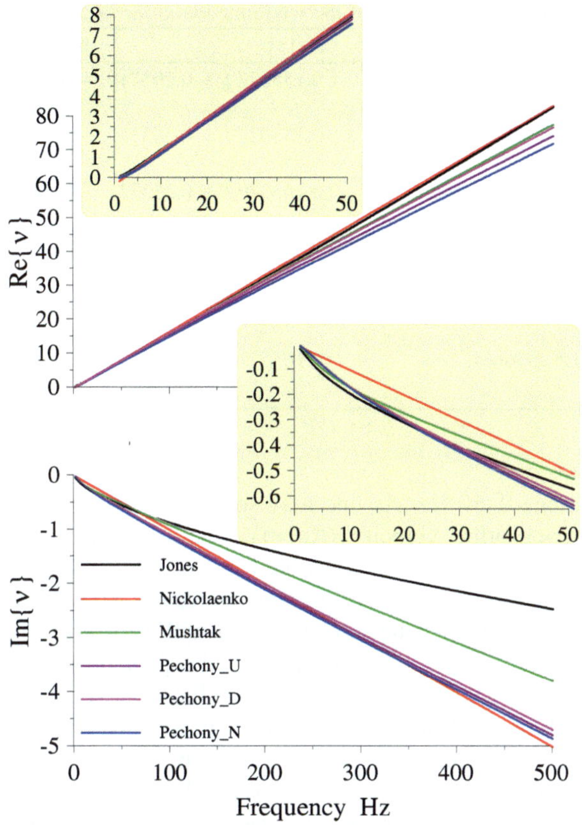

$$\nu(f) = -\frac{1}{2} + \sqrt{\frac{1}{4} + \left(ka\frac{H_M}{H_E}\right)^2}. \tag{14.11}$$

It should be noted that the upper sign of $\pm\sqrt{}$ expression is involved in the computation, otherwise the radio wave does not attenuate.

Equations (14.9–14.11) are similar to the relations by Greifinger and Greifinger (1978). The distinctions are in the details of obtaining the characteristics involved in the computations. Those authors of the works (Mushtak and Williams 2002; Greifinger et al. 2007) insist that these particulars provide a better correspondence of model spectra to the observation at least for the uniform distribution of thunderstorms over the globe (Williams et al. 2006).

The above parameters were used in the computations of dispersion curves that we show in Fig. 14.2. The abscissa in this figure depicts the frequency in Hz. The upper frame in Fig. 14.1 with insertion shows the real part of propagation constant, and the lower frame depicts the imaginary part (wave attenuation). Axes of the major frames cover the 0–500 Hz band, and the insets show variations in the SR

Table 14.2 Linear fit of the ν (f) function in the 50 Hz frequency band

Model name	Re{ν}	Im{ν}	ν
Knee model	$0.155077{\cdot}f + 0.295719$	$-\,0.00729324{\cdot}f - 0.171228$	$\nu(f) = \frac{f+1.907}{6.448} - i\frac{f+23.477}{137.113}$
PUK whole Day	$0.153023{\cdot}f + 0.503139$	$-\,0.00911865{\cdot}f - 0.163601$	$\nu(f) = \frac{f+3.288}{6.535} - i\frac{f+17.941}{109.665}$
PUK whole Night	$0.143379{\cdot}f + 0.55372$	$-\,0.00942982{\cdot}f - 0.183889$	$\nu(f) = \frac{f+3.862}{6.9745} - i\frac{f+19.501}{106.047}$
PUK Day and Night averaged	$0.149966{\cdot}f + 0.532781$	$-\,0.00931005{\cdot}f - 0.174449$	$\nu(f) = \frac{f+3.553}{6.668} - i\frac{f+18.378}{107.411}$
Linear model	$0.1666666{\cdot}f - 0.3333333$	$-\,0.01{\cdot}f$	$\nu(f) = \frac{f-2}{6} - i\frac{f}{100}$
Ishaq-Jones	$0.165571{\cdot}f + 0.862057$	$-\,0.00437689{\cdot}f - 0.414767$	$\nu(f) = \frac{f+5.207}{6.04} - i\frac{f+94.763}{228.473}$

range 0–50 Hz. As one may see, all models give very close results for the wave propagation velocity (Re{ν}). The wave attenuations (Im{ν}) show a greater deviation. As is demonstrated in Fig. 14.2, different models of Earth–ionosphere cavity give distinctive predictions for the ν (f) dependence. It would be interesting to see whether the model spectra will also seriously deviate from each other.

By applying the linear fit to the data of Fig. 14.2 in the frequency band of 50 Hz, we obtain the dependence listed in Table 14.2. One may see from the last column of Table 14.2 that the distance between resonance frequencies is different in the models. It is equal to 6.45; 6.54; 7.0; 6.67; 6, and 6.04 Hz for the knee, PUK-Day, PUK-Night, Linear, and Jones models respectively. Since the resonance frequency corresponds to the condition Re{$\nu(f)$} = n, the linear fit provides inexact values of the first mode frequencies.

To show the deviation among the models, we plot in Fig. 14.3 the power spectra of the vertical electric field component $|E(f)|^2$ computed for different $\nu(f)$ models. To abandon the influence of source–observer distance, we used the uniform distribution of random independent lightning strokes upon the globe. In this case, the power spectrum $|E(f)|^2$ is (Nickolaenko and Hayakawa 2002):

$$\left\langle |E(f)|^2 \right\rangle = \left| \frac{\nu(\nu+1)}{f} \right|^2 \sum_{n=0}^{\infty} \frac{(2n+1)}{|n(n+1) - \nu(f)[\nu(f)+1]|^2}. \tag{14.12}$$

We computed the power spectra by applying different ν (f) models listed in Table 14.2. The abscissa in Fig. 14.3 shows the signal frequency ranging from 6 to 28 Hz, i.e., the frequency band of the first four resonance modes. The logarithmic ordinate depicts the spectral power density in arbitrary units. Spectra of the models by Ishaq and Jones, linear model, knee model, PUK-Day, and PUK-Night models are shown by the black line with circles, red line, green line with stars, pink line with "bursts", and blue line with dots correspondingly.

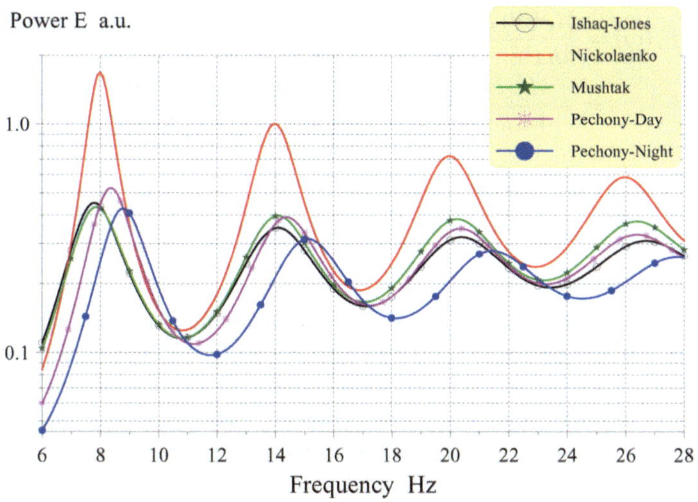

Fig. 14.3 Power spectra of vertical electric field component computed for typical models of ELF propagation constant and uniform distribution of lightning strokes upon the globe

One may observe that different radio propagation models result in deviating spectra: the peak frequencies and the Q-factors are different. Two models of the knee profile provide the resonance patterns rather close to the "sample" Ishaq and Jones ν (*f*) model: the Mushtak and the PUK-all-Day models. The linear and PUK-all-Night spectra deviate noticeably from the other curves.

We must remark here that a particular propagation model might be preferred depending on the personal opinion and experience of an investigator. The reason is that the spectra vary during the day, and their average outline depends on the observatory position with respect to global thunderstorm centers. Besides, the fine structure is always present in the observations, and it impedes the comparison with computations. As we demonstrate below, we expect reduction in the deviations between the $\nu(f)$ models reduce in the time domain, especially when a band-pass SR receiver is used in the measurements.

14.2 Model Q–Burst Spectra in the SR Band

As the next step, we compute the complex spectra of discrete pulses (Q-bursts). Only the TEM wave propagates in the Earth–ionosphere cavity at the frequency below 1 kHz. The following non-zero field components are present on the ground surface (e.g. Bliokh et al. 1977, 1980; Nickolaenko and Hayakawa 2002):

$$E_r(\omega) = \frac{M_C(\omega)}{4ha^2\varepsilon_0}\frac{iv(v+1)}{\omega}\frac{P_v[\cos(\pi-\theta)]}{\sin\pi v}$$
$$= -\frac{M_C(\omega)}{4\pi\varepsilon ha^2}\frac{iv(v+1)}{\omega}\sum_{n=0}^{\infty}\frac{(2n+1)P_n(\cos\theta)}{n(n+1)-v(v+1)}. \quad (14.13)$$

$$H_\varphi(\omega) = -\frac{M_C(\omega)}{4ha}\frac{P_v^1[\cos(\pi-\theta)]}{\sin\pi v} = \frac{M_C(\omega)}{4\pi ha}\sum_{n=1}^{\infty}\frac{(2n+1)}{n(n+1)-v(v+1)}P_n^1(\cos\theta). \quad (14.14)$$

Here a is the Earth's radius, h is the effective height of the ionosphere, $\varepsilon_0 = 1/(36\pi) \times 10^{-9}$ is the permeability of free space, $M_C(\omega)$ is the current moment of the vertical lightning stroke, θ is the source–observer angular distance, $v(\omega)$ is the propagation constant, and ω is the circular frequency, $P_v[\cos(\pi-\theta)]$ and $P_v^1[\cos(\pi-\theta)]$ are the Legendre and associated Legendre functions of complex order v. The following relation is valid $\frac{\partial}{\partial\theta}P_v[\cos(\pi-\theta)] = \frac{\partial}{\partial\theta}P_v^1[\cos(\pi-\theta)]$.

These formulas are regarded in the literature as the zero mode wave of the Earth–ionosphere duct. It is expanded into the zonal harmonics being the sum of resonance modes n (Gradstein and Ryzhik 1968; Bateman and Erdelyi 1953):

$$-\pi\frac{P_v[\cos(\pi-\theta)]}{\sin\pi v} = \sum_{n=0}^{\infty}\frac{(2n+1)P_n(\cos\theta)}{n(n+1)-v(v+1)}. \quad (14.15)$$

Zonal harmonic series (14.13) and (14.14) are simple in form, but they slowly converge. The reason is that the fields become infinite at the source point $\theta = 0$. Since every term in the sums of zonal harmonic series remains finite everywhere in the cavity, the series itself must diverge when $\theta \to 0$. There exist special procedures for accelerating the convergence, and these were described in the books by Bliokh et al. (1977, 1980) and Nickolaenko and Hayakawa (2002). We show the final results:

$$-\pi\frac{P_v(-x)}{\sin\pi v} = R_{\sum} + \sum_{n=0}^{\infty}a_n^{(4)}P_n(x). \quad (14.16)$$

where

$$R_{\sum} = 2R_1 + R_2 + 2[v(v+1)+1]R_3 + 3[3v(v+1)+2]R_4. \quad (14.17)$$

$$R_1 = \ln\frac{\sqrt{1-x}+\sqrt{2}}{\sqrt{1-x}} = \sum_{n=0}^{\infty}\frac{P_n(x)}{(n+1)}. \quad (14.18)$$

$$R_2 = 1 + (1-x)R_1 - \sqrt{2(1-x)} = \sum_{n=0}^{\infty}\frac{P_n(x)}{(n+1)(n+2)}. \quad (14.19)$$

$$R_3 = 1 + \frac{(1-x)(3x-1)}{4} R_1 - \frac{3}{4}\left(x + \sqrt{2(1-x)^3}\right)$$
$$= \sum_{n=0}^{\infty} \frac{P_n(x)}{(n+1)(n+2)(n+3)}. \tag{14.20}$$

$$R_4 = -\frac{(5x+1)(x-1)^2}{12} R_1 + \frac{15x^2 - 27x + 14}{36} + \frac{5(1-x^2)}{12}\sqrt{2(1-x)}$$
$$-\frac{11}{18}\sqrt{2(1-x)^3}. \tag{14.21}$$

and

$$a_n^4 = \frac{2n\left[v^2(v+1)^2 + 22v(v+1) + 12\right] + 17v^2(v+1)^2 + 74v(v+1) + 24}{(n-v)(n+v+1)(n+1)(n+2)(n+3)(n+4)}. \tag{14.22}$$

Similar expressions are valid for accelerated computations of magnetic field:

$$-\pi\frac{P_v^1(-x)}{\sin \pi v} = \frac{2v(v+1)}{\sqrt{1-x^2}}\left(2R_3 + 9R_4 + \sum_{n=0}^{\infty} b_n^4 P_n(x)\right). \tag{14.23}$$

where

$$b_n^4 = \frac{4n^3[v(v+1)+12] + 2n^2[19v(v+1)+78] - n[2v^2(v+1)^2 + 19v(v+1) + 66] - [17v^2(v+1)^2 - 34v(v+1) - 24]}{[v(v+1) - (n-1)n][v(v+1) - (n+1)(n+2)](n+1)(n+2)(n+3)(n+4)}. \tag{14.24}$$

The general term of the sum of Legendre polynomials decreases as $N^{-5.5}$ (N is the upper limit of summation) and the remainder of the sum behaves as $N^{-4.5}$ which is quite enough for obtaining accurate results: the relative error does not exceed 0.1 % when $N \geq 2\mathrm{Re}\{v(f)\}$. The only problem left is calculating the Legendre polynomials of high order, because we use the following recursion relation for the purpose: $P_{n+1}(x) = \frac{2n+1}{n+1} x P_n(x) - \frac{n}{n+1} P_{n-1}(x)$ (Erdelyi et al. 1953; Gradstein and Ryzhik 1968) starting from $P_0(x) = 1$ and $P_1(x) = x$.

An alternative set of formulas was suggested by Jones and Burke (1990):

$$-\pi\frac{P_v(-x)}{\sin \pi v} = \frac{64}{(1-x^2)^2}\sum_{n=0}^{\infty}\frac{N_3}{D_3}(2n+1)P_n(x). \tag{14.25}$$

and

$$-\pi\frac{P_v^1(-x)}{\sin \pi v} = \frac{16v(v+1)}{(\sqrt{1-x^2})^3}\sum_{n=0}^{\infty}\frac{N_2}{D_2}(2n+1)P_n(x). \tag{14.26}$$

where

$$N_2 = 2n(n+1) - (v-2)(v+3). \quad (14.27)$$

$$N_3 = n^2(n+1)^2 + 2n(n+1)(2v^2+2v-7) - (v-3)(v-1)(v+2)(v+4). \quad (14.28)$$

$$D_2 = [n(n+1) - (v-3)(v-2)][n(n+1) - (v-1)v][n(n+1) - (v+1)(v+2)][n(n+1) - (v+3)(v+4)]. \quad (14.29)$$

$$D_3 = [n(n+1) - (v-4)(v-3)][n(n+1) - (v-2)(v-1)][n(n+1) - v(v+1)][n(n+1) - (v+2)(v+3)][n(n+1) - (v+4)(v+5)]. \quad (14.30)$$

One can use any of the above presentations, and both of them provide the same results. Particular formulas (14.16–14.22) were used in the computations of distribution of the resonance field amplitude over the frequency–distance plane. The listing of the relevant FORTRAN procedure <zeromodes.for> is placed at the end of this chapter.

Distribution of the field amplitudes over the frequency–distance plane is shown in Fig. 14.4. Frequency is plotted along the abscissa in Hz, and the source–observer distance is shown on the ordinate in Mm. The amplitude (arbitrary units) of the field is shown by color inking. The left graph corresponds to the vertical electric field, while the right one depicts the amplitude of the horizontal magnetic field component.

We had used the linear frequency dependence of propagation constant in the computations. The maps illustrate the interaction of direct and antipodal waves in the spherical cavity. There is a broad maximum over the frequencies above one hundred Hertz when the source–observer distance is small. With an increase in the distance, the wave attenuation starts to play its role, and this peak rapidly disappears (Nickolaenko and Hayakawa 2002). Maps in Fig. 14.4 contain summits corresponding to SR modes. One may observe a general increase of the vertical electric field amplitude around the source and its antipode. Here, the well-resolved resonance peaks are seen. A wide valley is centered at the 10 Mm distance and separates these two 'ridges' at the small and great distances. This valley gradually descends to the right (higher frequencies) where the wave attenuation increases. The vertically aligned chains of peaks correspond to the spatial distribution of particular resonance modes.

The "side" trenches (blue zones) appear in the maps from interaction between the direct and antipodal wave. Each trench tends to connect the nodes of adjacent modes, so that "canyons" cut the relief and approach the source antipode when the frequency grows. The apparent condition $v(f)(\pi - \theta) \approx$ const is held for these structures describing a family of hyperbolic curves. In fact, this is the condition

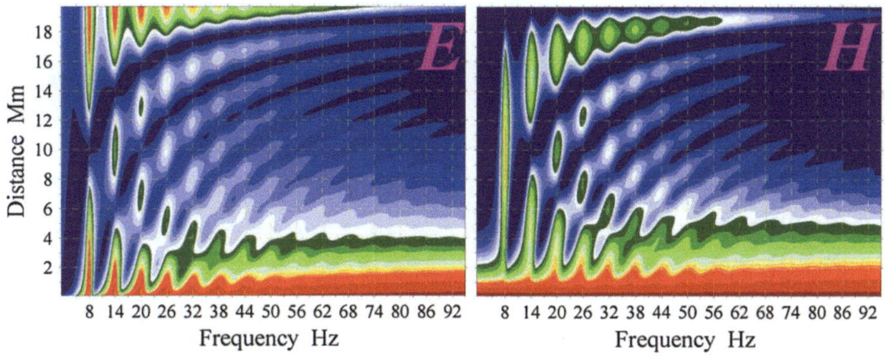

Fig. 14.4 SR spectra over the frequency–distance plane

that the mutual phase delay between the direct and antipodal waves is a constant. There exists a secondary system of canyons, which symmetrically approaches the source, however, this structure is not so lucid.

The right map in Fig. 14.4 corresponds to the amplitude of horizontal magnetic field. Interaction of direct and antipodal waves occurs here in a similar way. Still, there are obvious distinctions from the electric map. The most evident feature is absence of the magnetic field maximum at the source antipode because the magnetic field turns to zero at this particular point. Both the fields contain the broad "high frequency" peak when the source distance is small. A succession of equidistant saddles and peaks is seen at the fixed distance in both the maps. Say, at 10 Mm distance, the peaks of odd modes are absent in the E-field component, while the even peaks disappear in the H-field. This behavior shows that the H-amplitude map is kind of 'derivative' of the E map with respect to the distance. The 'antipodal effect' (e.g. Wait 1962) becomes visible at higher frequencies where the series of depressions appears becoming parallel to the source antipode.

The picturesque SR profile of Fig. 14.4 might be observed in the absence of finite band-pass receiver or in the case when its frequency band substantially exceeds 100 Hz (e.g. Ogawa and Komatsu 2007). Owing to the 50 Hz industrial interference, a typical SR receiver has the band-pass in the 4–40 Hz interval, so that natural ELF radio signals are distorted. We model below such deformations by introducing an idealistic receiver formed by the high-pass and the low-pass Butterworth filters of the 6th order. Second variant of the model 4–40 Hz receiver cuts the industrial interference by an additional 50 Hz notch filter of the second order that has the bandwidth of ±0.5 Hz. The relevant amplitude spectra are shown in Fig. 14.5. The abscissa in this figure depicts frequency in the 1–100 Hz range. The model spectra of a Q–burst (vertical electric and horizontal magnetic field components) were computed by using Eqs. (14.13–14.22). The source–observer distance was chosen (D = 10 Mm) and we used the linear frequency dependence $\nu(f)$.

The inset in Fig. 14.5 shows the amplitude frequency characteristic of the 4–40 Hz receiver with and without the notch filter. The upper frame shows the

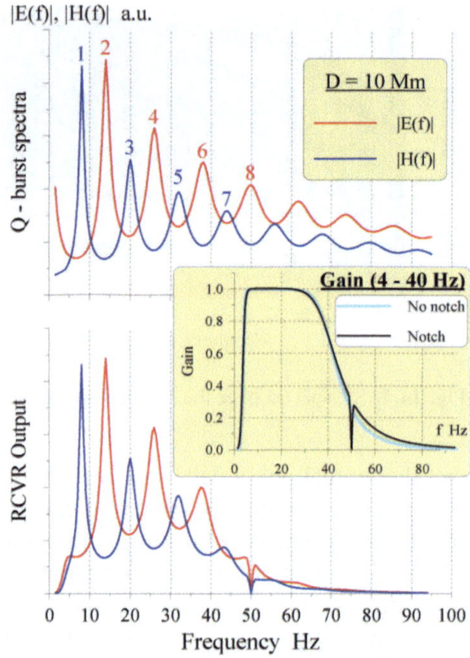

Fig. 14.5 SR spectra for the source—observer distance 10 Mm without a receiver (*upper frame*) and the output of a typical SR receiver (*lower frame*). Receivers are shown in the inserted plot of 4–40 Hz band-pass with and without notch filters

model amplitude spectra of the Q–burst arriving from the 10 Mm distance. The red curve corresponds to the vertical electric field component, which demonstrates that only even resonance modes are present in the pattern. The blue line depicts the amplitude spectra of the horizontal magnetic field containing only odd modes of SR. Both amplitudes are shown in arbitrary units. The color integer numbers in the upper plot mark the particular resonance peaks. As one may observe, the main portion of field is concentrated at the frequencies below 100 Hz: this is a result of natural filtering by the Earth–ionosphere cavity (Nickolaenko and Hayakawa 2002). Though the modes up to the 13th–14th might be resoled in the spectrum (Füllekrug 2005).

The lower frame in Fig. 14.5 demonstrates the SR pattern at the output of the typical SR receiver containing the 50 Hz notch filter. It demonstrates that only six resonance modes might be studied by such a hardware, and the last, sixth mode becomes slightly exaggerated by the roll-off of the receiver frequency response and by the notch filter. Spectral modifications inevitably modify the pulsed waveform of the ELF transients. We demonstrate such transformations in the following sub-section.

14.3 Waveforms of ELF Transients from the SR Receiver

First of all, we obtain the signal waveforms in an "ordinary" way, i.e., by using the procedure of Fast Fourier Transform (FFT). In the next sub-section, we will turn to the formal solution of the Q–burst waveform in the time domain. Here we confine ourselves to the 10 Mm source distance ($\theta_H = 90°$). Initially we compute the complex spectra of the fields, and the ideal waveforms are found then by using the FFT. When we take into account the bandwidth of the receiver, we multiply the complex field spectra by the complex gain of the receiver and apply FFT afterwards.

The frequency band and the frequency step must be chosen appropriately. For example, we compute the spectra in the band 1–512 Hz with the 1 Hz step. To correctly compute the relevant waveforms (the real functions of time), we must continue the complex spectra to the negative frequencies. The spectral densities at the negative frequencies are the complex conjugates of those at the positive frequencies, and they are zero at $f = 0$. The negative frequencies start at -511 Hz. In this case, the waveform is obtained in the time domain interval 1 s long with the time resolution of 1/1,024 s. The numbers mentioned provide reasonable sampling rate and the time interval.

Figure 14.6 demonstrates an impact of SR receiver on the detected pulsed signal. The time interval from 0 to 0.5 s is shown on the abscissa in all plots in this figure. The inset shows the time response function of two 4–40 Hz receivers: one does not contain the 50 Hz notch filter (black line) and the other performs filtering of the industrial interference. According to the definition, the time response function is the reaction of the device with a particular frequency response on the delta-pulse. One may observe the superposition of two characteristic periods in the response of band-pass receiver: about 1/4 and 1/40 s which corresponds to the cut-off frequencies. The presence of the node at 50 Hz adds distinctive "ringing" at 50 Hz (red curve). The upper frame in Fig. 14.6 depicts the waveforms of the pulse traveled the distance of 10 Mm from the positive stroke of lightning through the Earth–ionosphere cavity. Note that the pulse onset in the $E(t)$ dependence (red curve) is negative since the stroke is of positive polarity. Here we observe a characteristic "square" waveform in the magnetic field component (Nickolaenko and Hayakawa 2002) shown in the blue curve. Such a pattern appears due to the interaction of direct waveform and the antipodal pulse in the magnetic field. The polarities of pulse onsets are opposite since these arrive from opposite directions. "Square" wave is absent in the electric field, and all the pulses here have the negative onset.

The lower frame in Fig. 14.6 demonstrates the waveforms at the outputs of two receivers. These curves might be obtained by the convolution of the receiver time response function and the incident pulsed waveform. One may observe the signal modification by an ideal band-pass filter and by a receiver with notch filter. These two patterns are close to each other, though only the slight $\sim$50 Hz ringing is present at the output receiver with the notch filter. Computations explain why

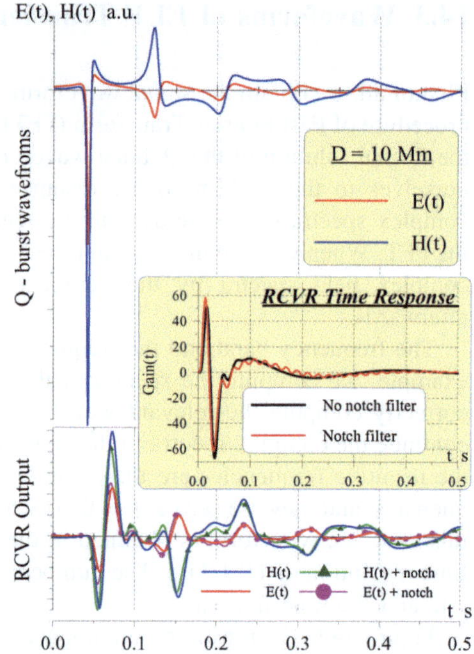

Fig. 14.6 Waveforms for the source–observer distance of 10 Mm. *Upper frame* shows pulses coming to the observer, and the *lower frame* shows the same pulses at the output of model receivers. Receiver time response functions are shown in the *inserted plot*

realistic waveforms were recorded only in the experiment by Ogawa and Komatsu (2007): they used an exceptional bandwidth of 11 kHz and applied no notch filters.

We must remark that the impractical assumption was made in our computations that the characteristics of electric and magnetic channels of the receiver were the same. These are different in practice, because frequency response of the electric and magnetic field sensors are different. Special efforts must be undertaken to match the channels (e.g. Belyaev et al. 1999); that is, the knowledge is strongly desirable of real through gains of receiving equipment both in the frequency and in the time domain.

Figure 14.7 demonstrates important features of the propagation models and the impact of SR receivers on the time domain record. The left frame in this figure shows waveforms of the vertical electric field component and the right one depicts the horizontal magnetic field. Temporal variations were computed for different $v(f)$ models. Data are shown by the black, red, magenta, blue, and green curves for the models by Ishaq and Jones, the knee model, the day and the night PUK models, and linear model correspondingly. The time domain transients were obtained as in Fig. 14.6, by applying the FFT procedure to relevant complex spectra of the fields. The upper plots in Fig. 14.7 depict the model waveforms in the absence of a receiver (the input waveform). All model pulses corresponding to the "realistic"

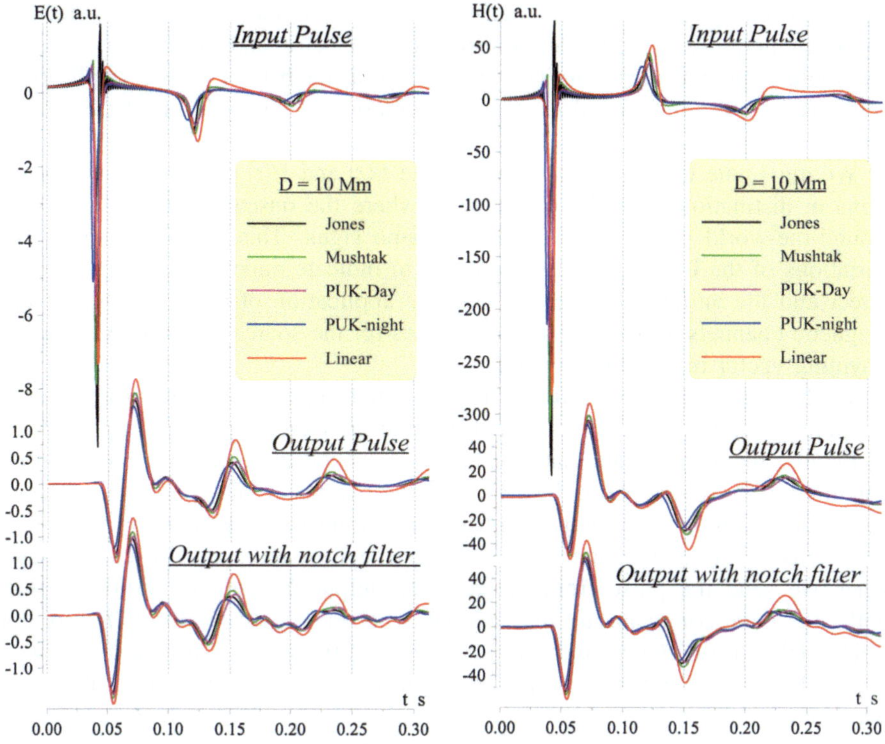

Fig. 14.7 Model waveforms for the source–observer distance 10 Mm

ionosphere models have the same characteristic feature—they are unstable and there is a "jitter" starting prior to the arrival of the pulse itself. One may observe that fast variations are superimposed on the regular pulse pattern similar to the predictions of linear $v(f)$ model. The fine structure (instability) in the time domain arises from the incorrect behavior of relevant propagation constants at high frequencies. This is why the fast variations vanish at the output of SR receivers. The collection of upper plots demonstrates that models have much in common, but deviate in details: There is slight deviation in the pulsed amplitudes and their position. We must also note that the "PUK whole night" model predicts an earlier arrival of pulse, which is explained by a systematic modification of the propagation velocity in the ambient night condition.

The lower plots in Fig. 14.7 depict the same pulses at the output of the model receivers shown by an inset in Fig. 14.5. One may observe the crucial impact of the signal filtering. First of all, the amplitude of all pulses has significantly reduced, approximately by an order of magnitude. All output pulses acquire close amplitudes regardless the propagation model, so that mutual deviations could hardly be detected in the experiment. All pulses are delayed against their real onset, which is not a novelty for those who worked with filters. The most dramatic alterations occurred in the signal pattern. For example, the "square" waveform of the magnetic pulse was completely lost, and the electric and magnetic waveforms became exceptionally similar. The presence of a notch filter adds a small 50 Hz component to the output pulses, but it definitely will be hidden in the SR background signal.

We must note that antipodal waves in the $E(t)$ and $H(t)$ patterns have opposite signs in distinction from the direct wave where the onsets have the same sign. Round-the-world pulses again have the same signs. This means that temporal variations of the Poynting vector continue to indicate opposite directions for the direct and the antipodal waves. Hence, the application of identical electric and magnetic channels in the SR receivers allows for the source location by using the Poynting vector in the time domain.

We hope that above waveforms demonstrated importance of measurements in the pure electromagnetic environment with no interference.

Finally, we demonstrate that crucial impact of the finite receiver bandwidth is conditioned by severe phase distortions in the output signal. Therefore, the compensation of phase perturbations substantially improves the situation with measurements of Q-burst parameters. The phase deformation will be compensated when we send the pulse from the receiver output through an additional filter with the gain conjugate to that of the receiver: $G_{AD}(f) = Gain^{*}(f)$, so that $\mathrm{Re}[G_{AD}(f) = \mathrm{Re}[Gain(f)]$ and $\mathrm{Im}[G_{AD}(f) = -\mathrm{Im}[Gain(f)]$. One may see from Fig. 14.8 that this kind of signal processing leads to a partial recovery of the initial waveform.

Figure 14.8 depicts two sets of waveforms. The upper plots show the $E(t)$ functions, and the lower plots $H(t)$. We confine ourselves to the linear $v(f)$ model in the illustration, because all other propagation models show the similar results. Time in seconds is shown along the abscissa in this figure. The source–observer distance is equal to 10 Mm. The black curves in the figure repeat the original waveforms that we have shown in Fig. 14.6. The red curves present the exaggerated pulsed waveforms at the output of the typical SR receiver with a notch filter at 50 Hz frequency. These functions also were shown in Figs. 14.6 and 14.7. The green lines depict the "recovered" pulses that we obtain after applying additional filtering compensating phase distortions. Simultaneously, the effective amplitude characteristic of the receiving equipment is not equal to $|G(f)|^2$, but $|G(f)|$.

As one can observe from Fig. 14.8, compensation of the phase distortions "reconstructs" the initial waveform. Of course, the pulse became wider since it passed through a band-pass and the notch filters. However, the characteristic elements of the waveforms returned to their true locations against the time axis. In particular, the pulse arrives at the correct time now. The "rectangular" waveform re-appears in the $H(t)$ record, and the obvious negative onset in the $E(t)$ waveform clearly indicates that the source was a positive cloud-to-ground stroke.

Applying this compensating technique is desirable when studying the sources of ELF transients. It might be readily implemented in a digital receiver. After sampling the Q-burst, one must calculate its complex spectrum and multiply the result

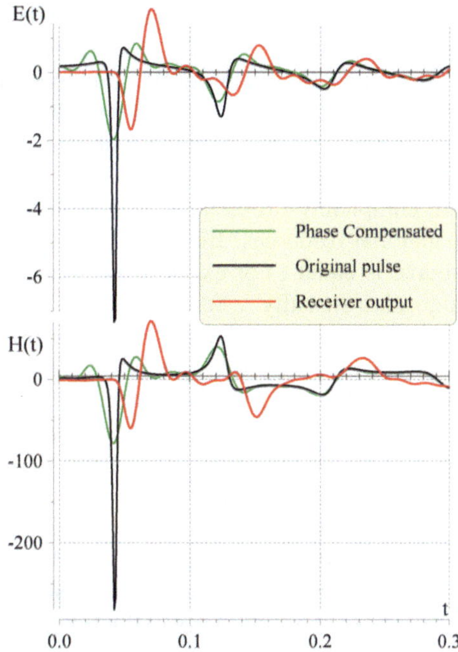

Fig. 14.8 Electric and magnetic field waveforms of a Q-burst arriving from the distance of 10 Mm

by the complex conjugate of the receiver gain. The product is transformed then into the time domain by using the FTT algorithm.

An alternative method exploits the direct convolution of the digitized signal with the receiver time response function *Gain*($-t$) recorded in the reversed time. The same result can also be achieved by the purely analog processing. To do this, one has to record the output pulse on a magnetic tape, for example. Then, the record must be played back in the reversed time (tape moves in the backward direction), and the signal obtained is fed to the receiving antenna. The output will show the compensated pulse.

All these tricks in the signal processing compensate the phase distortions caused by typical SR receiver. They are not a novelty and might be found almost in every textbook on the modern radiolocation and/or statistical processing of pulsed signals. It is a pity that such compensation was never used in the studies of natural ELF transients.

14.4 Time Domain Solution

We describe in this section the direct time domain solution (Nickolaenko and Rabinowicz 2000; Nickolaenko and Hayakawa 2002; Nickolaenko et al. 2004a, b). There are many numerical time domain solutions in terms of either finite element method (FEM) (Baba and Hayakawa 1995) or constructed with the finite difference time domain (FDTD) technique (e.g. Cummer 2000; Berenger 2002; Hayakawa and Otsuyama 2002; Simpson and Taflove 2002; Morente et al. 2003; Otsuyama et al. 2003; Yang and Pasko 2005, 2006; Yang et al. 2006). The waveforms obtained with the sophisticated modern digital techniques are similar to the direct time domain solution. This is an encouraging factor since the approaches are completely different. However, the new techniques require substantial computational resources. Besides, the method presumes that the minimal size of the source is equal to or exceeds the size of an elementary cell in the computational algorithm (usually, a few degrees). Therefore, we include here the compact, fast, and efficient time domain solution: it does not require anything unusual or sophisticated. The relevant listing of FORTRAN procedure <E_&_H(t).for> is placed at the end of this chapter.

The time domain solution is constructed as the formal Fourier transform of the frequency domain series (14.13) and (14.14). Since the spectral components of the field have the poles in the complex frequency plane, the time domain zonal harmonic representations are obtained for the fields by using the residual theorem (e.g. Wait 1962; Nickolaenko and Hayakawa 2002). Convergence of the time series is better than that in the frequency domain, however, it might be accelerated by applying the singularity extraction procedure (Kummer transform). The result is of the following form (Nickolaenko and Rabinowicz 2000; Nickolaenko and Hayakawa 2002; Nickolaenko et al. 2004a, b):

$$E(t) = E_A.Re\left\{A_\nu \cdot \tau^{-B_\nu}\left[Q_{-1} + (Q_0 - 1)B_\nu + \left(\frac{Q_1}{\tau} - 1\right)B_\nu(B_\nu + 1) + B_\nu(B_\nu + 1)^2\sum_{n=1}^{\infty}\frac{\tau^n P_n(x)}{(n - B_\nu)(n + 1)}\right]\right\}. \tag{14.31}$$

$$H(t) = H_A \cdot \mathrm{Im}\left\{\frac{\tau^{(1-B_\nu)}\sqrt{1 - x^2}}{\sqrt{(1 - 2x\tau + \tau^2)^3}}\right\}. \tag{14.32}$$

Here $E_A = \frac{M_C}{2\pi h a^2 \varepsilon_0}$ and $H_A = \frac{M_C}{2\pi a h c^2}$ are the electric and magnetic field amplitudes, $\tau = exp\left(i\frac{t}{A_\nu}\right)$ is the time factor, $x = \cos\,\theta$, the functions $Q_{-1} = \frac{1 - x\tau}{\sqrt{(1-2x\tau+\tau^2)^3}} - 1$ and $Q_1 = \ln\frac{\tau - x + \sqrt{1-2x\tau+\tau^2}}{1-x}$ are found from the generating function of Legendre polynomials $Q_0 = \frac{1}{\sqrt{1-2x\tau+\tau^2}}$. They are correspondingly its integral and the derivative with respect to the τ parameter.

The analytical time domain solutions (Eqs. 14.31 and 14.32) for the spherical Earth–ionosphere cavity imply the linear frequency dependence $\nu(\omega) = A_\nu\omega + B_\nu$, so that the complex frequency poles are positioned along the straight lines. The infinite series (14.31) for the electric field converges absolutely and uniformly everywhere in the cavity including the source point $\theta = 0$, provided that some time has passed from the pulse radiation, i.e., $t > 0$. Thus, we have a unique possibility of computing the waveform of a Q–burst returned to the point of the stroke.

The time domain solutions are expressed by the closed analytic form when the linear frequency dependence $\nu(\omega)$ is uniform, i.e., when $B_\nu = 0$, and $\nu(\omega) = A_\nu \omega$. In this case, the closed formulas describe the ELF transient:

$$E(t) = E_A \cdot \mathrm{Re}\left\{ A_\nu \cdot \left[\frac{1 - x\tau}{\sqrt{(1 - 2x\tau + \tau^2)^3}} - 1 \right] \right\}. \tag{14.33}$$

$$H(t) = H_A \cdot \mathrm{Im}\left\{ \frac{\tau\sqrt{1 - x^2}}{\sqrt{(1 - 2x\tau + \tau^2)^3}} \right\}. \tag{14.34}$$

These simplified expressions might be especially helpful when synthesizing the model time realization of the SR signal in a computer (Nickolaenko and Hayakawa 2002). Obtaining of analytical solution of any electrodynamic problem is a rarity. As far as we know, the explicit solution nearest in time for the low frequency radio propagation was obtained by Wait (1962) in the framework of the flat Earth–ionosphere waveguide.

The above time domain solution refers to the linear frequency dependence of the propagation constant. In the context of previous sub-section, this means that temporal variations will be stable. Propagation schemes implying ionosphere models result in the unstable time domain waveforms. However, the above time domain solution might be used for introducing a "more realistic" $\nu(f)$ function that deviates from the linear one in the limited frequency band.

Indeed, one may consider the situation when the $\nu(f)$ dependence deviates from the linear one in a finite number of SR modes. Accounting for such shifted poles is simple. The finite number of "linear" poles must be substituted by the novel "non-linear" eigen-values. Thus, there appear two finite correcting sub-sums : one to be extracted from the above result (the sum over "linear" poles) and the other to be added (the sum over the novel poles). The way of computing these sums is clear, so that we will not go deeper in this subject. Besides, it is obvious beforehand that the impact of such corrections cannot be great: changes in the solution appear due to the deviations of the poles of two similar complex functions. Most probably, such small alterations will be passed unnoticed in the experiment due to the permanent presence of the random SR background noise.

We have used formulas (14.31) and (14.32) when computing waveforms of Chap. 9. We use them to obtain the plots of Figs. 14.9 and 14.10. The corresponding listing of the FORTRAN procedure is placed below.

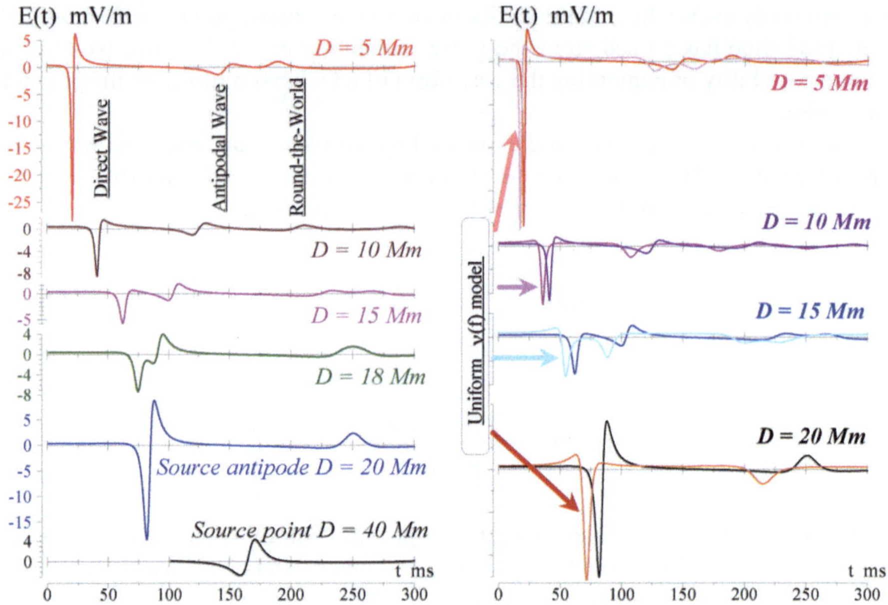

Fig. 14.9 Direct time domain solutions for a few characteristic distances

All frames in Fig. 14.9 depict the $E(t)$ functions corresponding to a few characteristic source distances computed for the linear (left frame) and uniform linear (right frame) $v(f)$ models. The time is shown along the horizontal axis in ms, and the vertical electric field is shown on the ordinate in mV/m. The vertically shifted plots depict pulses arriving from 5, 10, 15, 18, 20 and 40 Mm in the left frame and from 5, 10, 15, and 20 Mm in the right one. The 20 Mm distance corresponds to the point of observer antipode, and the distance of 40 Mm corresponds to the pulse returned back to the source point after the round-the-world travel. As one may observe, the outline of pulsed waveforms from the positive strokes at distances exceeding 15 Mm reminds us of the letter W: the W-type waveforms according to the classification introduced by Ogawa and Komatsu (2007).

The right frame of Fig. 14.9 compares pulses (in arbitrary units) obtained in the frameworks of linear and uniform linear models $v(f)$. The first model implies $B_v \neq 0$ and $B_v = 0$ in the second one. The color arrows indicate the pulses obtained in the uniform $v(f)$ model. One may see that this model has a higher propagation velocity of pulses, and the antipodal and round-the world waves are of different pattern in this model. The increase in the propagation velocity is conditioned by the denominator 7 since $\text{Re}\{A_v\} = 1/7$ in the uniform linear model instead of 6 ($\text{Re}\{A_v\} = 1/6$) in the linear model. Deviations in pattern might be explained if we recall the physical meaning of the B_v constant: it accounts for the spherical geometry of the cavity (Nickolaenko and Hayakawa 2002). Indeed, the eigen-values of a spherical cavity vary with the mode number as $\sqrt{n(n+1)}$, e.g.

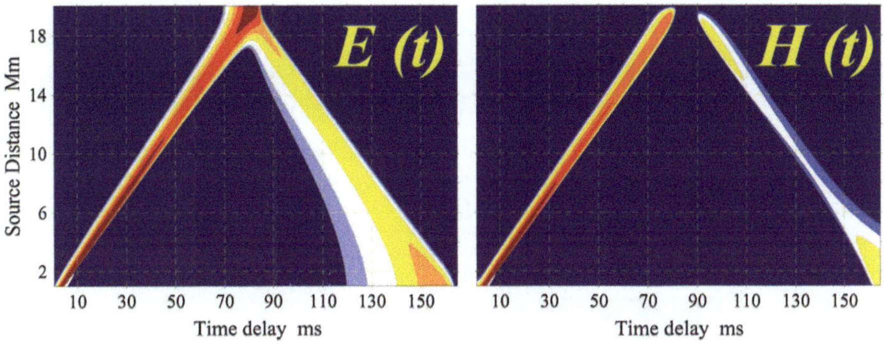

Fig. 14.10 ELF pulse over the time–distance plane

the classical Schumann formula. This square root is regarded as "the separation constant", and it is pertinent to the wave equation in the spherical coordinate system. The eigen-value grows linearly with the mode number: $f_n \sim n$ in the Cartesian (flat) geometry. As it was noted by Nickolaenko and Hayakawa (2002), the observed SR peak frequencies tend to increase with the mode number as $\sqrt{n(n+1)}$. Therefore, the credible linear approximation for the resonance frequencies must always have $B_\nu \neq 0$. At the moment when one suggests that the $\nu(f)$ model is the uniform linear or that $B_\nu = 0$, the geometry of the Earth–ionosphere cavity "transforms" into the "flat" one. This is why the antipodal and round-the-world waves become strongly distorted in such a model. These waves "cannot vanish" formally, as there is a succession of resonance frequencies hidden in the coefficient A_ν. Still, the pulsed pattern is modified, especially at the secondary and the higher order pulses. Apart from this point, the uniform linear model is compact and rather accurate.

We accepted the current moment of a powerful causative positive stroke $M_C = 3.8 \times 10^9$ A·m, which is equivalent to the 10 km stroke length and the peak current of 380 kA. As one may observe, the amplitude of the round-the-world pulse becomes rather small: about 3–4 mV/m. Such a decrease is conditioned by the wave attenuation in the Earth–ionosphere cavity, so that the focusing cannot compensate the losses in the round-the-world signal amplitude: it reduces dramatically even at the former source point. To conclude the discussion of this figure, we remark that the model data agree well with the unique wideband observations by Ogawa and Komatsu (2007) that we demonstrated in Chap. 9.

Individual pulsed waveforms might be combined into the 2D maps shown in Fig. 14.10 that reveal interesting properties of the pulse propagation in the spherical cavity. The significant feature is the "linear" time-distance dependence, as if the cavity has no sphericity. We already mentioned this feature: the reason is that circular wave fronts propagate between the source and its antipode. These fronts move with constant velocity of approximately 266 Mm/s in the Earth–ionosphere cavity. The left, rising rays in the maps represent direct waves while

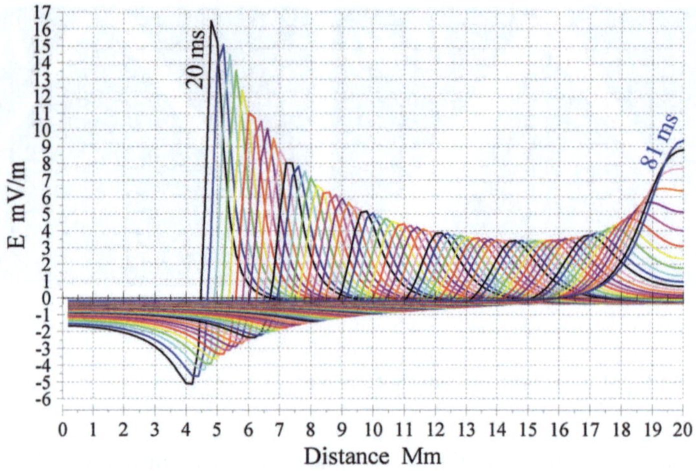

Fig. 14.11 Spatial distribution of the direct wave in the cavity for fixed moments of time

the descending wide rays are the antipodal pulses "spoiled" by the dispersion. The circular wave front converges at the source antipode, and we observe a peak in the electric field component and the node in the horizontal magnetic field. A gradual widening of the pulse takes place during the propagation, which is conditioned by the fast attenuation of high frequencies.

The time domain solution allows us to watch temporal evolution of the spatial field distribution. We confine ourselves to the vertical electric field component: its behavior at the source antipode and the source point is really remarkable. Figure 14.11 demonstrates the "motion" of the direct pulse in the cavity. The abscissa depicts the source–observer distance in Mm, and the ordinate shows the electric field computed by using formulas (14.31–14.32). Time is the parameter. We used the negative stroke polarity, because the positive pulse is more picturesque. Figure 14.11 illustrates the distance profile of the field for the time ranging from 20 to 81 ms after the onset of the negative lightning stroke. "Migration" of the pulsed envelope is obvious combined with the gradual changes in amplitude and the pulse width. The formation is clearly seen of the antipodal maximum in the field when the radio wave reaches the distance of 20 Mm. We stop at the moment of 81 ms when the positive antipodal pulse reaches its maximum.

The Cartesian coordinate system is not very convenient for showing the pulse traveling in the spherical cavity. Actually, there are circular fronts going away from the source point, and the single distance $D = \text{const}$ is relevant to the circle on the surface of the Earth. Therefore tops of the pulses in Fig. 14.11 correspond to the highest vertical electric field at a circle embracing the globe. Maybe, the 3D pictures of the pulse motion would be preferable, but its construction is a separate problem.

Pulses merging at the source antipode initially cause the positive "burst" that occurs on 81 ms. In the next 10 ms, the electric field remains equal to zero

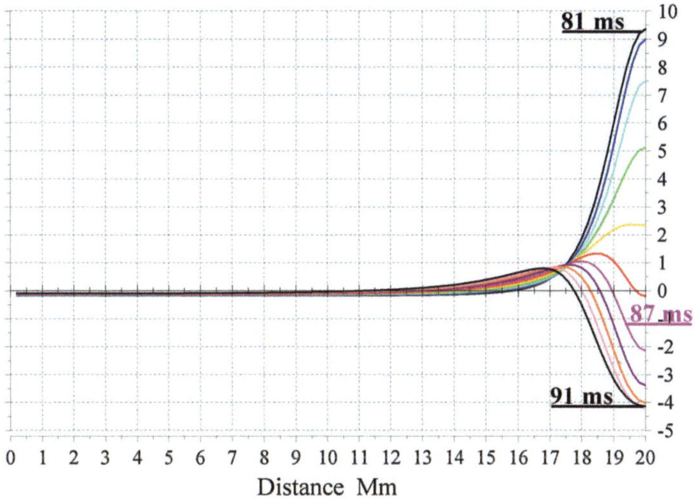

Fig. 14.12 Pulsed wave interaction around the antipode

everywhere in the cavity except the vicinity of the source antipode. It is gradually reformatted at the antipode, and the major excursion changes its sign prior to the backward travel toward the source. This fascinating behavior is shown in Fig. 14.12. In a way, the evolution of Fig. 14.12 reminds us of the waves in a pond arising from a stone thrown into the water.

The onset of the pulse leaving the source antipode is negative. The change of sign is in the strict accord with the description by J. R. Wait who specially noted that a monochromatic radio wave acquires the additional phase (180° jump of phase) when passing through the antipode (Wait 1962). Afterwards (see the film), the pulse amplitude becomes distributed over the whole length of the cavity: the wave excursions become small and even 'invisible'. With the progress of time, the circular wave front assembles around the former source point. The isolated pulse grows once more at the source, and its polarity is positive again. Figure 14.13 demonstrates the spatial field distribution during the arrival time of the round-the-world wave to the source point. Here again, the interplay is seen of the waves merging at the source point. Initially the field maximum is positive, but gradually the sign changes to negative prior to the wave leaving the focus. In this particular propagation model, the pulsed amplitudes were +9 and −4 mV/m at the antipode and +3 and −4 at the source point. Thus, the pulse dispersion (velocity variation with frequency) and the wave attenuation dramatically reduce the pulsed amplitude. The round-the-world pulse would hardly be seen in the real record: it will be hidden in the background noise.

If we look at the intermediate process (between the times of focusing) the field motion reminds us of the so-called 'seishi' or the waves on the surface of water in a big basin when the water comes from one end to another. On the other hand, distribution of the field during the intermediate times reminds us of the waves of a

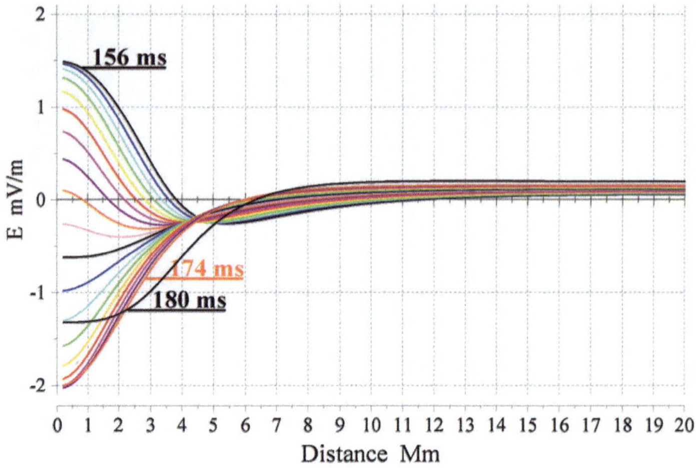

Fig. 14.13 Pulsed wave interaction around the source point

string or of the Legendre polynomials. This indicates indirectly that the higher order modes attenuate and so there remains only the basic mode finally. A poet could associate the splashes at the antipode and at the source with the picture of sea surf, the waves coming to a sea shore. Only, instead of the coast, we have the successive focusing of the wave fronts at two singular points of the sphere.

14.5 Formulas of Spherical Trigonometry

When performing computations, we often have to transform particular coordinates of the source and observer into conventional coordinates used in the computations: the source–observer distance and the source bearing from the observatory. To make such transformations, one has to apply formulas of spherical trigonometry. We place below the particular formulas. The geographic coordinates of the source and observer are: $\{a, \theta_S, \varphi_S\}$ and $\{a, \theta_0, \varphi_0\}$, where a is the Earth's radius, θ is the co-latitude and φ is the longitude of a point, both usually measured in degrees. The co-latitude is counted from the North Pole to the point, and the following relation is valid $\theta + \upsilon = 90°$ where υ is the latitude (positive to the North). The longitude is counted from the Greenwich meridian and is positive to the East. We must mention that all angular values such as co-latitude and longitude must be transformed to radians, and 180° correspond to $\pi \approx 3.14\ldots$ radians. The cosine of the source–observer angular distance θ_H is found from the relation:

$$\cos\,\theta_H = \cos\,\theta_0 \cdot \cos\theta_S + \sin\theta_0 \cdot \sin\theta_S \cdot \cos(\varphi_S - \varphi_0). \tag{14.35}$$

Hence the source–observer distance D in Mm is equal to:

$$D = \frac{20}{\pi} \cos^{-1} \theta_H. \tag{14.36}$$

By resolving the spherical triangles, one obtains the following expression for the source azimuth as seen from the observatory:

$$A_Z = \frac{20}{\pi} \cos^{-1}[\text{sign}(\varphi_S - \varphi_0) \cdot \cos(\psi_S)]. \tag{14.37}$$

where

$$\cos \psi_S = \frac{\cos \theta_S - \cos \theta_0 \cdot \cos \theta_H}{\sin \theta_0 \cdot \sin \theta_H}. \tag{14.38}$$

We must remind that the azimuth angle A_z is counted as the compass, i.e., from the direction to the North Pole through the east to the south. The above relations allow for computing all angles used in this book.

In the pulse propagation, the vectorial fields $\vec{E}, \vec{H}$ and the vector $\vec{k}$ of wave number (the propagation direction) form the right-hand set of three orthogonal vectors. This means that the vector $\vec{P} = [\vec{E} \times \vec{H}]$ is directed along the OY axis (to the North) when the positive electric field is vertical and the positive horizontal magnetic field is directed along the OX axis (to the east). Thus, the signs of the pulse onset in the $E(t)$ and $H_\phi(t)$ variations must be the same.

14.6 Caterpillar Procedure in the Processing of ELF Signal

To reveal the trend and periodic signal components hidden in a record, we multiply used the singular spectral analysis (SSA), in particular the “Caterpillar” algorithm (Danilov 1996; Danilov and Zhiglyavsky 1997; Golyandina et al. 2001; Troyan and Hayakawa 2002). In this subsection we list the stages of the data processing and demonstrate its application to the “raw” SR record. The algorithm is a realization of the SSA procedure; it is regarded also as Method of Principal Components (MPC). From the formal point of view, the “Caterpillar” algorithm is an automated version of the MUSIC (multiple signal classification) procedure (see e.g. Marple 1987).

At the first step of data processing, a 1D data series $x_k = x(t_k)$ is transformed into a matrix. The first L elements of the initial succession having the indices $k \in [1,\cdot L]$ are placed into the first line of the matrix. The parameter L is called the Caterpillar length, which is chosen at the beginning of processing. The elements from the second one form the second line, i.e., $k \in [2,\cdot L + 1]$. The third row contains the samples beginning from the third one: $k \in [3,\cdot L_G + 2]$, etc. The process continues

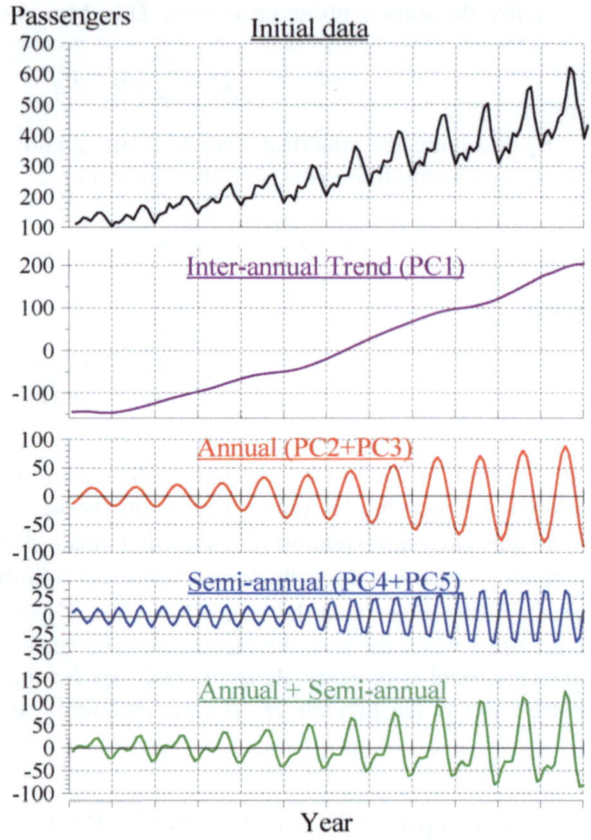

Fig. 14.14 Processing of the test data by the "Caterpillar" algorithm

until the rectangular matrix is filled. Then the code finds the eigen-vectors and the eigen-values of this matrix. The eigen-vectors form the "natural" orthogonal basis pertinent to the initial data, and no "outer" basis is used. The relevant eigen-values are the intensities corresponding to these eigen-vectors. The results of MPC are close to the signal processing with a bank of filters, but there are the following important distinctions:

1. The eigen-functions are not specified in advance, but instead they are found from the data themselves.
2. The processing efficiently separates the trends from the periodic components of the series.

The latter property was already demonstrated in Chap. 5 by using the standard "passengers" succession, which we reproduce in Fig. 14.14.

The test series of Fig. 14.14 imitates variations of the passenger number: the number of tickets sold by some transportation company. The upper plot shows

Fig. 14.15 Results of signal processing with the extraction of 50 Hz and its harmonics

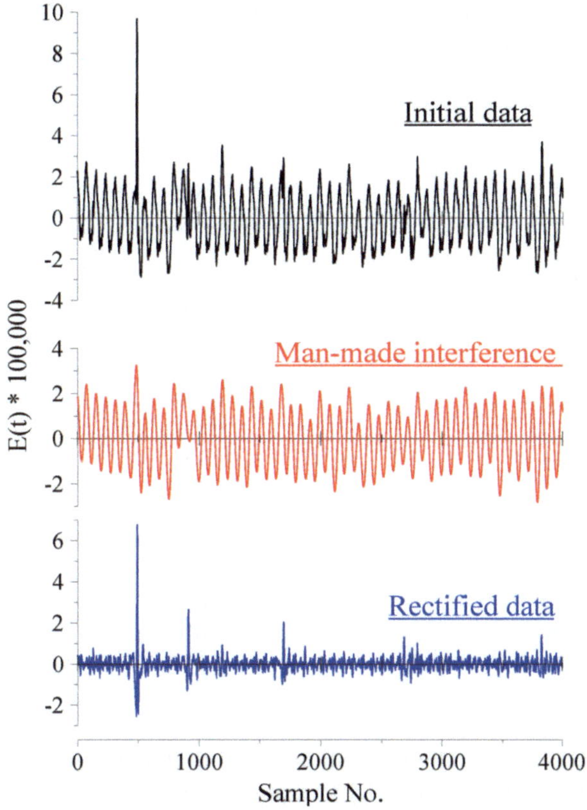

the initial data and the second line demonstrates the trend or PC1, retrieved by the Caterpillar algorithm when $L = 12$ months. The subsequent plots show the annual and semi-annual variations and, finally, their sum. Relevant numbers of principal components are printed beside the graphs. The advantage of such a processing is the reliable separation of the trend from periodic variations, and this cannot always might be done by conventional processing techniques.

One may see from Fig. 14.14 that amplitudes and frequencies of the principal components can vary slowly in time. If the initial series is an infinite periodic sequence, then the principal-component analysis provides the customary Fourier transform. Similarly to the Fourier-series expansion based on the sine and cosine functions of time, the principal-components are also grouped in pairs.

We demonstrate below the processing of real signal with the help of the Caterpillar algorithm. The initial data file was recorded by the vertical electric antenna at the Moshiri observatory. Since we only want to demonstrate the data processing, it is unimportant what particular component was registered and when the record

Table 14.3 Principal components found in a sample of Moshiri record

Component	Intensity %	What it is
Pc1	46.533	50 Hz interference
Pc2	46.480	50 Hz interference
Pc3	2.471	Trend
Pc4	0.525	150 Hz interference
Pc5	0.515	150 Hz interference
Pc6	0.348	Trend
Pc7	0.239	100 Hz interference
Pc8	0.239	100 Hz interference
Pc9	0.223	Trend
Pc10	0.219	Trend
Pc11	0.203	Trend
Pc12	0.172	Industrial
Pc13	0.172	Industrial
Pc14	0.132	Trend
Pc15	0.124	Trend
Pc16	0.091	Trend
Pc17	0.086	Trend
Pc18	0.081	Industrial
Pc19	0.080	Industrial
Pc20	0.070	Trend

was made. The realization of 1 s duration is shown in Fig. 14.15. The sampling frequency was 4 kHz, so that a period of 50 Hz interference occupies 80 samples. Hence, the lag L should be proportional to this number. We have chosen the $L = 160$ or two periods of the 50 Hz frequency. The obtained relative intensities of twenty principal components are listed in Table 14.3.

The result of signal 'rectification' is shown in Fig. 14.15. The initial record is shown in the top plot in black line. The industrial interference varying in time is depicted in the red curve in the middle plot. The lower blue line shows the signal after extraction of the man-made interference. It contains the series of distinct pulses. One may observe from this figure that:

1. The interference is not stationary, and it would hardly be removed by using the common Fourier transform or notch filters.
2. The interference amplitude is about ± 2 a.u., while the highest pulse is about 7 a.u. The relatively low level of interference indicates that the observatory position is rather good.
3. The rectified signal definitely contains the SR background pattern combined with individual pulses.

It would be interesting to obtain a spectrum of the signal "rectified" by the Caterpillar procedure.

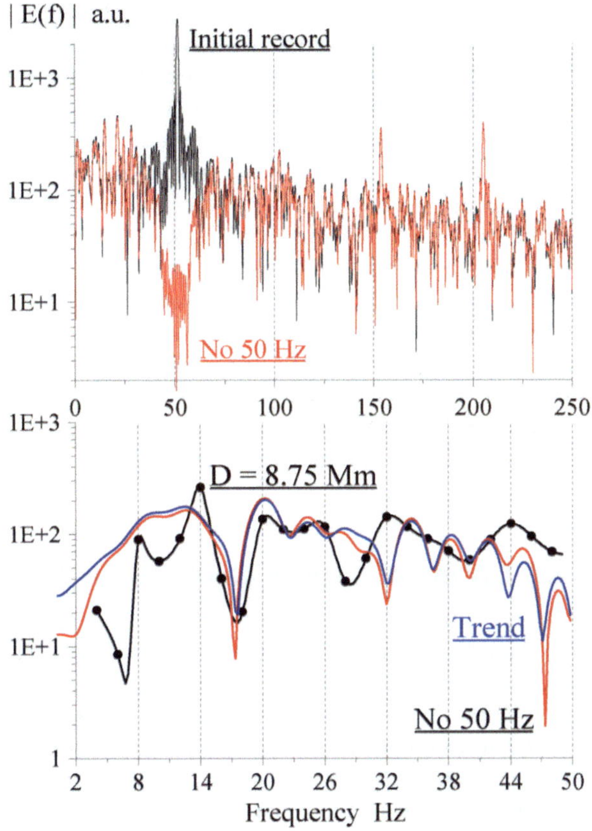

Fig. 14.16 Spectra of rectified data and model spectra

Results of the FFT application to the signal of Fig. 14.15 are shown in Fig. 14.16. Here we show the amplitude spectra of characteristic components of the signal. The black line in the upper frame of this figure depicts the spectrum of initial record. We picked the initial 0.5 s of the record (2,048 points) for the Fourier transform. The red line in the upper frame presents the spectrum of "rectified" data of Fig. 14.15.

Obviously, the spectrum of initial signal contains the large 50 Hz component and its less distinct harmonics. After extraction of periodic PC1 and PC2 components, we obtain a deep minimum in the vicinity of the 50 Hz frequency. The initial spectrum was not affected by the SSA processing at the frequency below 40 Hz. Thus, the SSA procedure worked as the notch filter in the 50 ± 5 Hz band uniformly reducing the industrial interference by about 50 dB. The data shown in the upper frame in Fig. 14.16 made us optimistic, and we performed the signal processing in the vicinity of the first pulse of rectified realization in Fig. 14.15.

To get rid of successive pulses, we pick initial 980 samples (about 0.25 s), center and zero pad the record. In other words, the realization was averaged over time, and this average was subtracted from the initial data. The centered realization thus obtained was continued by the zero samples up to 16,384 points and processed by the FFT algorithm afterwards. The resulting spectra are shown in the lower frame of Fig. 14.16 with ~0.25 Hz step. The red line demonstrates the spectrum with the subtraction of 50 Hz interference. The blue line is the composition of principal components relevant to trends listed in Table 14.3. Both the spectra are close to each other and contain the pronounced SR peaks around the first, the second, the third, and the fourth modes. The black line with dots is the model amplitude spectrum of a Q–burst arriving from the 8.75 Mm distance. Obviously, there is coincidence of many features of spectral patterns. Unfortunately, the closeness of successive pulses did not allow us to use the longer fragments of record, so that the spectral resolution of experimental spectra was not high. This particularity impeded a comparison with the model data.

Thus, we demonstrated usefulness of the SSA processing (Caterpillar procedure) for reducing the man-made interference in the SR records.

14.7 Listing of Typical Routines for Field Computations

The first listing presents the <ZeroMods.for> program, which computes the spectral components of the vertical electric *E*(*f*) and the horizontal magnetic $H_\phi(f)$ fields The propagation constant *ν*(*f*) is chosen by the user as a subroutine in the course of computations: the linear mode; Ishaq-Jones model; the knee model; PUK-day; and PUK-night model. Each field component is computed by a separate sub-routine based on the zonal harmonic series representation with the accelerated convergence (Nickolaenko–Rabinowicz algorithm). The particular program computes the spectra in the frequency band from 0.1 to 51.2 Hz ($dF = 0.1$ Hz) for distances from 0.1 to 19.7 Mm (dD = 0.1 Mm). The program can be adjusted for computations of the complex spectral components, so that the pulses in the time domain could be separately computed with the help of FFT.

```
1 ****************************************************
C  Program Name is ZeroMods.for  Zeroth mode spectra are
C  computed with Nick-Rabinowicz algorithm
C  \nu(f) are computed with sub-routines

    Parameter (N=190) ! For SR calculations
    Parameter (Nd=197)
    INTEGER N0
    Complex H1(N),E1(N)
    Real*8 dD, dF, D, Pi, Dr, Z
    Real*8 A, EPS0, C, F, Om, KA2
    Complex*16 J, Nu, B, Ever, Hphi, E_ver

     Pi=3.1415 92653 58979 32384 62643 D00
     J=(0.,1.)
        A=6.4e+06        ! Earth's  Radius in  meters
        EPS0=8.859e-12   ! Dielectric const., vacuum
        C=3.E+08         ! SPEED OF LIGHT meter/second
     dD=0.1000 00000 00000 00000 00000 D00

      dF=0.1

        Print*,'Specify NUMBER N0 of Desirable Nu model'
        Print*,'1 = Linear \nu(f) model'
        Print*,'2 = Ishaq-Jones \nu(f) model'
        Print*,'3 = Mushtak KNEE ionosphere \nu(f) model'
        Print*,'4 = Pechony Day KNEE \nu(f) model'
        Print*,'5 = Pechony Night KNEE \nu(f) model'
                Read*, N0

                     IF (N0.EQ.1) THEN
              open (7,file='LINEAR.dat')
c     write(7,*)'ComplE(f)_LINEAR      ComplH(f)_LINEAR\nu(f)'
     write(7,*)'F_Hz   D_Mm   |E(f)|_linear   |H(f)|_linear'
                     ENDIF
                  IF (N0.EQ.2) THEN
              open (7,file='JONES.dat')
c     write(7,*)'ComplE(f)_JONES      ComplH(f)_JONES\nu(f)'
     write(7,*)'F_Hz   D_Mm   |E(f)|_Jones   |H(f)|_Jones'
                  ENDIF
               IF (N0.EQ.3) THEN
              open (7,file='MUSHTAK.dat')
c     write(7,*)'ComplE(f)_Knee_MUSHTAK      ComplH(f)_Knee\nu(f)'
     write(7,*)'F_Hz   D_Mm   |E(f)|_knee   |H(f)|_knee'
               ENDIF
            IF (N0.EQ.4) THEN
              open (7,file='DAY.dat')
c     write(7,*)'ComplE(f)_PUK\nuDay      ComplH(f)_PUK\nuDay'
     write(7,*)'F_Hz   D_Mm   |E(f)|_PUK-Day   |H(f)|_PUK-Day'
          ENDIF
      IF (N0.EQ.5) THEN
               open (7,file='NIGHT.dat')
c     write(7,*)'ComplE(f)_PUK\nuNight      ComplH(f)_PUK\nuNight'
     write(7,*)'F_Hz   D_Mm   |E(f)|_PUK-Night   |H(f)|_PUK-Night'
      ENDIF

c     DO K =100, 100   !  D=10 Mm
    DO K =1, Nd   ! Distance variations AUTOMATIC  !!!!!
```

```
        D=dD*K
        Dr=Pi*D/20.      ! Source Distance in radians
        Z=cos(Dr)

        do J1=1, 512   !   FREQUENCY VARIATION
          F=DF*J1
          Om=2.*Pi*F      ! OMEGA - circular frequency
          KA2=(A*Om/C)*(A*Om/C)
                    IF (N0.EQ.1) THEN
           Call ANuN(F,Nu)     !LINEAR NICKOLAENKO \nu(f)= Nu
                    ENDIF
                  IF (N0.EQ.2) THEN
           Call ANuJ(F,Nu)    !ISHAQ-JONES model   \nu(f)= Nu
                  ENDIF
                IF (N0.EQ.3) THEN
           Call ANuKnee(F,Nu) !MUSHTAK KNEE model  \nu(f)= Nu
                ENDIF
              IF (N0.EQ.4) THEN
           Call PUKNuD(F,Nu)    !Part. Unif. KNEE model Day \nu(f)= Nu
              ENDIF
          IF (N0.EQ.5) THEN
           Call PUKNuN(F,Nu)    !Part. Unif. KNEE model Night \nu(f)= Nu
          ENDIF
             B=Nu*(Nu+1.)

          Call Ez(NU,Z,Ever) !CALCUL  vertical electric field Ever
          Call Hf(NU,Z,Hphi)!CALCUL  azimuthal magnetic field Hphi

           E_ver=(J*B/Om)*Ever
c      Array of output spectral data
           E1(j1)=E_ver
           H1(j1)=-Hphi
c          write(7,*) (E1(J1)), (H1(J1))
          write(7,*) F, D, Abs(E1(J1)), Abs(H1(J1))

                enddo  ! FREQUENCY VARIATION
         ENDDO  !  distance variations  AUTOMATIC !!!!!!
17      Format(6(G14.6))
        END

        subroutine ANuN(F,Nu)
c       Computes propagation constant \nu(f) LINEAR
        Complex*16 J, Nu
        Real*8 F
            J=(0.,1.)
            IF (F.LE.4) THEN
               Nu=F/12.-j*F/100.
                  ELSE
               Nu=(F-2.)/6.-j*F/100.  !  LINEAR Nu(f)
            ENDIF
          return
        END

        subroutine ANuJ(F,Nu)
c       Computes propagation constant \nu(f) by Ishaq and Jones
        Complex*16 J, SNU, S2, Nu
        Real*8 F, Pi, A, C, K, AKA, ST, SF, ALPH, CV
          Pi=3.1415 92653 58979 32384 62643 D00
          J=(0.,1.)
          A=6.4e+06        ! Earth's  Radius in  meters
          C=3.E+08         !       LIGHT Velocity  m/s
         K=2.*Pi*F/C       !       wave number
         AKA=K*A           !        ka
```

```
      ST=DLOG(F)                    !      Ln(F)
      SF=EXP(0.64*ST)               !      F^{0.64}
      ALPH=0.063*SF
      CV= 1.64 - 0.1759*ST + 0.01791*ST*ST    ! c/V
      SNU=CV - J*5.49*ALPH/F              ! COMPLEX Snu(f)
      S2 = SNU*SNU                     ! COMPLEX Snu^2
      Nu = SQRT(0.25 + AKA*AKA*S2) - 0.5     ! COMPLEX  Nu(f)
       return
      END

      subroutine ANuKNEE(F,Nu)
c     Computes propagation constant \nu(f) by MUSTAKI KNEE model
      Complex*16 J, HE, HM, S2, Nu
      Real*8 F, Pi, A,C,K,AKA, FKNEE,HKNEE,ZEB,ZEA,FMP,HMP,ZEMP,BM
        Pi=3.1415 92653 58979 32384 62643 D00
        J=(0.,1.)
        A=6.4e+06       ! Earth's  Radius in  meters
        C=3.E+08        !      LIGHT Velocity  m/s
       K=2.*Pi*F/C      !      wave number
       AKA=K*A           !      ka
      FKNEE=10.     !      KNEE Freq                    Hz
      HKNEE=55.     !      KNEE HEIGHT DAY-TIME  CAVITY     km
      ZEB=8.3       !      LOWER SCALE of PROFILE        km
      ZEA=2.9       !      UPPER SCALE of PROFILE       km
      FMP=8.        !     Star Freq                   Hz
      HMP=96.5      !      MAGNETIC HEIGHT                  km
      ZEMP=4.       !      MAGNETIC HEIGHT SCALE            km
      BM=20.        !      PARAMETER B                 km
      HE=HKNEE+ ZEA*DLOG(F/FKNEE)+0.5*(ZEA-ZEB)*DLOG(1.+(Fknee/F)**2)
     #  +J* (ZEA*Pi/2.-(ZEA-ZEB)*ATAN(Fknee/F) )
      ZEM=ZEMP + BM*(1./F - 1./FMP )
      HM=HMP-ZEM*DLOG(F/FMP) - J* (Pi/2.*ZEM)
      S2=HM/HE
       Nu = SQRT(0.25 + AKA*AKA*S2) - 0.5      ! COMPLEX  Nu(f)
       return
      END

      subroutine PUKNuD(F,Nu)
c     Computes propagation constant \nu(f) by Pechony Day KNEE model
      Complex*16 J, HE, HM, S2, Nu
      Real*8 F, Pi, A,C,K,AKA, FKNEE,HKNEE,ZEB,ZEA,FMP,HMP,ZEMP,BM
        Pi=3.1415 92653 58979 32384 62643 D00
        J=(0.,1.)
        A=6.4e+06       ! Earth's  Radius in  meters
        C=3.E+08        !      LIGHT Velocity  m/s
         K=2.*Pi*F/C      !     wave number
         AKA=K*A           !     ka
       FKNEE=13.     !      KNEE Freq                    Hz
       HKNEE=54.     !      KNEE HEIGHT DAY-TIME  CAVITY     km
       ZEB=7.5       !      LOWER DAY-TIME SCALE     km
       ZEA=2.7       !      UPPER DAY-TIME SCALE     km
       FMP=6.        !     Star Freq                   Hz
       HMP=97.5      ! MAGNETIC HEIGHT  in  DAY-TIME CAVITY  km
       ZEMP=3.7      ! MAGNETIC HEIGHT SCALE in DAY-TIME     km
       BM=5.0        !   PARAMETER B in DAY-TIME            km
      HE=HKNEE+ ZEA*DLOG(F/FKNEE)+0.5*(ZEA-ZEB)*DLOG(1.+(Fknee/F)**2)
     #  +J* (ZEA*Pi/2.-(ZEA-ZEB)*ATAN(Fknee/F) )
          ZEM=ZEMP + BM*(1./F - 1./FMP )
      HM=HMP-ZEM*DLOG(F/FMP) - J* (Pi/2.*ZEM)
          S2=HM/HE
          Nu = SQRT(0.25 + AKA*AKA*S2) - 0.5      ! COMPLEX  Nu(f)
       return
      END
```

```
      subroutine PUKNuN(F,Nu)
c     Computes propagation constant \nu(f) by Pechony NIGHT KNEE model
      Complex*16 J, HE, HM, S2, Nu
      Real*8 F, Pi, A,C,K,AKA, FKNEE,HKNEE,ZEB,ZEA,FMP,HMP,ZEMP,BM
        Pi=3.1415 92653 58979 32384 62643 D00
        J=(0.,1.)
        A=6.4e+06      ! Earth's  Radius in  meters
        C=3.E+08       !     LIGHT Velocity  m/s
          K=2.*Pi*F/C      !     wave number
          AKA=K*A          !     ka
       FKNEE=13.     !    KNEE Freq                    Hz
       HKNEE=60.     !    KNEE HEIGHT DAY-TIME  CAVITY     km
       ZEB=9.1      !    LOWER DAY-TIME SCALE     km
       ZEA=3.8      !    UPPER DAY-TIME SCALE     km
       FMP=6.       !   Star Freq                    Hz
       HMP=99.5      ! MAGNETIC HEIGHT  in  DAY-TIME CAVITY  km
       ZEMP=3.5      ! MAGNETIC HEIGHT SCALE in DAY-TIME     km
       BM=4.0       !   PARAMETER B in DAY-TIME             km
      HE=HKNEE+ ZEA*DLOG(F/FKNEE)+0.5*(ZEA-ZEB)*DLOG(1.+(Fknee/F)**2)
     #  +J* (ZEA*Pi/2.-(ZEA-ZEB)*ATAN(Fknee/F) )
      ZEM=ZEMP + BM*(1./F - 1./FMP )
      HM=HMP-ZEM*DLOG(F/FMP) - J* (Pi/2.*ZEM)
      S2=HM/HE
          Nu = SQRT(0.25 + AKA*AKA*S2) - 0.5       ! COMPLEX  Nu(f)
       return
      END

      subroutine Ez(Nu,x,E)
c     computes Ez with the complex Legendre function of index NU and
c     argument x=cos(theta). Nick-Rab formulae used for acceleration
      Complex*16 J,Nu,B,R,Zn,Ae,S,E
      Real*8 Pi,SiT,AN,EPS0,X,P,Q,O,R1,Sq,R2,R3,R4
      Integer UL
      J=(0.,1.)
      pi=3.1415 92653 58979 32384 62643 D00
      SiT=SQRT(1.-X**2) !                              sinus theta
      AN=Abs(Nu)
      B=Nu*(Nu+1.)
      EPS0=8.859e-12      !   Dielectric constant of the Vacuum
      S=(0.,0.)
      P=1.
      Q=1.
      O=.0
      UL=10+6*AINT(REAL(NU))        ! Upper limit of summation
      IF (1.-X.LT.EPS0**2) THEN    !           Avoiding SINGULARITY
        R1=log(1.+SQRT(2.)/EPS0)
          ELSE
        R1=log(1.+SQRT(2./(1.-X)))
      ENDIF
       Sq=SQRT((1.-X)*2.)
       R2=1.+(1.-X)*R1-Sq
       R3=1.+.25*(x-1.)*(3.*x-1.)*R1-.75*(x+(1.-x)*Sq)
       R4=-((1.-x)**2)*(1.+5.*x)*R1/12.+(15.*(x**2)-27.*x+14.)/36.
     #+5.*(1.-(x**2))*Sq/12.-11.*(1.-x)*Sq/18.
        R=2.*R1+R2+2.*(1.+B)*R3+3.*(2.+3.*B)*R4
C   summation of accelerated zonal harmonic series
          do N=0,UL   !            Zonal harmonic series summation
           Zn=(N*(N+1.)-B)*(N+1.)*(N+2.)*(N+3.)*(N+4.)
           Ae=(2.*N*((B**2)+22.*B+12.)+(B**2)*17.+B*74.+24.)/Zn
            IF (1.+X.LT.EPS0) THEN  !           Antipode
              S=S+Ae*((-1)**N)
              ELSE
```

```
          S=S+Ae*P
        ENDIF
        O=Q      ! recursion of Legendre polynomials
   Q=P
        P=x*Q*(N*2.+1.)/(1.+N)-O*N/(1.+N)
      enddo        ! End of ZHSR summation
    E=-(R+S)/Pi   !        P_nu(-X)/sin(Pi*Nn)  Rab-Nick
   return
   END

   Subroutine Hf(Nu,x,H)
c  Computes Hf with the complex associated Legendre function
c  of lower index NU, upper index 1, and argument x=cos(theta).
c  Nickolaenko-Rabinowicz formulae used for acceleration
C  Double precision
   Complex*16 J,Nu,B,A5,A6,Sh,Znh,Ah,H
   Real*8 Pi,x,SiT,AN,EPS0,P,Q,O,R1,R3,R4,Rh
   Integer UL
   J=(0.,1.)
c   pi= 3.1415926
   pi=3.1415 92653 58979 32384 62643 D00
   SiT=SQRT(1.-X**2) !                    sinus theta
   AN=Abs(Nu)
   B=Nu*(Nu+1.)
   A5=-2.*((B**2)-19.*B-66.)
   A6=-17.*(B**2)+34.*B+24.
   EPS0=8.859e-12      !   Dielectric constant of the Vacuum
   Sh=(0.,0.)
   P=1.
   Q=1.
   O=.0
   UL=10+6*AINT(REAL(NU))       ! Upper limit of summation
   IF (1.-X.LT.EPS0**2) THEN ! Avoiding SINGULARITY
     R1=log(1.+SQRT(2.)/EPS0)
          ELSE
     R1=log(1.+SQRT(2./(1.-X)))
   ENDIF
    Sq=SQRT((1.-X)*2.)
    R3=1.+.25*(x-1.)*(3.*x-1.)*R1-.75*(x+(1.-x)*Sq)
    R4=-((1.-x)**2)*(1.+5.*x)*R1/12.+(15.*(x**2)-27.*x+14.)/36.
  #+5.*(1.-(x**2))*Sq/12.-11.*(1.-x)*Sq/18.
    Rh=2.*R3+9.*R4
C summation of accelerated zonal harmonic series
    do N=0,UL       !    Zonal harmonic series summation
    Znh=(N*(N-1.)-B)*((N+1.)*(N+2.)-B)*(N+1.)*(N+2.)*(N+3.)*(N+4.)
    Ah=(4.*(N**3)*(B+12.)+2.*(N**2)*(19.*B+78.)+N*A5+A6)/Znh
    Sh=Sh+Ah*P
    O=Q                      ! recursion of Legendre polynomials
    Q=P
    P=x*Q*(N*2.+1.)/(1.+N)-O*N/(1.+N)
   enddo                   ! End of ZHSR summation
       IF (1.-ABS(X).LT.EPS0**2) THEN ! Avoiding SINGULARITY
         H=-(Rh+Sh)*B*2./(pi*EPS0)
           ELSE
   H=-(Rh+Sh)*B*2./(pi*SQRT(1.-(x**2))) !        P_'nu(-X)/sin(Pi*Nn)
       ENDIF
    return
   END
1 ****************************************************
```

The second listing presents the <E_&_H(t).for> program. It computes the vertical electric $E(t)$ and the horizontal magnetic $H_{\phi}(t)$ fields directly in the time domain. The linear model propagation constant $\nu(f)$ is used. The time series converges when the time $t > 0$. It starts from 1 ms and varies with the 0.5 ms step. Computations of the waveforms are made for any source distance specified by the user. The formulas are presented by Eqs. (14.31–14.32). The program can be adjusted to computations in the uniform $\nu(f)$ model, i.e., with the parameter $B_{\nu} = 0$.

```
2 *****************************************************
c  Program name is E_&_H(t).for
C  Direct ACCELERATED TIME DOMAIN formulas for the E(t) and H(t)
C  Linear \nu(f) dependence, Summed over 100 terms
C  Initial time should be > 1 ms

      PARAMETER (N=100, M=300)

      Complex j,ANu,Tau,ROOT,PsiM1,Psi0,Psi1,AS,S,EM1,E0,E1,Ever
      Real D, Pi, X, Y, dT, A, H, C, EPS0, Mc, EA, HA, B,T,P,Q,O,Hphi

      open (7,file='E_&_H(t).DAT')

      Print*,'Specify source distance D in Mm, please'
      Read*,D
      Print*,'Time  Em1  Em1+E0  Em1+E0+E1)  Ever  D= ',D, 'Mm'
      write(7,*)'T_ms  Ever_mV/m  Hphi_uA/m    D= ',D, 'Mm'

      j=(0.,1.)
      Pi= 3.1415926
      X=cos(Pi*D/20.)
      Y=sin(Pi*D/20.)
              Print*, X,  Y
      dT=5.0E-04            !     0.5 ms step
          A=6.366E+06       !              Earth's radius
          H=6.E+04          !        Ionosphere height 60 km
          C=2.997928E+05    !              Light Velocity
          EPS0=8.859E-12    !     DIELECTRIC CONSTANT FOR VACUUM
          Mc=1.0e+08        ! SOURCE CURRENT MOMENT (25 kA * 4 km)
        EA=Mc/(2.*Pi*EPS0*H*A*A)   ! ELECTRIC FIELD AMPLITUDE
        HA=Mc/(2.*Pi*H*C*C)        ! MAGNETIC FIELD AMPLITUDE
       FH=1.0e+06        !  Transform to uA/m

CC>>>>>>>>>  Linear \nu(\omega) dependence:          ANu and B
       ANu=(1./6.-j/100.) / (2.*Pi)  ! Linear \nu(\omega) dependence
       B=-1./3.                ! Linear \nu(\omega) dependence
CC>>>>>>>>>  UNIFORM Linear \nu(\omega) dependence:   ANu and B=0!
c      B=0                     !Uniform Linear \nu(\omega) dependence
c      ANu=(1./7.-j/100.) / (2.*Pi) !Uniform Linear \nu(\omega) dependence

      DO K=2, 600       ! Time VARIATIONS     from 1 ms to 300 ms
          T= K*dT
          S=0.     !            For recursion of Legendre Polynomials
          P=X      !            For recursion of Legendre Polynomials
          Q=1.     !            For recursion of Legendre Polynomials
        Tau=EXP( j*T/ANu )  !  Standard time factor
        ROOT=SQRT(1. - 2.*TAU*X + TAU*TAU )
          PsiM1=(1.-TAU*X)/(ROOT*ROOT*ROOT) - 1.
          Psi0=1./ ROOT
       IF ((1.-X).GT.(0.00001)) THEN
         Psi1=CLOG( (TAU - X + ROOT)/(1.-X) )
          ELSE
         Psi1=CLOG( 1./(1.-TAU) )
       ENDIF
              DO J1=1,N  ! E(t): ZHSR SUM left after acceleratrion
                AS= EXP(J1*j*T/ANu) / ((J1+1)*(J1-B ))
                S=S+As*P
                O=Q
                Q=P
```

```
            P=X*Q*(J1*2+1)/(1+J1)-O*J1/(1+J1)
          ENDDO    ! End of E(t) summation
      EM1=PSIM1*EXP(-j*T*B/ANu )
      E0=B*( Psi0-1)*EXP(-j*T*B/ANu )
      E1=B*( B+1)*( Psi1*EXP(-j*T/ANu ) - 1)*EXP(-j*T*B/ANu )
     Ever=EA*ANu*( EM1+E0+E1 + B*(B+1)*(B+1)*S*EXP(-j*T*B/ANu) )   ! mV/m
     Hphi=AIMAG( Y*EXP( j*T*(1-B)/ANu ) /ANu/ROOT/ROOT/ROOT)*HA*FH  ! uA/m
ccccccc>>>>> Fields should be NEGATIVE for positive stroke!!!!
      write(7,11) T*1000., -Real(Ever), -Hphi*fh
     ENDDO          ! Time VARIATIONS     from 1 ms to 300 ms
11   Format(F12.3, E15.6, E15.6, F12.3)

     END
2 ********************************************
```

References

Baba K, Hayakawa M (1995) The effect of localized ionospheric perturbations on subionospheric VLF propagation on the basis of finite element method. Radio Sci 30:1511–1517

Bateman H, Erdelyi A (1953) Higher transcendental functions, vol 1. McGraw-Hill Co., New York 296 pp

Belyaev GG, Schekotov AYu, Shvets AV, Nickolaenko AP (1999) Schumann resonances observed using poynting vector spectra. J Atmos Solar-Terr Phys 61:751–763

Berenger JP (2002) FDTD computation of VLF-LF propagation in the Earth-ionosphere waveguide. Ann Telecommun 5(11–12):1059–1070

Bliokh PV, Galyuk YuP, Hynninen EM, Nickolaenko AP, Rabinowicz LM (1977) On the resonance phenomena in the Earth-ionosphere cavity. Izvestiya Vuzov, Radiofizika 20:501–509 (in Russian)

Bliokh PV, Nickolaenko AP, Filippov YuF (1980) In: Jones DLl (ed) Schumann resonances in the Earth-ionosphere cavity. Peter Peregrinus, Oxford

Cummer SA (2000) Modeling electromagnetic propagation in the Earth-ionosphere waveguide. IEEE Trans Antennas Propag 48(9):1420–1429

Danilov DL (1996) Principal components in time series forecast. Proc Stat Comput Sect Am Stat Assoc, pp 56–160

Danilov DL, Zhiglyavsky AA (eds) (1997) Principal component of the time series: the Caterpillar method. St-Petersburg State University, St.-Petersburg, Russia, 307 pp (in Russian)

Erdelyi A, Magnus W, Oberhettinger F, Tricomi FG (1953) Higher transcendental functions, vol 1. McGraw-Hill, New York

Füllekrug M (2000) Dispersion relation for spherical electromagnetic resonances in the atmosphere. Phys Lett A 275:80–89

Füllekrug M (2005) Detection of thirteen resonances of radio waves from particularly intense lightning discharges. Geophys Res Lett 32:L13809. doi:10.1029/2005GL023028

Galejs J (1961) Terrestrial extremely low frequency noise spectrum in the presence of exponential ionospheric conductivity profiles. J Geophys Res 66:2789–2793

Galejs J (1970) Frequency variations of Schumann resonances. J Geophys Res 75:3237–3251

Galejs J (1972) Terrestrial propagation of long electromagnetic waves. Pergamon Press, New York

Golyandina N, Nerkutkin V, Zhiglyavsky AA (2001) Analysis of time series structure. Chapman and Hall, CRC, Boca Raton

Gradstein C, Ryzhik IM (1968) Tables of integrals, sums, series and products. Phys-Math Publ, Moscow
Greifinger C, Greifinger P (1978) Approximate method for determining ELF eigen-values in the Earth-ionosphere waveguide. Radio Sci 13:831–837
Greifinger PS, Mushtak VC, Williams ER (2007) On modeling the lower characteristic ELF altitude from aeronomical data. Radio Sci 42:RS2S12. doi:10.1029/2006RS003500
Hayakawa M, Otsuyama T (2002) FDTD analysis of ELF wave propagation in inhomogeneous subionospheric waveguide models. Appl Comput Electromagnet Soc J 17(3):239–244
Hynninen EM, Galyuk YuP (1972) Field of a vertical electric dipole above the spherical Earth and a non-uniform in height atmosphere. In: Problems of wave diffraction and propagation, vol. 11. Leningrad State University Press, pp 109–119 (in Russian)
Ishaq M, Jones DLl (1977) Method of obtaining radiowave propagation parameters for the Earth-ionosphere duct at ELF. Electron Lett 13:254–255
Jones DLl, Burke CP (1990) Zonal harmonic series expansions of Legendre functions and associated Legendre functions. J Phys A Math Gen 23:3159–3168
Jones DLl, Knott M (1999) Comparison of simplified and full-wave ELF propagation models, Abstracts of URSI General Assembly, Toronto, Ontario, Canada, 13–21 Aug 1999, section E6-10, p 295
Jones DLl, Joyce GS (1989) The computation of ELF radio wave fields in the Earth-ionosphere duct. J Atmos Terr Phys 51:233–239
Kirillov VV (1993) Parameters of the Earth-ionosphere waveguide at ELF. Prob Diffr Wave Propag 25:35–52 (in Russian)
Kirillov VV (1996) Two-dimension theory of propagation of ELF electromagnetic waves in the Earth-ionosphere cavity. Izvestiya Vuzov, Radiofizika 39:1103–1112 (in Russian)
Kirillov VV (2000) Two-dimension theory of propagation of electromagnetic waves of ELF-VLF bands in the waveguide Earth-ionosphere. Prob Wave Diffr Propag, St-Petersburg State Univ Press 28:184–203 (in Russian)
Kirillov VV, Kopeykin VN (2002) Solution of two dimension telegraph equation with anisotropic parameters. Izvestiya Vuzov, Radiofizika 45:1011–1024 (in Russian)
Kirillov VV, Kopeykin VN, Mushtak VC (1997) Electromagnetic waves of ELF band in the Earth-ionosphere waveguide. Geomagn Aeronomia 37:114–120 (in Russian)
Marple SL Jr (1987) Digital spectral analyses with applications. Prentice-Hall, Englewood Cliffs
Morente JA, Molina-Cuberos GJ, Portı JA, Schwingenschuh K, Besser BP (2003) A study of the propagation of electromagnetic waves in Titan's atmosphere with the TLM numerical method. Icarus 162:374–384. doi:10.1016/S0019-1035(03)00025-3
Mushtak VC, Williams ER (2002) ELF propagation parameters for uniform models of the Earth-ionosphere waveguide. J Atmos Solar-Terr Phys 64:1989–2001
Nickolaenko AP, Hayakawa M (2002) Resonance in the Earth-ionosphere cavity. Kluwer Academic Publishers, Dordrecht 380 pp
Nickolaenko AP, Rabinowicz LM (1982) Possible global electromagnetic resonances on the planets of the solar system. Translated from *Kosmicheskie Issledovaniya*, 20(1):82–88, Plenum Publishing Corporation
Nickolaenko AP, Rabinowicz LM (1987) Applicability of ultralow—frequency global resonances for investigating lightning activity on Venus. Translated from *Kosmicheskie Issledovaniya*, 25(2):301–306, Plenum Publishing Corporation
Nickolaenko AP, Rabinowicz LM (2000) Accelerating of the convergence of the time-domain solution for lightning stroke ELF pulses. Radiophys Electron 5:108–112 (in Russian)
Nickolaenko AP, Rabinowicz LM, Hayakawa M (2004a) Time domain presentation for ELF pulses with accelerated convergence. Geophys Res Lett 31:L05808. doi:10.1029/2003GL018700
Nickolaenko AP, Rabinowicz LM, Hayakawa M (2004b) Natural ELF pulses in the time domain: series with accelerated convergence. IEEJ Trans Fundam Mater 124(12):1210–1215
Ogawa T, Komatsu M (2007) Analysis of Q-burst waveforms. Radio Sci 42:RS2S18. doi:10.1029/2006RS003493

Otsuyama T, Sakuma D, Hayakawa M (2003) FTDT analysis of ELF wave propagation and Schumann resonance for a subionospheric waveguide model. Radio Sci 38(6):1103. doi:10.1029/2002RS002752

Pechony O (2007) Modeling and simulations of Schumann resonance parameters observed at the mitzpe ramon field station (Study of the day-night asymmetry influence on Schumann resonance amplitude records), Ph.D. Thesis, Tel-Aviv University, Israel, March, 92 pp

Pechony O, Price C (2004) Schumann resonance parameters calculated with a partially uniform knee model on Earth, Venus, Mars, and Titan. Radio Sci 39:RS5007. doi:10.1029/2004RS003056

Sentman DD (1990a) Approximate Schumann resonance parameters for a two-scale height ionosphere. J Atmos Terr Phys 52:35–46

Sentman DD (1990b) Electrical conductivity of Jupiter shallow interior and the formation of a resonant planetary—ionosphere cavity. Icarus 88:73–86

Simpson JA, Taflove A (2002) Two-dimensional FDTD model of antipodal ELF propagation and Schumann resonance of the Earth. Antennas Wirel Propag Lett 1(2):53–57

Troyan VN, Hayakawa M (2002) Inverse geophysical problems. TERRAPUB, Tokyo, 289 p

Wait JR (1962) Electromagnetic waves in stratified media. Pergamon Press, Oxford

Williams ER, Mushtak V, Nickolaenko AP (2006) Distinguishing ionospheric models using Schumann resonance spectra. J Geophys Res 111:D16107. doi:10.1029/2005JD006944

Yang H, Pasko VP (2005) Three-dimensional finite difference time domain modeling of the Earth-ionosphere cavity resonances. Geophys Res Lett 32:L03114. doi:10.1029/2004GL021343

Yang H, Pasko VP (2006) Three-dimensional finite difference time domain modeling of the diurnal and seasonal variations in Schumann resonance parameters. Radio Sci 41:RS2S14. doi:10.1029/2005RS003402

Yang H, Pasko VP, Yair Y (2006) Three-dimensional finite difference time domain modeling of the Schumann resonance parameters on Titan, Venus, and Mars. Radio Sci 41(2):203. doi:10.1029/2005RS003431

Index

A. Nickolaenko and M. Hayakawa, *Schumann Resonance for Tyros*,
Springer Geophysics, DOI: 10.1007/978-4-431-54358-9,

MIX
Papier aus verantwortungsvollen Quellen
Paper from responsible sources
FSC® C105338

If you have any concerns about our products,
you can contact us on
ProductSafety@springernature.com

In case Publisher is established outside the EU,
the EU authorized representative is:
Springer Nature Customer Service Center GmbH
Europaplatz 3, 69115 Heidelberg, Germany

Printed by Libri Plureos GmbH
in Hamburg, Germany